T0215764

CHUDLEY AND GREENO'S BUILDING CONSTRUCTION HANDBOOK

Twelfth edition

Roy Chudley, Roger Greeno and Karl Kovac

Routledge
Taylor & Francis Group

LONDON AND NEW YORK

Twelfth edition published 2020
by Routledge
2 Park Square, Milton Park, Abingdon, Oxon, OX14 4RN

and by Routledge
52 Vanderbilt Avenue, New York, NY 10017

Routledge is an imprint of the Taylor & Francis Group, an informa business

First edition published by Butterworth-Heinemann 1988
Eleventh edition published by Routledge 2016

British Library Cataloguing-in-Publication Data
A catalogue record for this book is available from the British Library

Library of Congress Cataloging-in-Publication Data
Names: Chudley, R., author. | Greeno, Roger, author. | Kovac, Karl, author.
Title: Chudley and Greeno's building construction handbook.
Other titles: Building construction handbook
Description: 12th edition. | Abingdon, Oxon ; New York,
NY : Routledge, 2019. | Includes index. | Identifiers: LCCN 2019027835
(print) | LCCN 2019027836 (ebook) | ISBN 9780367135423 (hardback) |
ISBN 9780367135430 (paperback) | ISBN 9780429027130 (ebook) |
ISBN 9780429651410 (adobe pdf) | ISBN 9780429646133 (mobi) | ISBN
9780429648779 (epub)
Subjects: LCSH: Building–Handbooks, manuals, etc. | Construction
industry–Handbooks, manuals, etc.
Classification: LCC TH151 .C52 2019 (print) | LCC TH151 (ebook) |
DDC 690–dc23
LC record available at https://lccn.loc.gov/2019027835
LC ebook record available at https://lccn.loc.gov/2019027836

ISBN: 978-0-367-13542-3 (hbk)
ISBN: 978-0-367-13543-0 (pbk)
ISBN: 978-0-429-02713-0 (ebk)

Typeset in Futura
by Swales & Willis, Exeter, Devon, UK

See the companion website at: www.routledge.com/9780367135430

CONTENTS

Contents

Contents

Contents

Contents

Figures

Figures

Figures

Figures

Figures

Figures

Figures

Figures

Tables

Tables

PREFACE

The *Building Construction Handbook* originated in 1982 as a series of four 'check-books' written and illustrated by Roy Chudley. In 1988 these successful study guides were consolidated into one volume. A second edition was published in 1995, and subsequently, Roger Greeno was enlisted to co-author the third edition, which published in 1998. Since then, Roger has been involved in revising no less than eight further editions, each time meticulously adding new sections on technological developments, new building regulations, updates to standards and filling in gaps in the book to make the *Handbook* the most comprehensive and up to date construction book on the market. 'Chudley', as it is affectionally known by many who have made use of the extensive coverage in their teaching and studies in the built environment over nearly four decades, remains one of the bestselling construction books in the UK. That is largely down to the hard graft and keen eye for detail that Roger Greeno has brought to his work on the book during his tenure as lead author. Many readers still don't believe that all the drawings, until this new edition, have been drafted by hand, either by Roy Chudley or by Roger. Such skill is rarely encountered in today's world of BIM and freely available 3D drawing software.

Since Taylor & Francis acquired the book in 2011, we have regularly consulted with readers of the *Handbook* to take in their views and have always remained cautious about changing the look and feel of such a long-standing and highly regarded book. However, in recent years voices have emerged that were keen to see a fresh layout and expressed concern about the sheer physical size of the book in an age where many readers prefer more portable ebook options. This feedback coincided with Roger's desire to reduce the amount of time he spent working on the book and increase the amount of time he spent with his grandchildren (and fishing). We are immensely grateful to Roger for his years of service and his skill and dedication in making the *Building Construction Handbook* as successful as it has been.

We have embarked on a search for a new custodian to take the reins and deliver a fully revised 12th edition, no easy feat for any author. We are extremely grateful to Karl Kovac for agreeing to take on this task and for his extensive review, editing and updating of the entire book.

The 12th edition of the book, renamed in honour of its two key contributors as *Chudley and Greeno's Building Construction Handbook*, has had the most extensive revision in its 37-year history. It retains its approach of using illustrations with supporting text to provide guidance on the practices, procedures, standards, legislations and regulations most relevant to the modern built environment professional and student. This edition has a completely redesigned and modernised internal layout and the content has been consolidated and restructured from 8 parts to 22 chapters which follow a logical sequence. In recognition of the importance and

Preface

popularity of the drawings in previous editions, we have archived those that were removed from the 11th edition online for anyone to download and enjoy.

To access these drawings please visit: https://www.routledge.com/978036 7135430

1 PRE-CONSTRUCTION

DESIGN CONSIDERATIONS
MODULAR COORDINATION
CONSTRUCTION CONTRACT DOCUMENTS
PLANNING LEGISLATION
BUILDING CONTROL
HEALTH AND SAFETY
BUILDING ORGANISATIONS
BUILDING INFORMATION MODELLING (BIM)
BUILDING INFORMATION CLASSIFICATION SYSTEMS

1.1 Design considerations

When considering any building development the design team will need to carefully consider the environmental and physical aspects of the site and proposed construction works.

Environmental considerations

1. Planning requirements.
2. Building regulations.
3. Land restrictions by vendor or lessor.
4. Availability of services.
5. Local amenities, including transport.
6. Subsoil conditions.
7. Levels and topography of land.
8. Adjoining buildings or land.
9. Use of building.
10. Daylight and view aspects.

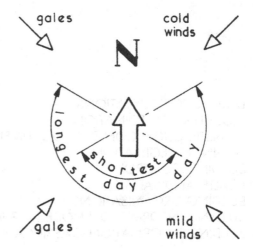

Figure 1.1 Orientation aspects

Physical considerations

1. Natural contours of land.
2. Natural vegetation and trees.
3. Size of land and/or proposed building.
4. Shape of land and/or proposed building.
5. Approach and access roads and footpaths.
6. Services available.
7. Natural waterways, lakes and ponds.
8. Restrictions such as rights of way, tree preservation and ancient buildings.
9. Climatic conditions created by surrounding properties, land or activities.
10. Proposed future developments.

1.2 Modular coordination

A module can be defined as a basic dimension which could, for example, form the basis of a planning grid in terms of multiples and submultiples of the standard module. In the diagram of a typical modular coordinated planning grid, let M = the standard module.

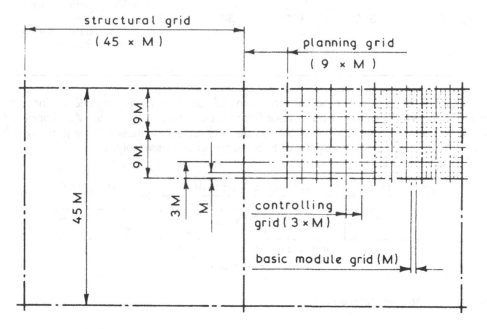

Figure 1.2 Basic module grid

Structural grid ~ used to locate structural components such as beams and columns.
Planning grid ~ based on any convenient modular multiple for regulating space requirements such as rooms.
Controlling grid ~ based on any convenient modular multiple for location of internal walls, partitions, etc.
Basic module grid ~ used for detail location of components and fittings.

All the above grids, being based on a basic module, are contained one within the other and are therefore interrelated. These grids can be used in both the horizontal and vertical planes, thus forming a three-dimensional grid system. If a first preference numerical value is given to M dimensional coordination is established.

BS 6750: *Specification for Modular Coordination in Building*

Dimensional coordination ~ the practical aims of this concept are to:

1. Size components so as to avoid the wasteful process of cutting and fitting on site.
2. Obtain maximum economy in the production of components.
3. Reduce the need for the manufacture of special sizes.
4. Increase the effective choice of components by the promotion of interchangeability.

Modular coordination

BS 6750 specifies the increments of size for coordinating dimensions of building components, as in Table 1.1.

Table 1.1 Coordinating dimensions

Preference	1st	2nd	3rd	4th	The 3rd and 4th preferences having a maximum of 300mm
Size (mm)	300	100	50	25	

Dimensional grids ~ the modular grid network as shown on page 2 defines the space into which dimensionally coordinated components must fit. An important factor is that the component must always be undersized to allow for the joint, which is sized by the obtainable degree of tolerance and site assembly.

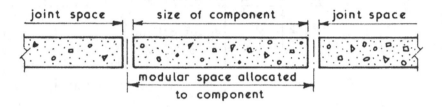

Figure 1.3 Modular tolerances

Controlling lines, zones and controlling dimensions ~ these terms can best be defined by an example.

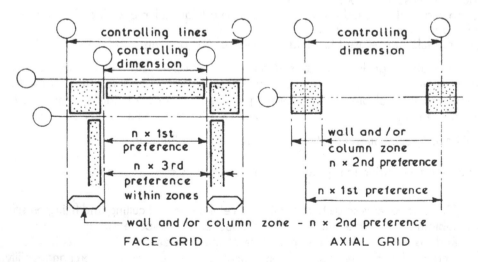

Figure 1.4 Controlling dimensions

1.3 Construction contract documents

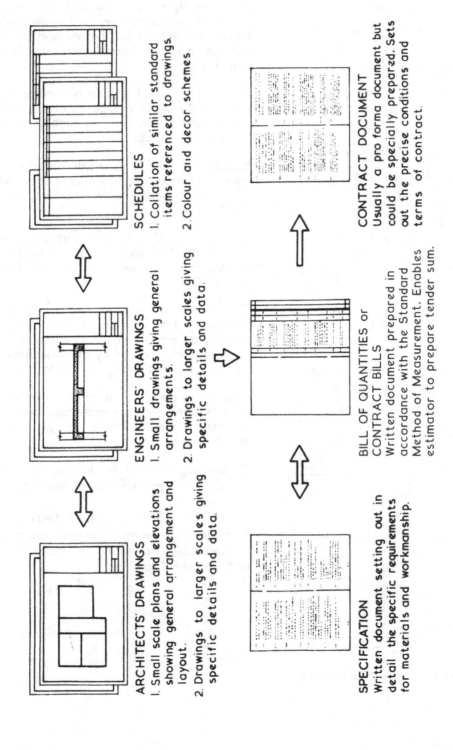

ARCHITECTS' DRAWINGS
1. Small scale plans and elevations showing general arrangement and layout.
2. Drawings to larger scales giving specific details and data.

ENGINEERS' DRAWINGS
1. Small drawings giving general arrangements.
2. Drawings to larger scales giving specific details and data.

SCHEDULES
1. Collation of similar standard items referenced to drawings.
2. Colour and decor schemes.

SPECIFICATION
Written document setting out in detail the specific requirements for materials and workmanship.

BILL OF QUANTITIES or CONTRACT BILLS
Written document prepared in accordance with the Standard Method of Measurement. Enables estimator to prepare tender sum.

CONTRACT DOCUMENT
Usually a pro forma document but could be specially prepared. Sets out the precise conditions and terms of contract.

Figure 1.5 Contract documents

Construction contract documents

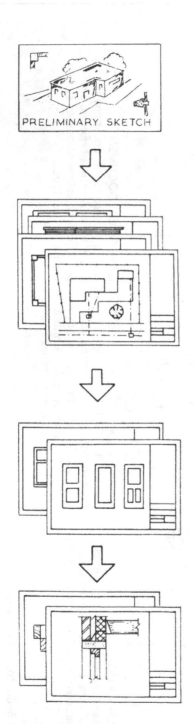

Figure 1.6 Construction drawings

Construction drawings

Site plans ~ used to locate site, buildings, define site levels, indicate services to buildings, identify parts of site such as roads, footpaths and boundaries and to give setting-out dimensions for the site and buildings as a whole. Suitable scale not less than 1:2500.

Floor plans ~ used to identify and set out parts of the building such as rooms, corridors, doors, windows, etc. Suitable scale not less than 1:100.

Elevations ~ used to show external appearance of all faces and to identify doors and windows. Suitable scale not less than 1:100.

Sections ~ used to provide vertical views through the building to show method of construction. Suitable scale not less than 1:50.

Component drawings ~ used to identify and supply data for components to be supplied by a manufacturer or for components not completely covered by assembly drawings. Suitable scale range 1:100 to 1:1.

Assembly drawings ~ used to show how items fit together or are assembled to form elements. Suitable scale range 1:20 to 1:5.

All drawings should be fully annotated, fully dimensioned and cross-referenced.

EN ISO 7519: Technical Drawings. Construction Drawings. General Principles of Presentation for General Arrangement and Assembly Drawings.

Construction drawings – orthographic projections

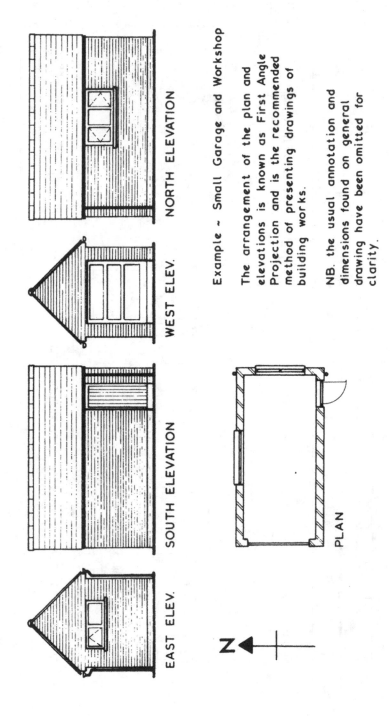

NORTH ELEVATION

WEST ELEV.

SOUTH ELEVATION

EAST ELEV.

PLAN

N

Example ~ Small Garage and Workshop

The arrangement of the plan and elevations is known as First Angle Projection and is the recommended method of presenting drawings of building works.

NB. the usual annotation and dimensions found on general drawing have been omitted for clarity.

Figure 1.7 Orthographic projection ~ a means of drawing independent views of a solid object on a plane surface

Construction drawings – floor plans and elevations

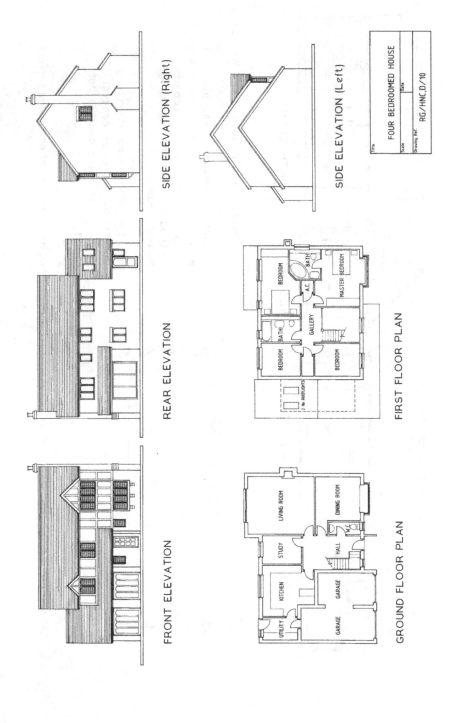

Figure 1.8 Floor plans and elevations

Construction drawings – location plans and site plans

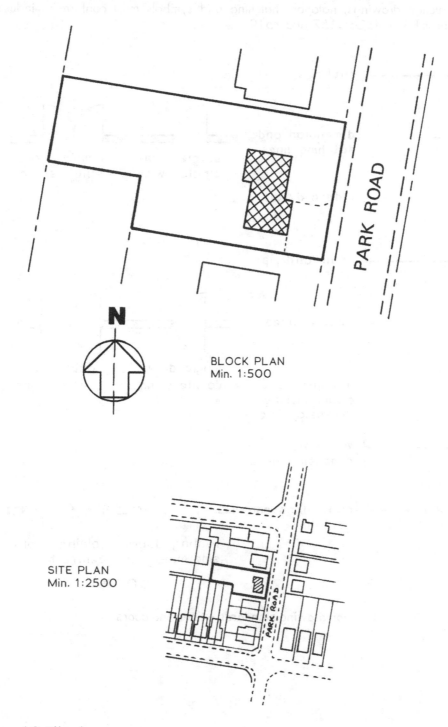

BLOCK PLAN
Min. 1:500

SITE PLAN
Min. 1:2500

Figure 1.9 Site plans

Construction contract documents

Construction drawings – standards: lines and doors

Construction drawings, notation, hatching and symbols must conform to industry standards BS EN ISOs 4157 and 7519.

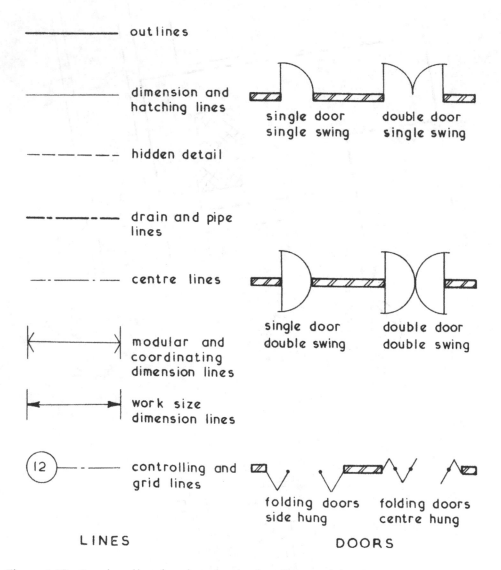

LINES   DOORS

Figure 1.10 Construction drawing standards – lines and doors

Construction drawings – standards: hatching and windows

Hatchings ~ the main objective is to differentiate between the materials being used, thus enabling rapid recognition and location. Whichever hatchings are chosen they must be used consistently throughout the whole set of drawings. In large areas it is not always necessary to hatch the whole area.

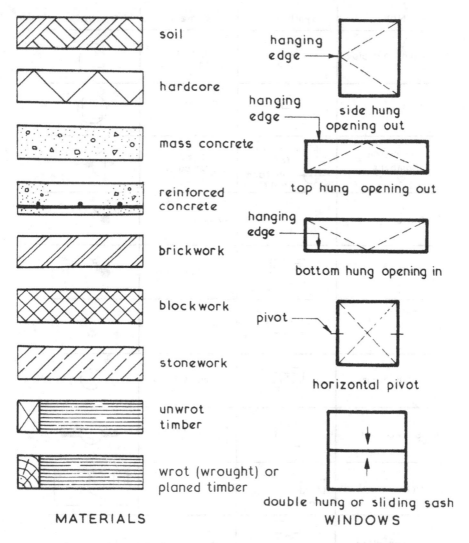

MATERIALS   WINDOWS

Figure 1.11 Construction drawing standards – hatching and windows

Construction contract documents

Construction drawings – standards: electrical components

Symbols ~ these are graphical representations and should, wherever possible, be drawn to scale. Above all they must be consistent for the whole set of drawings and clearly drawn.

Name	Symbol	Name	Symbol
Rainwater pipe	◯ R W P	Distribution board	☐
Gully	☐ G	Electricity meter	⊚
Inspection chambers	IC ⊡ soil or foul / IC ⊙ surface water	Switched socket outlet	⊳
Boiler	☐ B	Switch	●
Sink	▭ S •	Two-way switch	●
Bath	▭ •	Pendant switch	●
Wash basin	▭ W B	Filament lamp	◯
Shower unit	☐ S	Fluorescent lamp	━◯━
Urinal	stall bowl	Bed	▭
Water closet	🚽	Table and chairs	▦

Figure 1.12 Typical component, fitment and electrical symbols

Construction drawings – standards: electrical layout

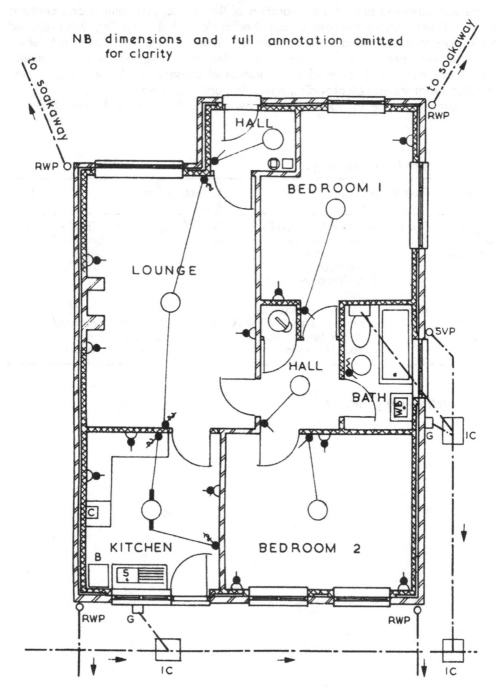

Figure 1.13 Typical plan of a two–bedroom bungalow

Construction contract documents

Method statements

A method statement precedes preparation of the project programme and contains the detail necessary for the construction of each element of a building. It is prepared from information contained in the contract documents. It also functions as a brief for site staff and operatives in sequencing activities, indicating resource requirements and determining the duration of each element of construction. It complements construction programming by providing a detailed analysis of each activity.

A typical example for foundation excavation could take the format shown in Table 1.2.

Table 1.2 Method statement examples

Activity	Quantity	Method	Output/ hour	Labour	Plant	Days
Strip site for excavation	300m^2	Exc. to reduced level over construction area – JCB-4CX face shovel/loader. Topsoil retained on site.	50m^2/ hr	Exc. driver +2 labourers	JCB-4CX backhoe/ loader	0.75
Excavate for foundations	60m^3	Excavate foundation trench to required depth – JCB-4CX backhoe. Surplus spoil removed from site.	15m^3/ hr	Exc. driver +2 labourers. Truck driver	JCB-4CX backhoe/ loader. Tipper truck	0.50

Construction programme

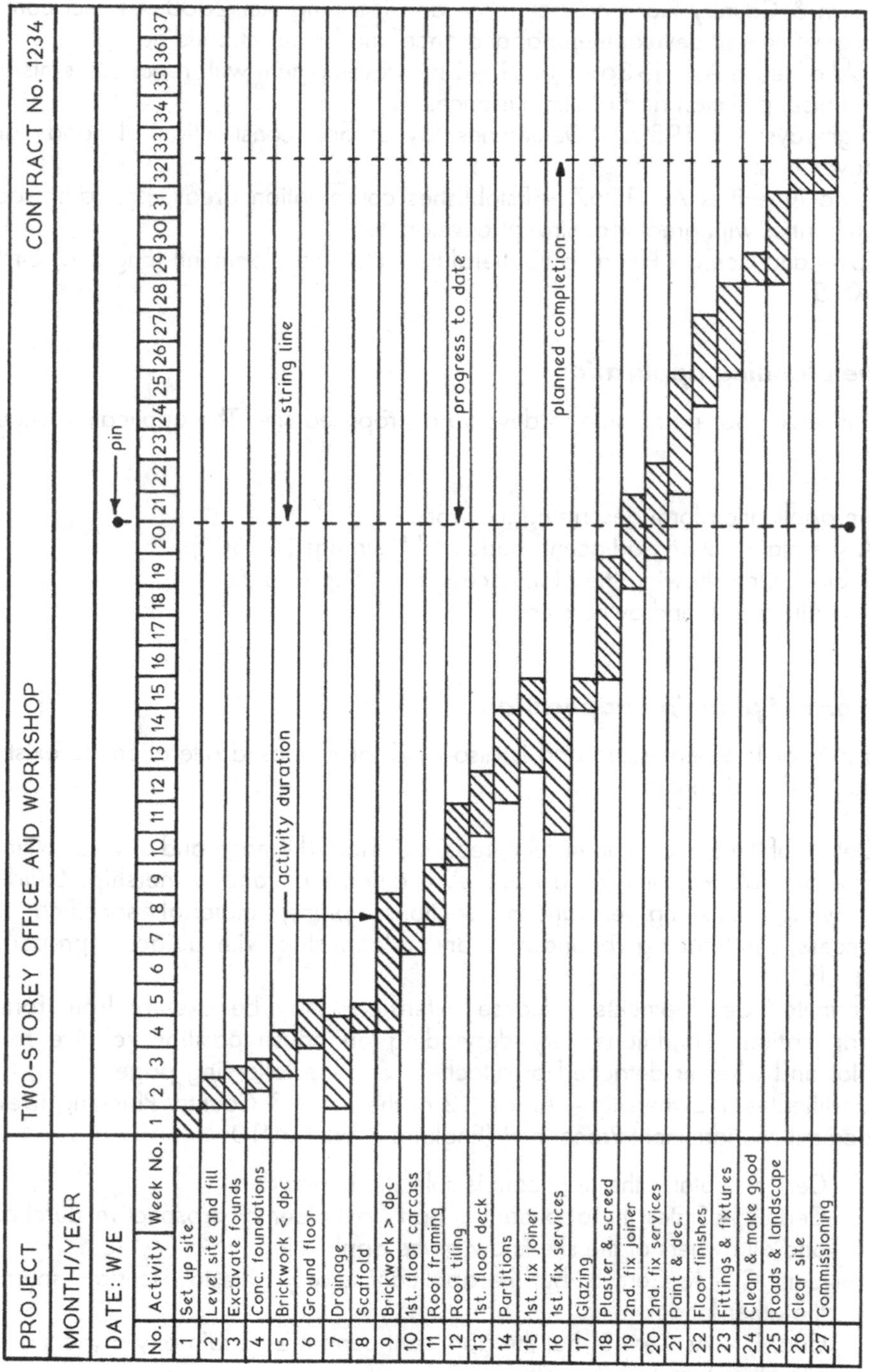
PROJECT
TWO-STOREY OFFICE AND WORKSHOP
CONTRACT No. 1234
MONTH/YEAR
DATE: W/E
No.
Activity
Week No.
1 2 3 4 5 6 7 8 9 10 11 12 13 14 15 16 17 18 19 20 21 22 23 24 25 26 27 28 29 30 31 32 33 34 35 36 37
1 Set up site
2 Level site and fill
3 Excavate founds
4 Conc. foundations
5 Brickwork < dpc
6 Ground floor
7 Drainage
8 Scaffold
9 Brickwork > dpc
10 1st. floor carcass
11 Roof framing
12 Roof tiling
13 1st. floor deck
14 Partitions
15 1st. fix joiner
16 1st. fix services
17 Glazing
18 Plaster & screed
19 2nd. fix joiner
20 2nd. fix services
21 Paint & dec.
22 Floor finishes
23 Fittings & fixtures
24 Clean & make good
25 Roads & landscape
26 Clear site
27 Commissioning
pin
string line
activity duration
progress to date
planned completion

Figure 1.14 Basic construction programme

1.4 Planning legislation

- Town & Country Planning Act 1990 and Planning Act 2008 ~ Effect control over volume of development, appearance and layout of buildings.
- Public Health Acts 1936 to 1961 ~ Limit development with regard to emission of noise, pollution and public nuisance.
- Highways Act 1980 ~ Determines layout and construction of roads and pavements.
- Civic Amenities Act 1967 ~ Establishes conservation areas, providing local authorities with greater control of development.
- Town and Country Planning (General Permitted Development) (England) Order 2015

Outline planning application

This is necessary for permission to develop a proposed site. The application should contain:

- An application form describing the work.
- A site plan showing adjacent roads and buildings (1:2500).
- A block plan showing the plot, access and siting (1:500).
- A certificate of land ownership.

Detail or full planning application

This follows outline permission and is also used for proposed alterations to existing buildings. It should contain:

- Details of the proposal, to include trees, materials, drainage and any demolition.
- Site and block plans (as above). A certificate of land ownership. Building drawings showing elevations, sections, plans, material specifications, access, landscaping, boundaries and relationship with adjacent properties (1: 100).
- Permitted developments – House extensions may be exempt from formal application. Conditions vary depending on house position relative to its plot and whether detached or attached. See the following pages.
- Certificates of ownership – Article 12 of the Town & Country Planning (Development Management Procedure) (England) Order 2010:
  - Cert. A – States the applicant is sole site freeholder.
  - Cert. B – States the applicant is part freeholder or prospective purchaser and all owners of the site know of the application.
  - Cert. C – As Cert. B, but the applicant is only able to ascertain some of the other landowners.
  - Cert. D – As Cert. B, but the applicant cannot ascertain any owners of the site other than him/herself.

Permitted development

Permitted development applies specifically to extensions and alterations to houses, but not flats. Houses in conservation areas and those listed for historical interest may be excluded. The Local Planning Authority (LPA) should be consulted for clarification on all proposals, as planning departments will interpret the 'Order' with regard to their locality. Exemption from the formal planning process does not include exemption from Building Regulation approval. Most extensions and some alterations will still require this.

Tree preservation orders (TPO)

Local planning authorities have powers under the Town and Country Planning Act and the Town and Country Planning (Tree Preservation) (England) Regulations to protect trees by making tree preservation orders (TPO). A TPO may be applied if the LPA consider that it is 'expedient in the interests of amenity to make provision for the preservation of trees or woodlands in their area' (Section 198[1] of the Town and Country Planning Act).

Before cutting down, uprooting, severing roots, topping off, lopping, damaging or destroying a tree, a formal application must be submitted to the LPA for consent. Contravention of such an order can lead to a substantial fine and a compulsion to replace any protected tree which has been removed or destroyed. Trees on building sites that are covered by a tree preservation order should be protected by a suitable fence.

Listed buildings

Buildings of special historic or architectural interest are protected by provisions in the Planning (Listed Buildings and Conservation Areas) Act. Historic England, funded by the Department for Culture, Media and Sport and from donations and commercial activities, is responsible for safeguarding and protecting the character of buildings that could otherwise be lost through demolition or unsympathetic alterations, extensions, modifications, refurbishment or inadequate maintenance.

Buildings considered to be a national asset and worthy of preservation are listed. Statutory listing applies to about half-a-million buildings. This status places responsibility on their owners to keep them in good order. These buildings are legally protected, therefore proposals for development on the site that they occupy, as well as proposals for both internal and external alterations, are subject to a listed building consent being obtained from the local planning authority.

Examples of the type of work may include the following:

* Extensions and any demolition.
* Removal of internal walls and floors.
* Changes to room layout.
* Window and door replacement.
* Painting of unpainted surfaces.

Planning legislation

- Exposed plumbing and electrical installations.
- Alterations to internal features, e.g. doors, panelling, fireplaces.
- Changes to existing materials/colour specifications.
- Removal of finishes to expose brickwork and structural timber.

The LPA should be consulted about all proposed work. It is a fineable offence to alter the character of listed buildings without the necessary consent and an order can be imposed on the building owner to rectify matters at their own expense.

Categories of listing

- Grade I: Buildings of exceptional interest.
- Grade II*: Particularly important buildings of more than special interest.
- Grade II: Buildings of special interest, warranting preservation; 90% of listed buildings are in this category.

The grading applies in England and Wales. Similar provisions exist for Northern Ireland and Scotland.

1.5 Building control

Building Act 1984 consolidated various other building legislation and enabled the making of Building Regulations, which are a statutory instrument for enforcing minimum material and design standards.

The regulations are supported by approved documents which give guidance on how to achieve the required performance standards. The relationship of these and other documents is set out below:

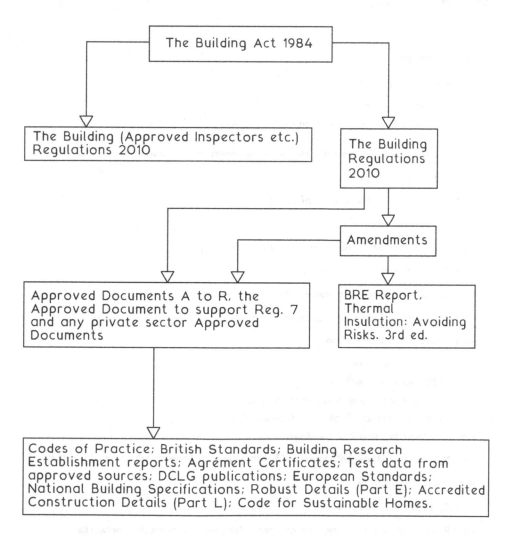

Figure 1.15 Building control legislation

Building Regulations apply throughout the UK. Specific requirements for England, Wales, Scotland and Northern Ireland are administered by their regional governments.

Building control

Approved documents

These publications support the Building Regulations. They are prepared by the Ministry of Housing, Communities and Local Government, approved by the Secretary of State and issued by the Stationery Office. The approved documents (ADs) have been compiled to give practical *guidance* to comply with the performance standards set out in the various regulations. They are not mandatory but show compliance with the requirements of the Building Regulations. If other solutions are used to satisfy the requirements of the regulations, proving compliance rests with the applicant or designer is necessary.

Part A: Structure

Part B: Fire safety

    Volume 1 Dwelling houses
    Volume 2 Buildings other than dwelling houses

Part C: Site preparation and resistance to contaminants and moisture

Part D: Toxic substances

Part E: Resistance to the passage of sound

Part F: Ventilation

Part G: Sanitation, hot water safety and water efficiency

Part H: Drainage and waste disposal

Part J: Heat producing appliances and Fuel storage system

Part K: Protection from falling, collision and impact

Part L: Conservation of fuel and power

    L1A – New dwellings
    L1B – Existing dwellings
    L2A – New buildings other than dwellings
    L2B – Existing buildings other than dwellings

Part M: Access to and use of buildings

Part N: Glazing – Safety (*withdrawn 2013 functional requirements subsumed into Part K*)

Part P: Electrical safety

Part Q: Security – Dwellings

Part R: Physical infrastructure for high-speed electronic communication networks

Regulation 7: Materials and workmanship

Figure 1.16 Approved documents

Building control options

There are two routes a builder can take to achieve Building Control Approval.

- Local authority building control.
- Approved inspector (private company).

Local authority building control

This is a public service administered by borough and unitary councils through their building control departments. The local authority has a duty to provide building control services to any client.

Approved inspectors

These are a private sector building control alternative – they can pick and choose which clients they work for. Approved inspectors may be suitably qualified individuals or corporate bodies employing suitably qualified people, e.g. National House Building Council (NHBC Ltd).

Unless the applicant has opted for control by a private approved inspector under The Building (Approved Inspectors, etc.) Regulations 2010, the control of building works in the context of the Building Regulations is vested in the local authority. There are two systems of control: namely the Building Notice and the Deposit of Plans. The sequence of systems is shown on the next page.

Building control

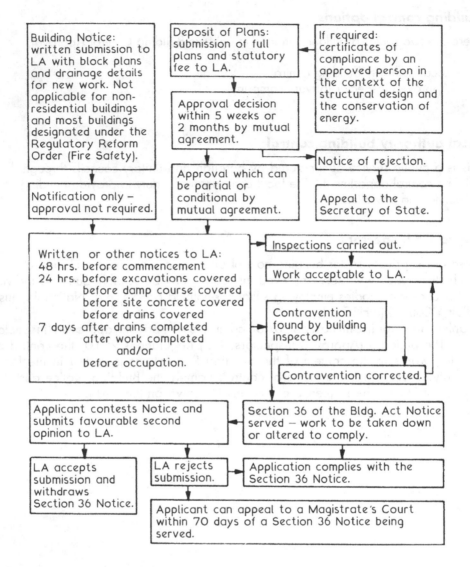

Building Notice: written submission to LA with block plans and drainage details for new work. Not applicable for non-residential buildings and most buildings designated under the Regulatory Reform Order (Fire Safety).

Deposit of Plans: submission of full plans and statutory fee to LA.

If required: certificates of compliance by an approved person in the context of the structural design and the conservation of energy.

Approval decision within 5 weeks or 2 months by mutual agreement.

Notice of rejection.

Notification only – approval not required.

Approval which can be partial or conditional by mutual agreement.

Appeal to the Secretary of State.

Written or other notices to LA:
48 hrs. before commencement
24 hrs. before excavations covered
before damp course covered
before site concrete covered
before drains covered
7 days after drains completed
after work completed
and/or
before occupation.

Inspections carried out.

Work acceptable to LA.

Contravention found by building inspector.

Contravention corrected.

Applicant contests Notice and submits favourable second opinion to LA.

Section 36 of the Bldg. Act Notice served – work to be taken down or altered to comply.

LA accepts submission and withdraws Section 36 Notice.

LA rejects submission.

Application complies with the Section 36 Notice.

Applicant can appeal to a Magistrate's Court within 70 days of a Section 36 Notice being served.

Figure 1.17 Building control processes

Note: In some stages of the sequence statutory fees are payable as set out in The Building (Local Authority Charges) Regulations 2010.

Building Regulations approval

This is required if 'Building Work' as defined in Regulation 3 of the Building Regulations is proposed. This includes:

- Construction or extension of a building.
- Alterations to an existing building that would bring into effect any of the complying regulations.

- Installing replacement windows where the installer is not known to the local Building Control Authority as being a 'competent' registered installer, e.g. FENSA (**FEN**estration **S**elf-**A**ssessment) scheme.
- Alteration or installation of building services and fittings that bring into effect any of the complying regulations.
- Installation of cavity wall insulation.
- Underpinning of a building's foundations.
- Change of purpose or use of a building.

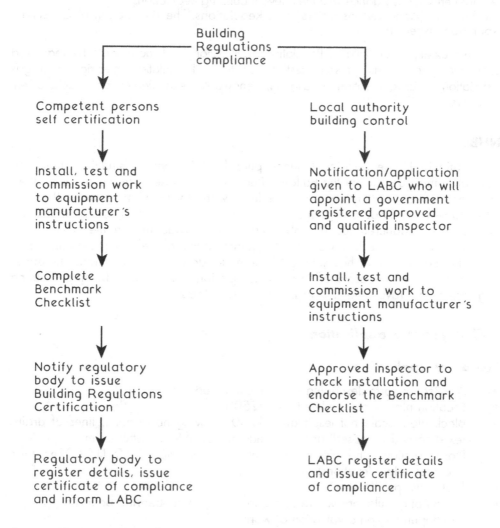

Figure 1.18 Comparison of LA and AI processes

'Competent' persons are appropriately qualified and experienced to the satisfaction of a relevant scheme organiser. For example, Capita Group's 'Gas Safe Register' of engineers for gas installation and maintenance services. They can

Building control

'self-certify' that their work complies with Building Regulations, thereby removing the need for further inspection.

The Building (Approved Inspectors, etc.) Regulations The Association of Consultant Approved Inspectors.

Benchmark checklist ~ an initiative that places responsibilities on manufacturers and installers to provide equipment of an appropriate standard for the situation. Further requirements are that it is installed, commissioned and serviced to the manufacturer's requirements in accordance with the relevant Building Regulations.
The Building (Approved Inspectors, etc.) Regulations. The Association of Consultant Approved Inspectors.

Some examples of Building Regulations notification work acceptable by registered competent persons: air pressure testing, cavity wall insulation, electrical and gas installation, micro-generation installation, renewable technologies and replacement windows.

NHBC

The National House Building Council publish their own construction rules and standards that supplement the Building Regulations. These form the basis for their own independent quality control procedures whereby their inspectors will undertake stage and periodic examinations of work in progress to ensure that these standards are adhered to. The objective is to provide new home buyers with a quality assured product warranted against structural defects (10 years), provided the house builder has satisfied certain standards for registration. Therefore, the buyer should be provided with a completion certificate indicating Building Regulations approval and a warranty against defects.

Building control application

Full plans application

- Application form describing the proposed work.
- Location plan, scale not less than 1:2500.
- Block plan, scale not less than 1:1250 showing north point, lines of drains (existing and proposed) and size and species of trees within 30m.
- Plans, sections and elevations, scale not less than 1:50 (1:100 may be acceptable for elevations).
- Materials specification.
- Structural calculations where appropriate, e.g. load-bearing beams.
- Fee depending on a valuation of work.

The appointed inspector examines the application and, subject to any necessary amendments, an approval is issued. This procedure ensures that work on site is conducted in accordance with the approved plans. Also, where the work is being financed by a loan, the lender will often insist the work is only to a Full Plans approval.

Building notice

- A simplified application form.
- Block plan as described above.
- Construction details, materials specification and structural calculations if considered necessary by the inspector.
- Fee depending on a valuation of work.

This procedure is only really appropriate for minor work, for example, extensions to existing small buildings such as houses. Building control/inspection occurs as each element of the work proceeds. Any Building Regulation contravention will have to be removed or altered to attain an acceptable standard.

Regularisation

- Application form.
- Structural calculations if relevant.
- A proportionally higher fee.

Regularisation applies to unauthorised work undertaken since November 1985. In effect, it is a retrospective application that will involve a detailed inspection of the work. Rectification may be necessary before approval is granted.

Accredited construction details

A UK Department for Communities and Local Government (DCLG) publication contains a series of construction details that are applied to five different building techniques. The fully detailed illustrations are for relatively light construction appropriate to dwellings. The details and supplementary support notes concentrate on continuity of thermal insulation with a regard for thermal (cold) bridging and on quality of construction to maintain airtightness.

Publication sections and details

Section 1: Introduction relating mainly to continuity of insulation and airtightness.
Section 2: Detailed drawings and checklists for constructing the external envelope.

The five types of construction are:

- Externally insulated masonry solid walls.
- Part and fully filled cavity insulated masonry walls.
- Internally insulated masonry walls.
- Timber framed walls.
- Lightweight steel framed walls.

Note: All five construction practice details include the critical areas of junctions and interfaces between wall and roof, ground and intermediate floors. Treatment at door and window openings is also included with specific applications where appropriate.

The guidance shown indicates the categories of buildings that do not normally require submission of a Building Notice or Deposit of Plans for approval by the Building Control Section of the Local Authority. However, they may still require planning permission – see page 16

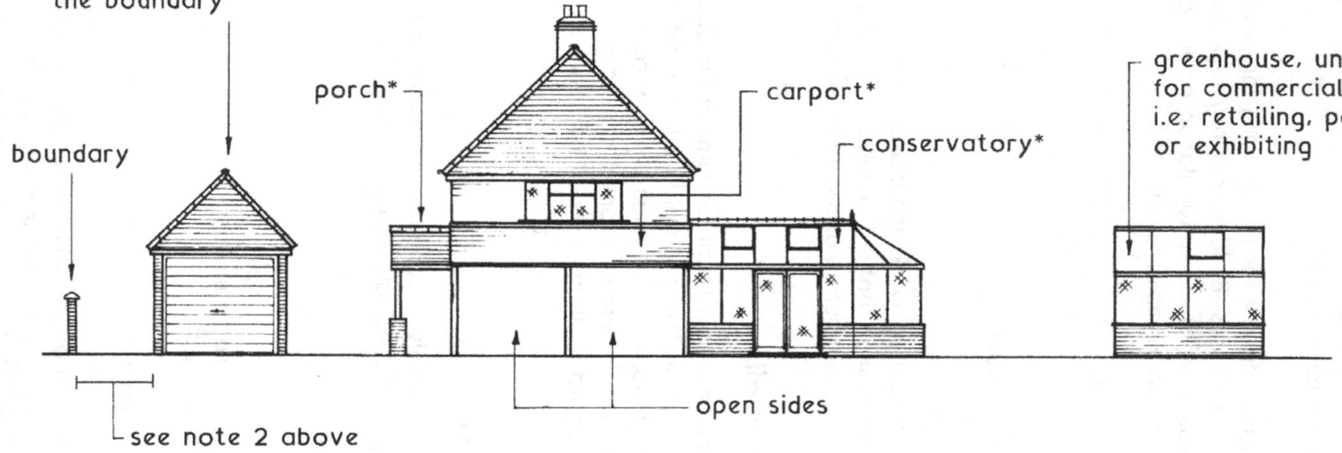

Small detached buildings:-
1. floor area < 15m² not containing sleeping accommodation, or
2. floor area < 30m² not containing sleeping accommodation, and either:
   • constructed substantially from non-combustible materials, or
   • located in excess of 1m from the boundary

boundary

porch*

carport*

conservatory*

greenhouse, unless for commercial use, i.e. retailing, packing or exhibiting

open sides

see note 2 above

* Single storey ground level additions which are not open to the house and are < 30m² floor area, to include carport, covered yard, conservatory and porch.

Notes: 1. a carport must be open on at least two sides.
2. conservatories must have fully glazed laminated or toughened glass, or translucent plastic roofs.
3. glazed doors and windows in a porch or conservatory as defined in Buildings Regulations Approved Document K – see page 564
4. single storey additions should not impede escape from an upper floor window specified for emergency egress. Building Regulations, A.D.: B1 Section 2

Figure 1.19 Single storey additions

BUILDING REGULATIONS APPLICATION

		APPLICATION No
Use this form to give notice of intention to erect, extend, or alter a building, install fittings or make a material change of use of the building.	Unless specified differently overleaf, Please return:- • 2 copies of the Form • 4 copies of the Plans • the correct fee	
		DATE RECEIVED

1. NAME AND ADDRESS OF APPLICANT
Applicant will be invoiced on commencement of work.

Post Code _____

Tel. No. _____

2. NAME AND ADDRESS OF AGENT (If Used)

Post Code _____

Tel. No. _____

3. ADDRESS OR LOCATION OF PROPOSED WORK

4. DESCRIPTION OF PROPOSED WORKS

5. IF NEW BUILDING OR EXTENSION PLEASE STATE PROPOSED USE

6. IF EXISTING BUILDING PLEASE STATE PRESENT USE

7. DRAINAGE

Please state means of:-

Water Supply _____

Foul Water Disposal _____

Storm Water Disposal _____

8. CONDITIONS

Do you consent to the Plans being passed subject to conditions where appropriate?   Yes ☐   No ☐

Do you agree to an extension of time if this is required by the Council?   Yes ☐   No ☐

9. COMPLETION CERTIFICATE

Do you wish the Council to issue a Completion Certificate upon satisfactory completion of the work?

Yes ☐   No ☐

10. REGULATORY REFORM ORDER (Fire Safety) 2005

Is the building intended for any other purpose than occupation as a domestic living unit by one family group?

Yes ☐   No ☐

11. FEE

Please state estimated cost of the work (at current market value) £.................... Amount of Fee submitted £....................

Has Planning Permission been sought?   Yes ☐   No ☐   If Yes, please give Application No _____

12. PLEASE SIGN AND DATE THIS FORM BEFORE SUBMITTING

I/We hereby give notice of intention to carry out the work set out above and deposit the attached drawings and documents in accordance with the requirements of Regulations 12 (2) (b). Also enclosed is the appropriate Plan Fee and I understand that a further Fee will be payable when the first inspection of work on site is made by the Local Authority.

Signed _____ Date _____ On behalf of (if agent) _____

Figure 1.20 Building regulations application form

Building control

The Building Regulations 2010 and associated Approved Document L: Conservation of Fuel and Power.

Accredited Construction Details, Communities and Local Government publications.

Limiting U-values for energy loss through the enclosing fabric are shown in Table 1.3.

Table 1.3 Limiting U values for external elements

Element of construction	Limiting area weighted average U-value (W/m²K)	Objective or target design values (W/m²K)
External wall	0.35	0.23
Roof	0.25	0.15
Floor	0.25	0.20
Windows	2.20	1.50
Doors	2.20	1.50
Curtain walling	2.20	1.50

Note: Building airtightness $\leq 5m^3$ per hour per m^2 at 50Pa pressure.

Energy Performance Certificates (EPC)

Applications include:

- Construction of new buildings.
- Extensions to existing buildings.
- Alterations to existing buildings to provide an increase or a reduction in the number of separate occupancies, e.g. a house conversion into flats or vice versa.
- Refurbishment or modification to include provision or extension of fixed energy consuming building services for hot water, heating, air conditioning or mechanical ventilation (applies to buildings with a floor area exceeding 1000 m^2 but can also be required for smaller buildings depending on specification of installation).
- Part of the marketing particulars when selling or letting a new or existing property.

The above applications relate quite simply to buildings that are roofed, have enclosing walls and use energy consuming appliances to condition the internal space. Some building types are exempted an EPC, these include the following:

- Buildings listed under the Planning (Listed Buildings and Conservation Areas) Act.*
- Buildings within a conservation area as determined under Section 69 of the Act.*

- Structures included in the monuments schedule under Section 1 of the Ancient Monuments and Archaeological Areas Act.*
- Churches and other buildings designated primarily for worship.
- Temporary buildings with less than two years' expected use.
- Industrial buildings, workshops and non-residential agricultural buildings/barns with low demands on energy.
- Detached buildings other than dwellings with usable floor area of less than 50m^2.

* The objective is to preserve the character and appearance of specific buildings that would otherwise be altered or spoilt by applying contemporary energy efficiency requirements.

The Energy Performance of Buildings (Certificates and Inspections) (England and Wales) Regulations.

An Energy Performance Certificate (EPC) provides a rating for fuel use efficiency in a building. This rating relates to the amount of carbon dioxide (CO_2) emitted by the energy producing appliances.

Asset rating

An estimate of the fuel energy required to meet the needs of a building during normal occupancy. A performance rating based on a building's age, location/exposure, size, glazing system, materials, insulation, general condition, fuel use controls and fixed appliance efficiency, e.g. boiler. Rating is alphabetical from A to G–A the highest grade for energy efficiency with lowest impact on environmental damage in terms of CO_2 emissions. An EU-type energy rating label is part of the certification documents. The alphabetic rating relates directly to SAP numerical ratings as shown in Table 1.4.

Table 1.4 EU type energy rating categories

A (92–100)	B (81–91)	C (69–80)	D (55–68)
E (39–54)	F (21–38)	G (1–20)	

EPC asset rating (SAP rating)

Operational rating ~ an alternative to asset rating, using the numerical scale for energy consumed over a period of time. This could be presented monthly or seasonally to indicate varying demands.

CO_2 emission rate calculations for new-build and refurbishment work ~ *Before* work commences the local building control authority (LA) to be provided with the following for approval:

- Target CO_2 emission rate by calculation (TER).
- Dwelling CO_2 emission rate by calculation and design (DER).
- Building design specification relative to calculated CO_2 emissions.

Building control

After (within five days of completion), LA to be provided with certification confirming:

- Target CO_2 emissions.
- Calculated CO_2 emissions as constructed.
- Confirmation that the building design specification is adhered to. If not, details of variations to be provided.

Note: TER and DER energy performance requirements are expressed in mass of CO_2 in units of kg per m^2 floor area per year.

EPC content:

- Address and location of the building assessed.
- Activity/function of the building, e.g. dwelling house.
- Date of construction, approximate if very old.
- Construction, e.g. solid walls, cavity walls, etc.
- Materials of construction.
- Heat energy source, system type and fuel used.
- Electrical energy source, lighting and power provision.
- Energy efficiency asset rating.
- Environmental impact rating (CO_2 emissions).
- Recommendations for improvements.
- Date of issue (valid 10 years unless significant changes occur).
- Reference/registration number.
- Assessor's name, accreditation number and scheme number.

EPC assessor/surveyor ~ an appropriately qualified energy assessment member of an accredited scheme approved by the Secretary of State for the Department for Communities and Local Government (DCLG), as defined in the Energy Performance of Buildings (Certificates and Inspections) (England and Wales) Regulations. Within five days of work completion, the assessor must provide the building owner with an EPC and the local authority is to be informed of the details of the EPC reference as entered in the register maintained under Regulation 31.

Recommendations for improvements ~ in addition to an energy assessment survey and rating, the assessor is required to provide a report identifying areas that could improve the energy performance of a building. Examples may include cost-effective recommendations for cavity wall insulation, increased insulation in the roof space, provision of a central-heating room temperature control thermostat, double/secondary glazing, etc. These recommendations to include a cost analysis of capital expenditure relative to potential savings over time and enhanced asset rating that the building could attain.

Building Regulation 29 – Declaration of giving an EPC
Building Regulations Approved Document L.
Standard Assessment Procedure.

Energy Performance of Buildings Directive – 2010/31/EU.
Recast of the Energy Performance of Buildings Directive: Impact Assessment –

DCLG 1051. Typical EPC content is shown in Table 1.5.

Table 1.5   EPC building details

Dwelling type............	Assessment date.........
Certificate date..........	Ref. No....................
Total floor area	Type of assessment

This certificate can be used to:

- Compare the energy efficiency with other properties/dwellings.
- Determine the potential economies of energy saving installations.

Estimated energy costs over three years: £3,750
Energy saving potential over three years: £1,330

Estimated energy costs of this dwelling over three years are shown in Table 1.6.
　Figures do not include costs of running subsidiary appliances such as TV, fridge, cooker, etc.

Table 1.6   EPC energy assessment

	Current cost	Potential cost	Potential savings
Lighting	£380	£190	£190
Heating	£2,820	£1,980	£840
Hot water	£550	£250	£300
Totals	£3,750	£2,420	£1,330

Energy efficiency rating

Figure 1.21 shows current efficiency rating. The higher the rating, the lower the fuel costs. Potential rating includes recommendations indicated within the figure. Average rating for a dwelling in England and Wales is band D (60). Recommended measures are shown in Table 1.7.

Building control

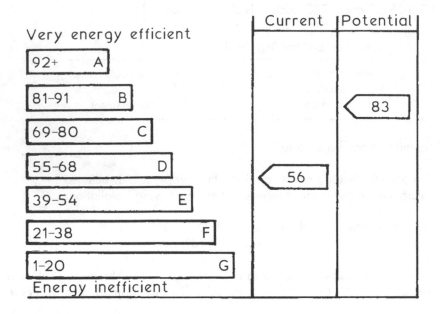

Figure 1.21 Energy efficiency band rating

Table 1.7 EPC recommended energy improvements

	Indicative capital cost	Typical savings over 3 years
Increase loft insulation	£150–£350	£125
Solar photovoltaic panels	£9,000–£14,000	£700
Low energy lighting throughout	£120	£150

British Standards (BS)

British Standards ~ these are publications issued by the British Standards Institution which give recommended minimum standards for materials, components, design and construction practices. These recommendations are not legally enforceable but some of the Building Regulations refer directly to specific British Standards and accept them as deemed to satisfy provisions. All materials and components complying with a particular British Standard are marked with the British Standards kitemark (shown in Figure 1.22) together with the appropriate BS number.

Figure 1.22   British Standards kitemark

This symbol assures the user that the product so marked has been produced and tested in accordance with the recommendations set out in that specific standard.

There are over 1,500 British Standards which are directly related to the construction industry:

1. British Standards – these give recommendations for the minimum standard of quality and testing for materials and components. Each standard number is prefixed BS.
2. Codes of Practice – these give recommendations for good practice relative to design, manufacture, construction, installation and maintenance with the main objectives of safety, quality, economy and fitness for the intended purpose. Each code of practice number is prefixed CP or BS.

European Standards (EN)

European Standards are prepared under the auspices of Comité Européen de Normalisation (CEN), of which the BSI is a member. European Standards that the BSI have not recognised or adopted are prefixed EN. These are Euro norms and will need revision for national acceptance.

For the time being, British Standards will continue and, where similarity with other countries' standards and ENs can be identified, they will run side by side until harmonisation is complete and approved by CEN.

Some products which satisfy the European requirements for safety, durability and energy efficiency carry the CE mark. This is not to be assumed as a mark of performance and is not intended to show equivalence to the BS kitemark. However, the BSI is recognised as a notified body by the EU and as such is authorised to provide testing and certification in support of the CE mark.

International Standards (ISO)

These are prepared by the International Organisation for Standardisation and are prefixed ISO. Many are compatible with, complement and have superseded BSs, e.g. ISO 9001 Quality Management Systems and BS 5750: Quality Systems.

Building control

Construction Products Regulation (CPR)

For manufacturers' products to be compatible and uniformly acceptable in the European market, there exists a process for harmonising technical specifications. These specifications are known as harmonised European product standards (hENs), produced and administered by the Comité Européen de Normalisation (CEN). European Technical Approvals (ETAs) are also acceptable where issued by the European Organisation for Technical Approvals (EOTA).

CPR harmonises the following compliance requirements:

- Energy economy and heat retention.
- Hygiene, health and environment.
- Mechanical resistance and stability.
- Protection against noise.
- Safety and accessibility in use.
- Safety in case of fire.
- Sustainable use of natural resources.

Member states of the European Economic Area (EEA) are legally required to ensure their construction products satisfy the above basic criteria.

UK attestation accredited bodies include: BBA, BRE and BSI.

CE mark ~ a marking or labelling for conforming products. A 'passport' permitting a product to be legally marketed in any EEA. It is not a quality mark, e.g. the BS kitemark, but where appropriate this may appear with the CE marking.

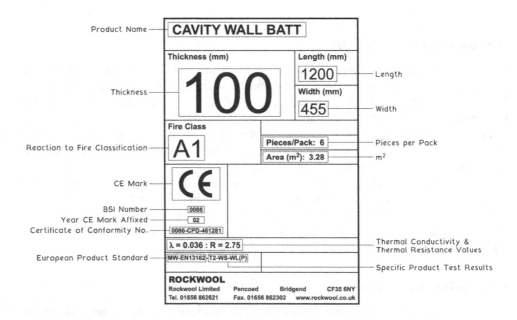

Figure 1.23 CE marking – reproduced with kind permission of Rockwool Ltd

1.6 Health and Safety

These are Statutory Instruments made under the Factories Acts of 1937 and 1961. They are now largely superseded by the Health and Safety at Work, etc. Act 1974, but still have relevance to aspects of hazardous work on construction sites. The requirements contained within these documents must therefore be taken into account when planning construction operations and during the actual construction period. Reference should be made to the relevant document for specific requirements but the broad areas covered can be shown.

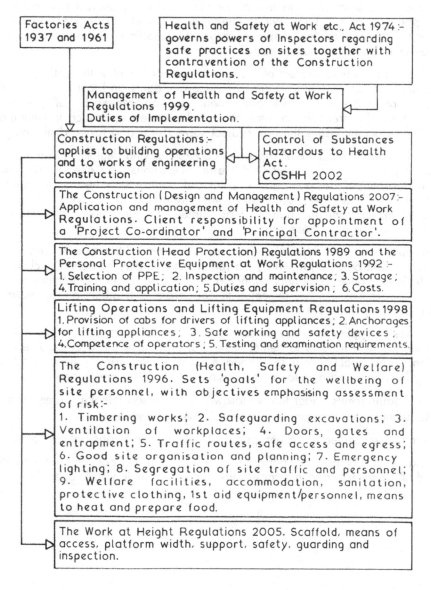

Figure 1.24 Health and safety regulations

Health and Safety

Construction (Design and Management Regulations 2015)

The Construction Design and Management Regulations were introduced in 1994 following European Construction Sites Directive, reference 92/57/EEC. Since their introduction the regulations have been nominally amended several times, with new and updated editions issued in 2007 and 2015. As regulations they are secondary legislation or a Statutory Instrument effected under an Act of Parliament. In this instance the primary legislation or Act is the Health and Safety at Work etc. 1974.

Administering body ~ construction division of the Health and Safety Executive (HSE).

Notification ~ project client to inform the HSE using a standard F10 notification form. Applies to all building, civil engineering and works of engineering construction involving more than 500 employee days or exceeding 30 days with more than 20 operatives working simultaneously.

Objective ~ to improve project planning and management with particular regard for standards of site safety in order to reduce accidents and lost production time. This is to be achieved by risk assessment and awareness by all persons on a building site by creating an integrated and planned approach to anticipation of health and safety issues through project supervision and forward planning.

Scope ~ intended to embrace all aspects of construction and demolition with the exception of very minor works, typically domestic alterations. The client is required to appoint a principal designer and a principal contractor where more than one contractor is on site. Contractor being interpreted as trade. Very few projects will be excepted from nomination of appointees but on smaller projects it may be possible to combine these roles.

Responsibilities ~ CDM Regulations apportion responsibilities to everyone involved in a construction project. Known by the regulations as duty holders, i.e. client, principal designer, other designers, principal contractor, other contractors and any other site operatives. All are expected to cooperate with others and conform with health and safety directives.

Capabilities ~ applies to every person involved on a construction project. It is an obligation for duty holders to demonstrate that they have the necessary resources, experience and qualifications to undertake the work. In terms of health and safety this may be expressed as information, instruction, training and supervision.

Note: Requirements under the withdrawn Construction (Health, Safety and Welfare) Regulations are incorporated into the CDM Regulations.

Project management team ~ client, principal designer and principal contractor.

Duty holders ~ client, principal designer, designers, principal contractor, trade contractors and all others engaged in a construction project, including trade operatives.

Client role:

- Notify the project to the HSE.
- Check competence of proposed principal designer, other designers, principal contractor and other contractor appointments.
- Appoint principal designer and principal contractor.
- Ensure that adequate welfare facilities are provided.
- Provide sufficient time and resources for all stages of design and construction.
- Ensure that sufficient pre-construction information is available to the principal contractor for production of a construction phase plan.
- Ensure that a health and safety information file and plan is available.

Principal designer role:

- Identify potential hazards and apply effective means for eliminating them.
- Determine that client and principal contractor are fully briefed of their responsibilities.
- Facilitate effective communications procedures.
- Provide information appropriate for the compilation of a health and safety file. Maintain and coordinate use of the file with other administrative duty holders.
- Ensure compliance with workplace and facilities regulations.
- Consult, advise, support and liaise with the client and principal contractor in the execution of their duties.
- Determine competence of client's nominated appointments and advise accordingly.
- Contribute to the pre-construction information.
- Plan, manage and monitor the pre-construction phase.
- Liaise with principal contractor and client on work progression and ongoing design information.

Principal contractor role:

- Develop, manage and administer the construction phase plan.
- Manage health and safety procedures on site with definition of employee duties.
- Identify hazards on site, assess risks and effect safety measures.
- Convey relevant parts of the construction phase plan to other contracted parties.
- Determine competence, capabilities and qualifications of specialist contractors, e.g. scaffolders registration.
- Ensure that site operatives/sub-contractors are suitably briefed and made aware of specific site procedures and client directives.
- Direct site operations by liaison with the various trade line managers. Plan and coordinate activities accordingly.
- Assess the need for site security and ensure that adequate and suitable provisions are in place.

Health and Safety

Sub-contractors role:

- Assess, plan and manage their work without risk to their own health and safety and that of others.
- Convey to all employees and other appointees details of site health and safety procedures.
- Ensure adequate provision of welfare facilities, including personal protective equipment.
- Check client's appointment of principal designer and principal contractor and that HSE are notified.
- Cooperate with principal contractor with regard to planning, effecting and managing work.
- Provide information required for health and safety file.
- Assess risks and communicate findings to the principal contractor to eliminate potential hazards.
- Advise principal contractor on foreseeable problems that may affect progressing the construction phase plan.
- Document any accidents, hazards and risk appraisals. Provide copy for the principal contractor to file.

Note: The preceding listings are not a one-fits-all-projects, they are intended as an outline for guidance only. Every project is an individual workplace and will require a health and safety assessment before and during the construction phase. Administrative policy should be applied accordingly.

Credibility and qualification

All parties to a building contract must verify their own credibility; demonstrate capability and competence before undertaking the work. Evidence of membership of a recognised trade association, professional body and possession of certificated up-to-date instruction and training qualifications are an expectation. Cooperation with others is essential, particularly with communications relating to issues of health and safety, not least in identifying and ensuring awareness of potential hazards and risks. Negligence of responsibilities is unacceptable and in contravention of the CDM Regulations.

Construction phase plan

A requirement for all projects with the exception of very minor works. Construction phase planning is usually coordinated by the principal contractor. Its purpose is for planning and managing health and safety procedures, site rules and any special measures that will become apparent during the construction phase of a building project. The phase plan must be in place prior to site work commencing. It is more than a work in progress chart: it should identify personnel on site, their appointment and responsibilities, any anticipated as well as possible hazards and risks, in addition to the management procedures with which the project is controlled and progressed.

Inevitably there will be some overlap of responsibilities. Where this occurs it should be perceived as a useful double check and back up. Liaison between client, principal designer and principal contractor is an essential requirement and can lead to simplification by agreeing apportioning and delegation of duties. It is for individual duty holders to recognise this and to make certain that effective measures are in place to fulfil their obligations.

Under these regulations, employers are required to provide and maintain health and safety signs conforming to European Directive 92/58 EEC:

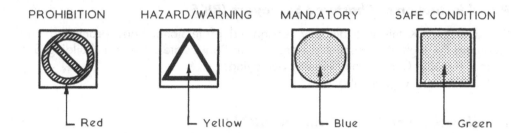

PROHIBITION	HAZARD/WARNING	MANDATORY	SAFE CONDITION
Red	Yellow	Blue	Green

Figure 1.25 Hazard symbols

In addition, employers' obligations include conducting a risk assessment, providing and maintaining safety signs where there is a risk to health and safety, e.g. obstacles, and training staff to comprehend safety signs.

BS ISO 3864–1: *Graphical Symbols. Safety Colours and Safety Signs. Design Principles for Safety Signs and Safety Markings.*

BS EN ISO 7010: *Graphical symbols. Safety Colours and Safety Signs. Registered Safety Signs.*

1.7 Building organisations

Chartered Institute of Building (CIOB)

This is the professional body for construction management and leadership, its Royal Charter is to promote science and practice of building for the benefit of society.
Ref: www.ciob.org/about

Royal Institution of Chartered Surveyors (RICS)

Based in the UK this is a globally recognised professional body promoting and enforcing the highest international standards in Real Estate surveying, Building Surveying, Building Control and Quantity Surveying.
Ref: www.rics.org/eu/about-rics/

Royal Institute of British Architects (RIBA)

The RIBA is a global professional membership body driving excellence in architecture. It aims to deliver better buildings and places, stronger communities and a sustainable environment, and is inclusive, ethical, environmentally aware and collaborative.
Ref: www.architecture.com/about

Chartered Association of Building Engineers (CABE)

CABE was founded to promote and advance knowledge, study and practice of all the arts and sciences concerned with building technology, planning, design, construction maintenance and repair of the built environment.
www.cbuilde.com/the-cabe/

Chartered Institution of Building Services Engineers (CIBSE)

This professional body exists to support science, art and practice of Building Services Engineering. They provide internationally recognized codes of practice and guidance in: refrigeration, air conditioning, space heating plumbing, drainage and electrical and mechanical engineering.
www.cibse.org/about-cibse/what-we-do

Building Research Establishment (BRE)

The BRE was founded as a UK government agency in 1921. It was known initially as the Building Research Department and thereafter, until the early 1970s, as the Building Research Station.

The BRE incorporates and works with other specialised research and material testing organisations. It is accredited under the United Kingdom Accreditation Service (UKAS) as a testing laboratory authorised to issue approvals and certifications such as CE product marking. Certification of products, materials and applications is effected through BRE Certification Ltd.

British Board of Agrément (BBA)

This is an approvals authority established in 1966, known then as the Agrément Board. It was based on the French Government's system of product approval. In 1982 it was renamed. Accredited by UKAS and a UK representative in EOTA.

The BBA's UK premises are at the BRE in Garston, near Watford, a convenient location for access to the BRE's research and testing facilities. It is an independent organisation with the purpose of impartially assessing materials, systems, practices, new market products and existing products being used in a new or innovative way. The objective is to evaluate these where an existing British Standard, Eurostandard or similar quality benchmark does not exist. Agrément certification is a quality assurance standard for products and innovations not covered by a CE mark and/or a BS kitemark. Once established, an agrément certificate may be used to attain CE marking or for development into a new BS.

The Construction Industry Research and Information Association (CIRIA)

Main areas of activity:

- Research – industry wide covering construction and the environment.
- Publications – a catalogue of over 600 titles.
- Training – short courses, seminars and conferences.
- Networks – support processes at training events, seminars and workshops.
- Information services – newsletters, bulletins and a biannual magazine.
- National Building Specifications

    o NBS (National Building Specification) ~ The UK national standard specification for building work, part of RIBA Enterprises Ltd.
    o NES (National Engineering Specification) ~ This is for building services projects. Standard, concise specifications endorsed by CIBSE, Chartered Institution of Building Services Engineers and BSRIA, Building Services Research and Information Association.

- RIBA Plan of Work – established by the Royal Institute of British Architects as a methodical procedure for the design team to apply to building projects. Its 'work stages' and 'key tasks' have become universally accepted by the other building professions as a planning model, as guidance for progressing projects and for processing contract procurement. Subsequent revisions, notably in 1998, 2007 and 2011, have retained its relevance.

1.8 Building Information Modelling (BIM)

In March 2016 the UK Government published its Construction Strategy 2016–2020, which is part of a suite of documents seeking to improve the delivery, efficiency and performance of construction and infrastructure projects in the public, private and regulated sectors. A key element of the strategy is developing the digital capability in design and construction, and with a target of procuring all government assets using Building Information Modelling (BIM) Level 2 from April 2016 (Table 1.8).

Building Information Modelling (BIM) is the framework of managing building information/data from initial design, construction phase, handover and maintenance through to de-commission/demolition.

Table 1.8   BIM levels of information management

Level 0	Unmanaged 2D CAD drawings
Level 1	Managed CAD design in 2D or 3D, electronic sharing of data from a common Data Environment (CDE)
Level 2	Managed 3D CAD model environment with collaborative working, data in IFC or COBie formats

Standards

A range of standards have been developed by BSI for the implementation of BIM Level 2 for the construction industry.

- PAS 1192-3:2014
- BS 1192-4:2014
- PAS 1192-5:2015
- BS 8536-1:2015
- BS 8536-2:2016
- ISO 19650-1:2018
- ISO 19650-2:2018

Common Data Environment (CDE)

Cloud-based virtual environment for collaborative working of digital information, avoiding duplication and retaining ownership.

BIM Execution Plan (BEP)

This plan identifies the responsible parties and defines how the project's information management will be carried out by the delivery team.

Construction Operation Building Information Exchange (COBIE)

This is a data spreadsheet containing all the digital information about maintainable assets of the building in a pre-defined structure that is used to both store and index information transferred within the CDE.

1.9 Building information classification systems

CI/SFB system of coding

This is a coded filing system for the classification and storage of building information and data. The system consists of five sections, called tables, that are subdivided by a series of letters or numbers which are listed in the CI/SfB index book, and to which reference should always be made in the first instance to enable an item to be correctly filed or retrieved.

Table 0 – Physical environment
Table 1 – Elements
Table 2 – Construction form
Table 3 – Materials
Table 4 – Activities and requirements

Other classification systems include:

- CPI System of Coding
- Uniclass
- EPIC

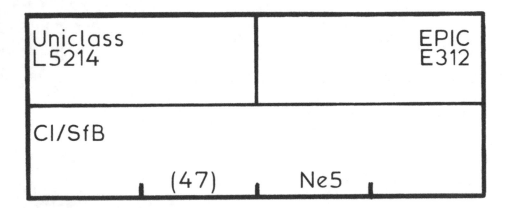

Figure 1.26 Product literature label showing CI/SfB, Uniclass and EPIC notation

2 SITE INVESTIGATION

OPENING REMARKS
DESK STUDY
FIELD STUDY
SITE SURVEY AND GROUND INVESTIGATION
SOIL ANALYSIS
SITE SOIL TESTING TECHNIQUES

Opening remarks

For any new build development it is essential to conduct a thorough site investigation to collect and record all the necessary data that will be needed, or will help in the design and construction processes of the proposed work, including anything on adjacent sites that may affect the proposed works or, conversely, anything appertaining to the proposed works that may affect an adjacent site and so should also be recorded.
Procedures ~
1. Desk study.
2. Field study or walk-over survey.
3. Site survey.
4. Ground Investigation.

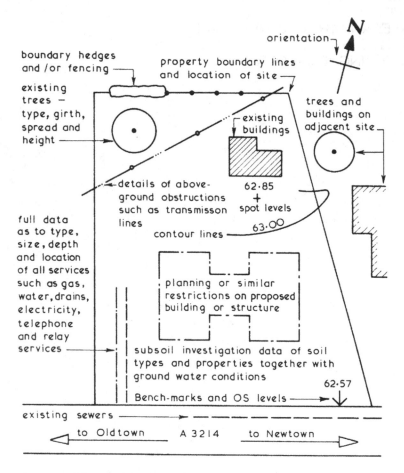

Figure 2.1 Proposed development details

2.1 Desk study

A desk study and collection of known data about the site including:

- Ordnance Survey maps – historical and modern, note grid reference.
- Rights of way.
- Geological maps – subsoil types, radon risk.
- Site history – green-field/brown-field.
- Previous planning applications/approvals of the site.
- Current planning applications in the area.
- Development restrictions – conservation orders.
- Utilities – location of services on and near the site and Wayleaves.
- Aerial photographs.
- Ecology factors – protected wildlife.
- Proximity of local land-fill sites – methane risk.
- Flood risk and history.
- Underground mining and local subsidence history.
- Any restrictions made with the deeds of the land.
- Archaeological findings.

2.2 Field study

Also known as a walk-over survey, this is a visual check of the site to corroborate the findings or deviation from the desk study and to assess the current physical conditions on site, including:

- Checking the site boundaries match the deeds.
- Looking for potential hazards to health and safety including tipping of hazardous waste.
- Assessing the general condition of any existing structures.
- Assessing the general surface conditions including ground slope/contours.
- Looking for invasive vegetation species such as Japanese Knot Weed.
- Confirming Tree Preservation Orders.
- Assessing adjacent structures and property that might be adversely affected by any development works.

2.3 Site survey and ground investigation

A site survey is a digital measured survey of the land, including ground levels. This is usually necessary at this stage to establish the location of ground investigations.

Ground investigation

The purpose of this is to identify the types, properties and depths of subsoil strata on the site and the level of the water table to enable the design of economic foundations. Generally, a series of samples extracted at the intersection points of a 10–20m^2 grid pattern should be adequate for most cases. Alternatively, representative samples are taken from locations close to, but not interfering with, the proposed works.

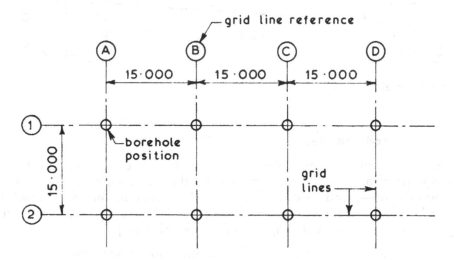

Figure 2.2 Location plan of boreholes

Two principal methods are used to obtain soil samples, which are then tested on site or in a laboratory: trial pits and boreholes. The test results of soil samples are usually shown on a drawing which gives the location of each sample and the test results in the form of a hatched legend or section.

Disturbed soil samples

These are soil samples obtained from boreholes and trial pits. The method of extraction disturbs the natural structure of the subsoil but such samples are suitable for visual grading, establishing the moisture content and some laboratory tests. Disturbed soil samples should be stored in labelled airtight jars.

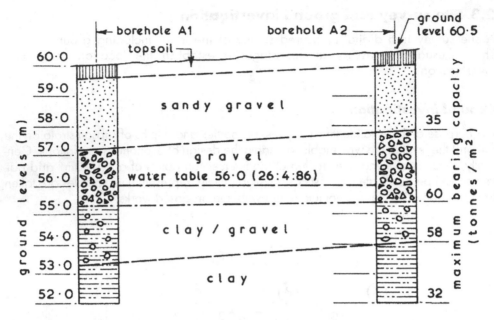

Figure 2.3 Borehole data

Undisturbed soil samples

These are soil samples obtained using coring tools that preserve the natural structure and properties of the subsoil. The extracted undisturbed soil samples are labelled and laid in wooden boxes for dispatch to a laboratory for testing. This method of obtaining soil samples is suitable for rock and clay subsoils, but difficulties can be experienced in trying to obtain undisturbed soil samples in other types of subsoil.

BS EN 1997–2: *Geotechnical design. Ground investigation and testing.*

Soil investigation methods

Method chosen will depend on several factors:

1. Size of contract.
2. Type of proposed foundation.
3. Type of sample required.
4. Types of subsoils which may be encountered.

As a general guide the most suitable methods in terms of investigation depth are:

1. Foundations up to 3.000 deep – trial pits.
2. Foundations up to 30.000 deep – borings.
3. Foundations over 30.000 deep – deep borings and in-situ examinations from tunnels and/or deep pits.

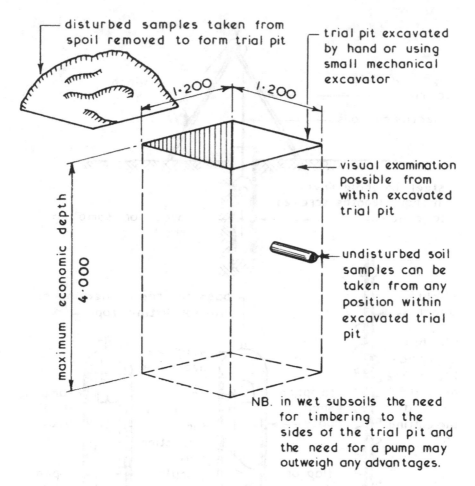

disturbed samples taken from spoil removed to form trial pit

trial pit excavated by hand or using small mechanical excavator

1·200   1·200

maximum economic depth

4·000

visual examination possible from within excavated trial pit

undisturbed soil samples can be taken from any position within excavated trial pit

NB. in wet subsoils the need for timbering to the sides of the trial pit and the need for a pump may outweigh any advantages.

Figure 2.4 Typical trial pit details

Trial pits

These are excavations up to 3.0m deep made by machine in dry ground, which require little or no temporary support to the sides of the excavation. These can also be used to expose and/or locate underground services. Subsoil can be visually examined in situ – both disturbed and undisturbed samples can be obtained. Trial pits are generally used for low rise buildings where shallow foundations are deemed adequate.

Boreholes

A cheaper and simpler method of obtaining subsoil samples than the trial pit method.

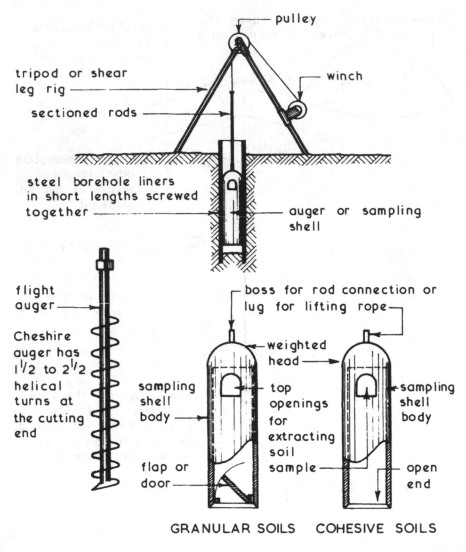

Figure 2.5 Typical bored sample details

Boring methods to obtain disturbed soil samples:

1. Hand or mechanical auger – suitable for depths up to 3.0m using a 150 or 200mm diameter flight auger.
2. Mechanical auger – suitable for depths over 3.0m using a flight or Cheshire auger; a liner or casing is required for most granular soils and may be required for other types of subsoil.

3. Sampling shells – suitable for shallow to medium depth borings in all subsoils except rock.

Borehole data ~ the information obtained from trial pits or boreholes can be recorded on a pro forma sheet or on a drawing showing the position and data from each trial pit or borehole. Boreholes can be taken on a 15.000 to 20.000 grid covering the whole site or in isolated positions relevant to the proposed foundation(s).

2.4 Soil analysis

Soils can be classified in many ways, such as geological origin, physical properties, chemical composition and particle size. It has been found that the particle size and physical properties of a soil are closely linked and are therefore of particular importance and interest to a designer.

Particle size distribution

This is the percentages of the various particle sizes present in a soil sample as determined by sieving or sedimentation. BS 1377 divides particle sizes into groups as follows:

Gravel particles – over 2mm
Sand particles – between 2mm and 0.06mm
Silt particles – between 0.06mm and 0.002mm
Clay particles – less than 0.002mm

The sand and silt classifications can be further divided, as in Table 2.1.

Table 2.1 Soil particle sizes

CLAY	SILT			SAND			GRAVEL
	fine	medium	coarse	fine	medium	coarse	
0.002	0.006	0.02	0.06	0.2	0–6	2	>2

The results of a sieve analysis can be plotted as a grading curve, as in Figure 2.6.

Soil classification

A general classification of soils composed predominantly from clay, sand and silt can be made using a soil triangle chart. Each side of the triangle represents a percentage of a material component. Following laboratory analysis, a sample's properties can be graphically plotted on the chart and classed

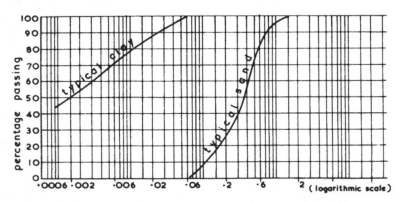

Figure 2.6 Soil classification

accordingly. E.g. sand – 70%, clay – 10% and silt – 20% = sandy loam. Note: Silt is very fine particles of sand, easily suspended in water. Loam is very fine particles of clay, easily dissolved in water.

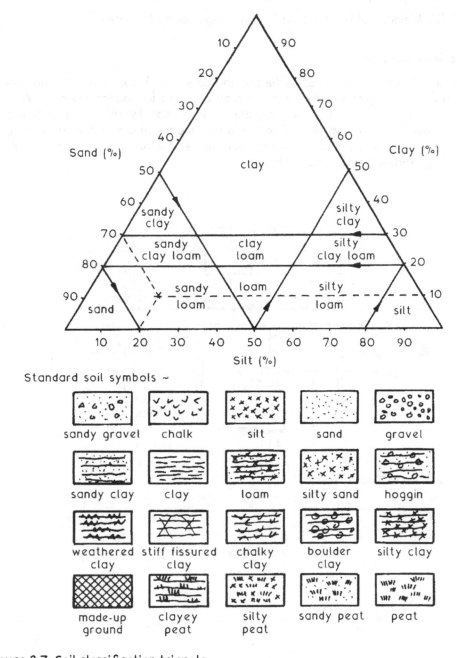

Figure 2.7 Soil classification triangle

2.5 Site soil testing techniques

These tests are designed to evaluate the bearing capacity, density or shear strength of soils and are very valuable since they do not disturb the soil under test.

BS 1377: *Methods of Test for Soils for Civil Engineering Purposes.*

Plate loading test

This is a rudimentary test to assess the bearing capacity of unknown soil conditions.

A trial hole is excavated to the anticipated depth of foundation bearing. Within the excavation is placed a steel test plate of proportionally smaller area than the finished foundation. This plate is then loaded by means of a jack and a Kentledge weight to exert pressure into the ground, and the settlement is monitored to establish a building load/soil settlement relationship.

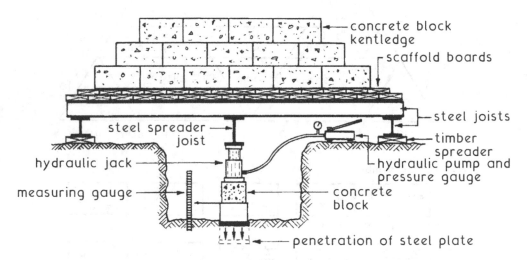

Figure 2.8 Plate loading test

Plate loading test results

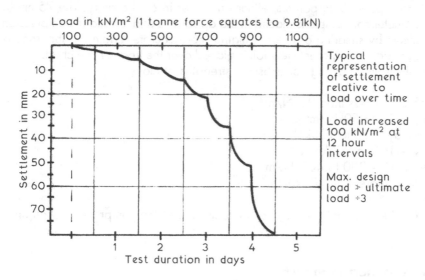

Load in kN/m² (1 tonne force equates to 9.81kN)

Typical representation of settlement relative to load over time

Load increased 100 kN/m² at 12 hour intervals

Max. design load ≯ ultimate load ÷3

Figure 2.9 Plate test results

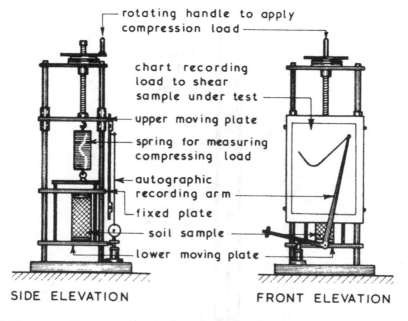

rotating handle to apply compression load

chart recording load to shear sample under test

upper moving plate

spring for measuring compressing load

autographic recording arm

fixed plate

soil sample

lower moving plate

SIDE ELEVATION   FRONT ELEVATION

Figure 2.10 Unconfined compression test equipment

Site soil testing techniques

Unconfined compression test (Fig. 2.10)

This test can be used to establish the shear strength of a non-fissured cohesive soil sample using portable apparatus, either on site or in a laboratory. The 75mm long × 38mm diameter soil sample is placed in the apparatus and loaded in compression until failure occurs by shearing or lateral bulging. For accurate reading of the trace on the recording chart a transparent view foil is placed over the trace on the chart.

Typical results, showing compression strengths of clays:

Very soft clay – less than 25kN/m^2
Soft clay – 25 to 50kN/m^2
Medium clay – 50 to 100kN/m^2
Stiff clay – 100 to 200kN/m^2
Very stiff clay – 200 to 400kN/m^2
Hard clay – more than 400kN/m^2

Note: The shear strength of clay soils is only half of the compression strength values given above.

Standard penetration test

This test measures the resistance of a soil to the penetration of a split spoon or split barrel sampler driven into the bottom of a borehole. The sampler is driven into the

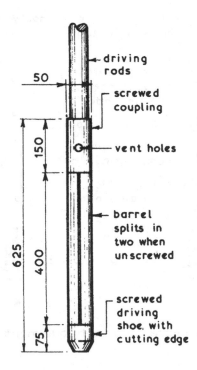

Figure 2.11 Typical split barrel sampler

soil to a depth of 150mm by a falling standard weight of 65kg falling through a distance of 760mm. The sampler is then driven into the soil a further 300mm and the number of blows counted up to a maximum of 50 blows. This test establishes the relative density of the soil.

Typical results for non-cohesive soils are shown in Table 2.2.

Table 2.2 Penetration test classifications for non-cohesive soils

No. of blows	Relative density
0 to 4	very loose
4 to 10	loose
10 to 30	medium
30 to 50	dense
50+	very dense

Typical results for cohesive soils are shown in Table 2.3.

Table 2.3 Penetration test classifications for cohesive soils

No. of blows	Relative density
0 to 2	very soft
2 to 4	soft
4 to 8	medium
8 to 15	stiff
15 to 30	very stiff
30+	hard

The results of this test in terms of the number of blows and amount of penetration will need expert interpretation.

Vane test

This test measures the shear strength of soft cohesive soils. The steel vane is pushed into the soft clay soil and rotated by hand at a constant rate. The amount of torque necessary for rotation is measured and the soil shear strength calculated, as shown on page 61.

This test can be carried out within a lined borehole where the vane is pushed into the soil below the base of the borehole for a distance equal to three times the vane diameter before rotation commences. Alternatively, the vane can be driven or jacked to the required depth, the vane being protected within a special protection

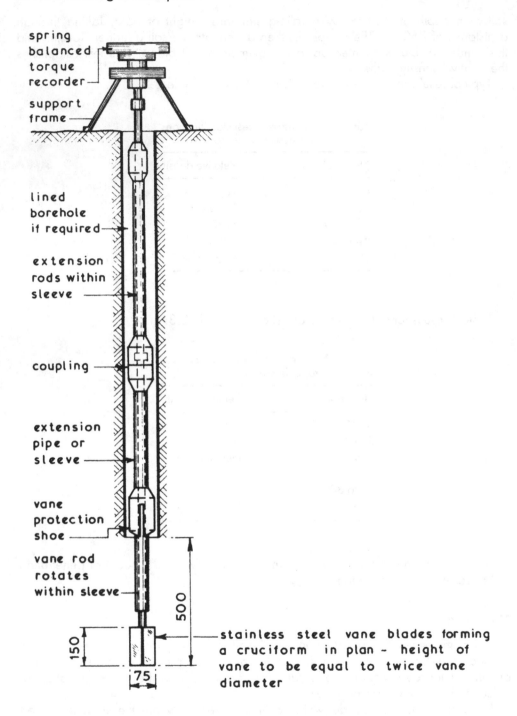

spring
balanced
torque
recorder

support
frame

lined
borehole
if required

extension
rods within
sleeve

coupling

extension
pipe or
sleeve

vane
protection
shoe

vane rod
rotates
within sleeve

500

150

75

stainless steel vane blades forming
a cruciform in plan - height of
vane to be equal to twice vane
diameter

Figure 2.12 Typical vane test apparatus

shoe. The vane is then driven or jacked a further 500mm before rotation commences.

Calculation of shear strength:

$$\text{Formula}: \quad S = \frac{M}{K}$$

where S = shear value in kN/m^2
    M = torque required to shear soil
    K = constant for vane
      = 3.66 D^3 × 10^{-6}
    D = vane diameter

Laboratory testing

Tests for identifying and classifying soils with regard to moisture content, liquid limit, plastic limit, particle size distribution and bulk density are given in BS 1377. (For moisture content see also BS EN ISO 17892–1: *Geotechnical Investigation and Testing. Laboratory Testing of Soil. Determination of Water Content.*)

Bulk density

This is the mass per unit volume, which includes mass of air or water in the voids and is essential information required for the design of retaining structures where the weight of the retained earth is an important factor.

Shear strength

This soil property can be used to establish its bearing capacity and also the pressure being exerted on the supports in an excavation. The most popular method to establish the shear strength of cohesive soils is the Triaxial Compression Test. In principle this test consists of subjecting a cylindrical sample of undisturbed soil (75mm long × 38mm diameter) to a lateral hydraulic pressure in addition to a vertical load. Three tests are carried out on three samples (all cut from the same large sample), each being subjected to a higher hydraulic pressure before axial loading is applied. The results are plotted in the form of Mohr's circles.

Consolidation of soil

This property is very important in calculating the movement of a soil under a foundation. The laboratory testing apparatus is called an oedometer.

Site soil testing techniques

Shrinkable soils

Soils subject to volume changes due to variation in moisture content.

SHRINKABLE SOIL CLASSIFICATION

- \> 35% fine particles.
- Plasticity index (PI) ≥ 10%.

Fine particles ~ minute portions of clay (loam) or silt having a nominal diameter of < 6 μm (0.06mm).

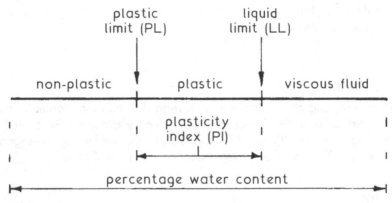

Note: See below and next page for PL and LL

Figure 2.13 Plasticity index
Note: PI = Liquid limit (LL) – Plastic limit (PL).

Plasticity index (PI) ~ a numerical measure of the potential for a soil to change in volume. Can be determined by the Atterberg limits test on fine particles < 425μm (0.425mm). Particles greater than this are removed by sieving.

Soils with a high PI tend to be mainly composed of clay, those with a low PI, of silt (see Table 2.4).

Plastic limit

A simple test established by hand rolling a sample of soil on a flat, non-porous surface. Soil is defined as plastic when the sample retains its shape down to a very small diameter. The plastic limit occurs when the slender thread of soil breaks apart at a diameter of about 3mm. It is considered non-plastic if the thread of soil cannot be rolled down to 3mm at any moisture content. This is a measure of the relatively low water content at which a soil changes from a plastic state to a solid state.

Table 2.4 Soil plasticity index

Plasticity index (PI)	Characteristic plastic quality
0	None
1–5	Slight
5–10	Low
10–20	Medium
20–40	High
> 40	Very high

Liquid limit

The water or moisture content when a soil sample changes from plastic to liquid behavioural characteristics. It can also be regarded as the relatively high water content at which soil changes from a liquid to a plastic state.

3 SITE WORKS

GENERAL CONSIDERATIONS
SETTING OUT
LEVELLING
ROAD AND PAVING
LANDSCAPING
PERSONAL PROTECTIVE EQUIPMENT (PPE)
DEMOLITION

3.1 General considerations

Before any specific considerations and decisions can be made regarding site layout, a general appreciation should be obtained by conducting a thorough site investigation at the pre-tender stage and examining, in detail, the drawings, specification and Bill of Quantities to formulate proposals of how the contract will be carried out if the tender is successful. This will involve a preliminary assessment of plant, materials and manpower requirements plotted against the proposed time scale.

Site security

Hoardings ~ under the Highways Act 1980 a close-boarded fence hoarding (see Figure 3.1) must be erected prior to the commencement of building operations if such operations are adjacent to a public footpath or highway. The hoarding needs to be adequately constructed to provide protection for the public, resist impact damage, resist anticipated wind pressures and be adequately lit at night. Before a hoarding can be erected a licence or permit must be obtained from the local authority who will usually require 10 to 20 days' notice. The licence will set out the minimum local authority requirements for hoardings and define the time limit period of the licence.

Sheet metal or plywood hoarding 2.4m high is commonly used. For smaller sites, 2.0m high panels of galvanised mesh, known as Heras fencing, are commonly used and these can easily be hired.

Traffic management plans

Traffic routes to and from the site must be checked as to the suitability for transporting all the plant and materials necessary for the proposed works and within the site. City centre developments often present significant challenges in this respect, requiring detailed traffic management plans. Access around the site for mobile plant should be within fenced off traffic routes, with separate pedestrian routes for safety.

Material storage

This requirement varies with the construction and procurement strategy of the project and the types of material to be stored. Storage of materials should be designed to reduce double handling to a minimum without impeding the general site circulation and/or works in progress. Just in Time delivery of materials reduces the need for storage space on site and reduces the risk of damage.

Accommodation

This requirement depends upon the size of the project and the number and type of site staff to be accommodated, along with associated welfare requirements. Modular steel buildings and converted container units are used either

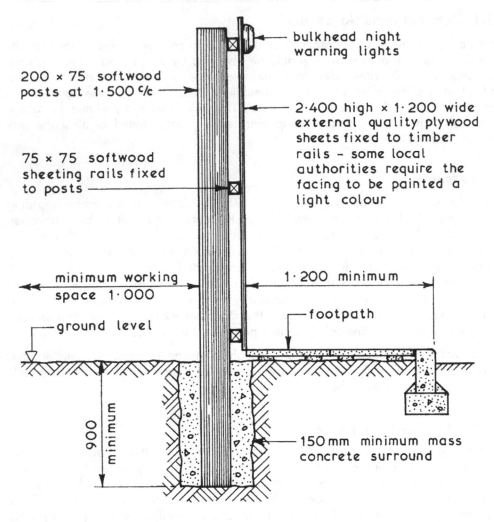

200 × 75 softwood posts at 1·500 c/c

75 × 75 softwood sheeting rails fixed to posts

bulkhead night warning lights

2·400 high × 1·200 wide external quality plywood sheets fixed to timber rails – some local authorities require the facing to be painted a light colour

minimum working space 1·000

1·200 minimum

ground level

footpath

900 minimum

150 mm minimum mass concrete surround

Figure 3.1 Typical hoarding details

singly or in multiples and often stacked to minimise use of the available site area. Cabins can be fitted out as offices/meeting rooms/toilets/showers/canteens/drying rooms/etc. Select siting for offices to give easy and quick access for visitors but at the same time with a reasonable view of the site. Select siting for mess room and toilets to reduce walking time to a minimum without impeding the general site circulation and/or works in progress.

The minimum requirements of such accommodation are governed by the Offices, Shops and Railway Premises Act 1963 unless they are:

1. Mobile units in use for not more than six months.
2. Fixed units in use for not more than six weeks.

3. Any type of unit in use for not more than 21 man hours per week.
4. Office for exclusive use of self-employed person.
5. Office used by main contractor-only staff.

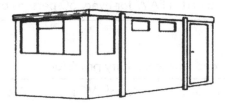

Figure 3.2   Portable site cabin

Fire prevention on construction sites

The joint code of practice on the protection from fire of construction sites and buildings undergoing renovation, published by Construction Industry Publications and The Fire Protection Association, is applicable where the work is notifiable to the HSE, i.e. work extends to more than 30 construction days with 20 or more persons working simultaneously or it involves more than 500 person days.

First aid

*   Box readily accessible and distinctly marked.
*   Contents sufficient for the number of persons on site.
*   Person appointed with responsibility for first aid facilities and calls to emergency ambulance service (not necessarily a qualified first aider).
*   Information displayed confirming name of appointed person.
*   Person trained (first aider) in first aid at work holding an HSE recognised FAW qualification or an emergency first aid at work EFAW qualification (see also Table 3.1).

Sanitary facilities

Toilets: separate male and female if possible, with lockable doors. Minimum number where men and women are employed is shown in Table 3.2. Thereafter, an additional WC and basin for every 25 persons. Minimum requirement where only men are employed is shown in Table 3.3.

For portable toilets, used where there is no plumbing and drainage, at least one per seven persons, emptied at least once a week, is required.

General considerations

Table 3.1   First aid personnel requirements

No. of site personnel	Minimum requirements
<5	One appointed person preferably trained in first aid
5–50	One trained in EFAW or FAW depending on assessment of possible injuries
> 50	One trained in FAW for every 50 persons or part thereof

Table 3.2   Male and female sanitary provision

No. of persons on site	No. of WCs	No. of washbasins
≤5	1	1
6–25	2	2

Table 3.3   Male-only sanitary provision

No. of men on site	No. of WCs	No. of urinals
1–15	1	1
16–30	2	1
31–45	2	2
46–60	3	2
61–75	3	3
76–90	4	3
91–100	4	4

Washing facilities

- Next to toilets and changing areas.
- Hot and cold water or mixed warm water.
- Soap, towels or hot air dryer.
- Adequate ventilation and lighting.
- Washbasins large enough to wash face, hands and forearms.
- Showers for particularly dirty work.

Drinking water

Wholesome supply direct from the mains.

- Bottled water acceptable if mains supply unavailable.
- Cups or other drinking vessels at outlets or a drinking fountain.

Accommodation for rest, shelter, changing and eating

- Separate provisions if men and women are on site.
- Located close to washing facilities.
- Heated place for shelter from inclement weather.
- Space for changing with security for personal clothing, etc.
- Lockers and/or place to hang clothes.
- Place for wet clothing to be dried.
- Rest facilities with tables and raised-back chairs.
- Ventilation and lighting.
- Means for heating water and warming food, unless a separate provision is made for providing meals.

Note: All facilities to be cleaned regularly and serviced with soap, paper, towels, etc.
Health and Safety at Work, etc. Act 1974.
Construction (Design and Management) Regulations 2015
[incorporating the Construction (Health, Safety and Welfare)Regulations of 1996].
Health and Safety (First Aid) Regulations 1981.
Workplace (Health, Safety and Welfare) Regulations 1992.
Management of Health and Safety at Work Regulations 1999.

Temporary services

Electrical supply to building sites ~ a supply of electricity is usually required at an early stage in the contract to provide light and power to the units of accommodation. As the work progresses power could also be required for site lighting, hand-held power tools and large items of plant. The supply of electricity to a building site is the subject of a contract between the contractor and the local area electricity company who will want to know the date when supply is required; site address together with a block plan of the site; final load demand of proposed building; and an estimate of the maximum load demand in kilowatts for the construction period. The latter can be estimated by allowing $10W/m^2$ of the total floor area of the proposed building plus an allowance for high load equipment such as cranes. The installation should be undertaken by a competent electrical contractor to ensure that it complies with all the statutory rules and regulations for the supply of electricity to building sites.

The units must be strong, durable and resistant to rain penetration with adequate weather seals to all access panels and doors. All plug and socket outlets should be colour coded: 400V – red; 230V – blue; 110V – yellow.

Having permanent electrical services installed at an early stage and making temporary connections for site use during the construction period is the optimal solution. Alternatively, temporary electrical services may be provided by portable generators. Toilet units can be temporarily connected to existing drains where present.

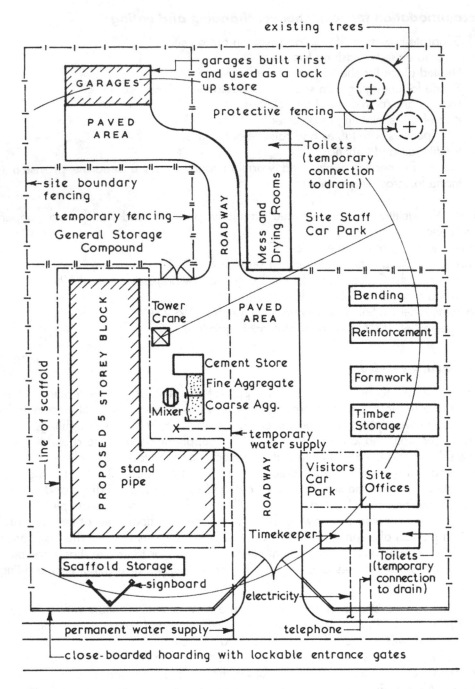

Figure 3.3 Example of a site set-up with office accommodation and storage facilities

Plant considerations

This includes the siting of fixed cranes and movement of mobile plant around the site. Fixed cranes should be selected and positioned for maximum efficiency of moving material direct from lorries or storage areas to where they are to be used.

Site lighting

To plan an adequate system of site lighting the types of activity must be defined and given an illumination target value which is quoted in lux (lx). Recommended minimum target values for building activities are shown in the diagram.

External lighting	–	general circulation	10 lx
		materials handling	
Internal lighting	–	general circulation	5 lx
		general working areas	15 lx
		concreting activities	50 lx
		carpentry and joinery	100 lx
		bricklaying	
		plastering	
		painting and decorating	200 lx
		site offices	
		drawing board positions	300 lx

Figure 3.4 Site lighting requirements

Such target values do not take into account deterioration, dirt or abnormal conditions; therefore, it is usual to plan for at least twice the recommended target values. Generally the manufacturers will provide guidance as to the best arrangement to use in any particular situation

Fixed lighting around the site is a deterrent to would-be trespassers and helps site safety. Walkway and local lighting to illuminate the general circulation routes bulkhead and/or festoon lighting could be used either on a standard mains voltage of 230V or on a reduced voltage of 110V. For local lighting at the place of work, hand lamps with trailing leads or lamp fittings on stands can be used and positioned to give the maximum amount of illumination without unacceptable shadow cast.

General considerations

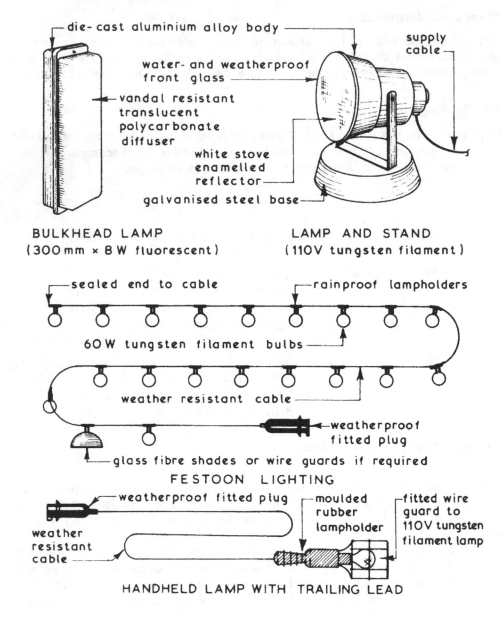

die-cast aluminium alloy body

water- and weatherproof front glass

vandal resistant translucent polycarbonate diffuser

white stove enamelled reflector

galvanised steel base

supply cable

BULKHEAD LAMP
(300 mm × 8 W fluorescent)

LAMP AND STAND
(110V tungsten filament)

sealed end to cable

rainproof lampholders

60 W tungsten filament bulbs

weather resistant cable

weatherproof fitted plug

glass fibre shades or wire guards if required

FESTOON LIGHTING

weatherproof fitted plug

weather resistant cable

moulded rubber lampholder

fitted wire guard to 110V tungsten filament lamp

HANDHELD LAMP WITH TRAILING LEAD

Figure 3.5 Site lighting equipment

3.2 Setting out

Reduced level excavations

The overall outline of the reduced level area can be set out using a theodolite, ranging rods, tape and pegs working from a baseline. To control the depth of excavation, sight rails are set up at a convenient height and at positions that will enable a traveller to be used.

1. Setting up sight rails:

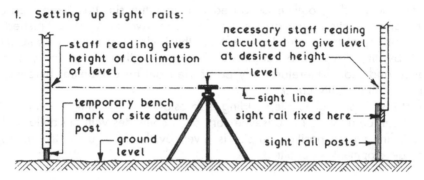

2. Controlling excavation depth:

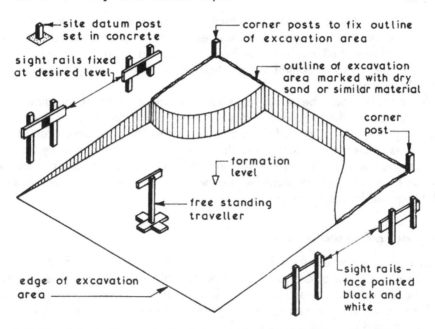

height of traveller = desired level of sight rail - formation level

Figure 3.6 Levelling and setting out for excavations

Setting out

This will involve the removal of topsoil together with any vegetation, scraping and grading the required area down to formation level, plus the formation of any cuttings or embankments. The soil immediately below the formation level is called the subgrade, whose strength will generally decrease as its moisture content rises. Therefore, if it is to be left exposed for any length of time, protection may be required. To preserve the strength and durability of the subgrade it may be necessary to install cut-off subsoil drains alongside the proposed road.

Buildings

Setting out the building outline is carried out after the site has been cleared of all debris or obstructions and any reduced-level excavation work is finished. It is the responsibility of the contractor to set out the building(s) using the information provided by the designer or architect. Accurate setting out is of paramount importance and should therefore only be carried out by competent persons, with all their work thoroughly checked.

The first task in setting out the building is to establish a baseline to which all the setting out can be related. The baseline very often coincides with the building line – which is a line whose position on site is given by the local authority, in front of which no development is permitted.

Trenches

The objective of this task is twofold. First, it must establish the excavation size, shape and direction; and second, it must establish the width and position of the walls. The outline of building will have been set out and using this outline the profile boards can be set up to control the position, width and possibly the depth of the proposed trenches.

Profile boards should be set up at least 2.0m clear of trench positions so that they do not obstruct the excavation work. The level of the profile cross board should be related to the site datum and fixed at a convenient height above ground level if a traveller is to be used to control the depth of the trench. Alternatively, the trench depth can be controlled using a level and staff related to site datum. The trench width can be marked on the profile with either nails or saw cuts and with a painted band if required for identification.

Setting out a framed building

Framed buildings are usually related to a grid, the intersections of the grid lines being the centre point of an isolated or pad foundation. The grid is usually set out from a baseline (which does not always form part of the grid). Setting out dimensions for locating the grid can either be given on a drawing or they will have to be accurately scaled off a general layout plan.

The grid is established using a theodolite and marking the grid line intersections with stout pegs. Once the grid has been set out, offset pegs or profiles can be fixed clear of any subsequent excavation work. Control of excavation depth can be by means of a traveller sighted between sight rails or by level and staff related to site datum.

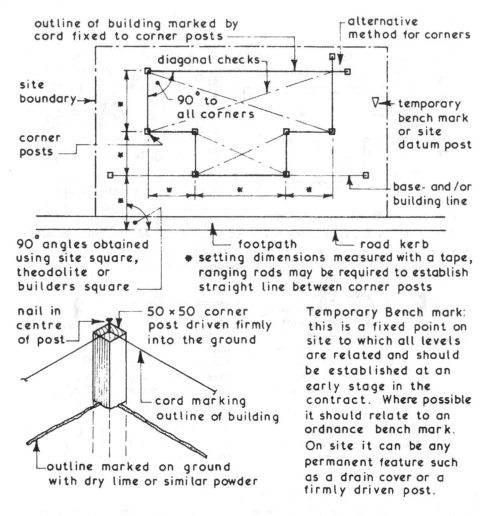

outline of building marked by
cord fixed to corner posts

alternative
method for corners

diagonal checks

site
boundary

90° to
all corners

corner
posts

temporary
bench mark
or site
datum post

base- and /or
building line

90° angles obtained
using site square,
theodolite or
builders square

footpath

road kerb

setting dimensions measured with a tape,
ranging rods may be required to establish
straight line between corner posts

nail in
centre
of post

50 x 50 corner
post driven firmly
into the ground

cord marking
outline of building

outline marked on ground
with dry lime or similar powder

Temporary Bench mark:
this is a fixed point on
site to which all levels
are related and should
be established at an
early stage in the
contract. Where possible
it should relate to an
ordnance bench mark.
On site it can be any
permanent feature such
as a drain cover or a
firmly driven post.

Figure 3.7 Setting out building lines

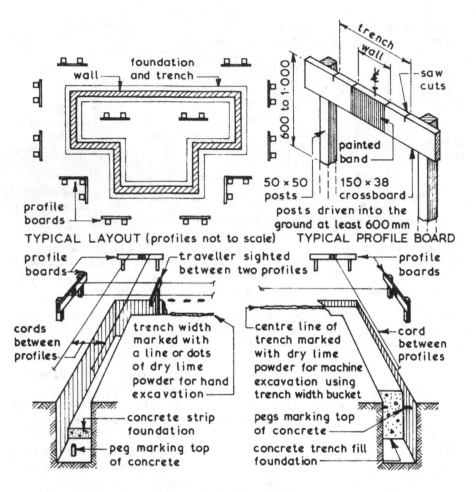

Figure 3.8 Setting out profile boards for foundations

Note: Corners of walls transferred from intersecting cord lines to mortar spots on concrete foundations using a spirit level.

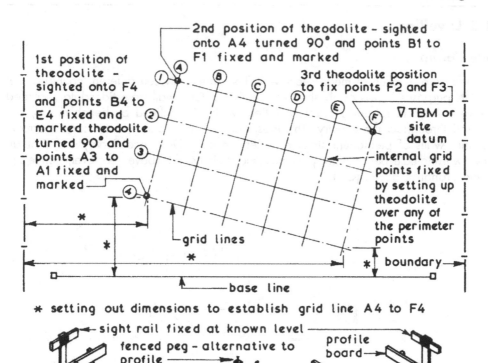

2nd position of theodolite - sighted onto A4 turned 90° and points B1 to F1 fixed and marked

1st position of theodolite - sighted onto F4 and points B4 to E4 fixed and marked theodolite turned 90° and points A3 to A1 fixed and marked

3rd theodolite position to fix points F2 and F3

▽ TBM or site datum

internal grid points fixed by setting up theodolite over any of the perimeter points

grid lines

boundary

base line

✳ setting out dimensions to establish grid line A4 to F4

sight rail fixed at known level
fenced peg - alternative to profile
profile board
cords
profile board
pad template
excavation
grid setting out peg
profile board
traveller sighted between sight rails to control depth

1. Pad template positioned with cords between profiles and pad outline marked with dry lime or similar powder.

2. Pad pits excavated using traveller sighted between sight rails fixed at a level related to site datum.

Figure 3.9 Setting out pad foundations on structural grid

77

3.3 Levelling

Site datum

Altitude zero is taken at mean sea level. This varies between different countries, but for UK purposes it was established at Newlyn in Cornwall from tide data recorded between May 1915 and April 1921. Relative levels, defined by bench marks, are located throughout the country. The most common are identified as carved arrows cut into walls of permanent structures. Reference to Ordnance Survey maps of an area will indicate bench mark positions and their height above sea level, hence the name ordnance datum (OD).

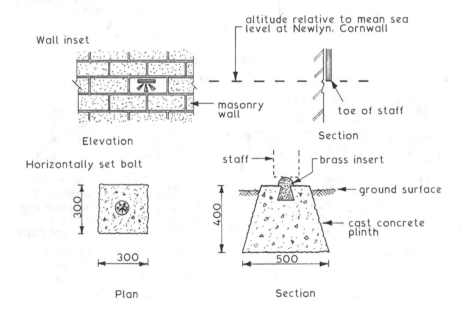

Figure 3.10 Ordnance benchmarks and temporary benchmarks

Ordnance datum benchmark (OD)

If the nearest OD bench mark is impractical to access, the alternative is to establish a datum or temporary bench mark (TBM) from a fixture such as a manhole cover. Otherwise, a fixed position for a TBM could be a robust post set in concrete or a cast concrete plinth set in the ground to one side of ongoing work.

Instruments consist of a level (optical or laser) and a staff. Cross hairs of horizontal and vertical lines indicate image sharpness on an extending staff of 3, 4 or 5m in length. Staff graduations are in 10mm intervals on an 'E' pattern, estimates are taken to the nearest millimetre.

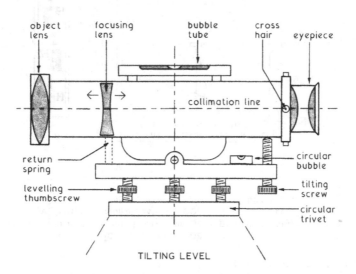

Figure 3.11 Optical level principles

Taking levels

An ordnance datum bench mark or a temporary bench mark is located. The levelling instrument and tripod are positioned on firm ground and sighted to the bench mark. Further staff height readings are taken at established positions around the site or at measured intervals corresponding to a grid pattern at convenient intervals, typically 10.0m. Each intersection of the grid lines represents a staff position, otherwise known as a station. Levels taken from the staff readings (four, taken from the grid corners) are computed with the plan area calculations ($100m^2$ for a 10m grid). From this the volume of site excavation or cut and fill required to level the site can be calculated.

Levelling

Application ~ methods to determine differences in ground levels for calculation of site excavation volumes and costs.

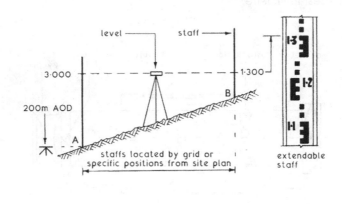

Rise and fall:

Staff reading at A = 3·00m, B = 1·30m
Ground level at A = 200m above ordnance datum (AOD)
Therefore level at B = 200m + rise (– fall if declining)
So level at B = 200 + (3·00 – 1·30) = 201·7m

Height of collimation (HC):

HC at A = Reduced level (RL) + staff reading
= 200 m + 3·00m = 203m AOD
Level at B = HC at A – staff reading at B
= 203 – 1·30 = 201·7m

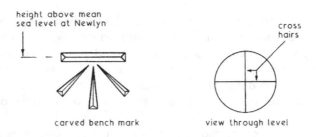

Figure 3.12 Levelling procedures

The theodolite in principle

Theodolite ~ a tripod mounted instrument designed to measure angles in the horizontal or vertical plane.

Measurement: a telescope provides for focal location between instrument and subject. Position of the scope is defined by an index of angles. The scale and presentation of angles varies from traditional micrometer readings to computer-compatible crystal displays. Angles are measured in degrees, minutes and seconds, e.g. 165° 53′ 30″.

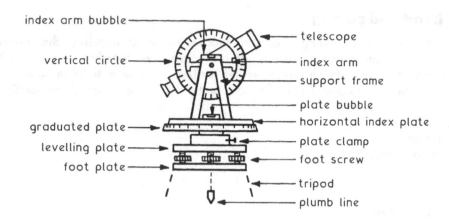

Figure 3.13 Theodolite details

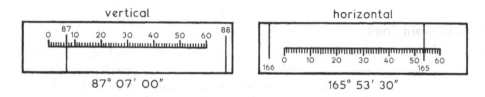

Figure 3.14 Taking angular readings

Direct reading micrometer scale

Application: at least two sightings are taken and the readings averaged. After the first sighting, the horizontal plate is rotated through 180° and the scope is also rotated 180° through the vertical to return the instrument to its original alignment for the second reading. This process will move the vertical circle from right face to left face, or vice versa. It is important to note the readings against the facing.

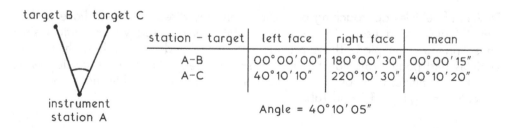

station – target	left face	right face	mean
A–B	00°00′00″	180°00′30″	00°00′15″
A–C	40°10′10″	220°10′30″	40°10′20″

Angle = 40°10′05″

Figure 3.15 Angular measurement

3.4 Road and paving

Within the context of building operations roadworks usually consist of the construction of small estate roads, access roads and driveways, together with temporary roads laid to define site circulation routes and/or provide a suitable surface for plant movement. The construction of roads can be considered under three headings.

1. Setting out.
2. Earthworks.
3. Paving construction.

Setting out roads

This activity is usually carried out after the topsoil has been removed using the dimensions given on the layout drawing(s). The layout could include straight length junctions, hammer heads, turning bays and intersecting curves.

Vertical sight lines

Vertical in height and horizontal or near horizontal for driver and pedestrian visibility. These clear sight lines are established between a driver's eye height of between 1.050m and 2.000m to allow for varying vehicle heights, to an object height of between 0.600m and 2.000m over the horizontal plane. Straight road lengths are usually set out from centre lines that have been established by traditional means.

Lane width

This is established by assessment of the amount of traffic flow and speed restriction. For convenience of road users, the minimum width can be based on a vehicle width not exceeding 2.500m plus an allowance for clearance of at least 0.500m between vehicles. This will provide an overall dimension of 5.500m for two-way traffic. One-way traffic will require a lane width of at least 3.000m.

Road junctions

Drivers of vehicles approaching a junction from any direction should have a clear view of other road users. Unobstructed visibility is required within vertical sight lines, triangular on plan. These provide a distance and area within which other vehicles and pedestrians can be seen at specific heights above the carriageway. No street furniture, trees or other obstructions are permitted within these zones, as indicated in Figure 3.17 and Table 3.4.

Horizontal sight lines

These should be provided and maintained thereafter with a clear view, to prevent the possibility of danger from vehicle drivers having obstructed outlook when

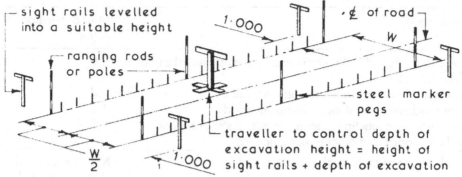

sight rails levelled into a suitable height

ranging rods or poles

1·000

₵ of road

W

steel marker pegs

traveller to control depth of excavation height = height of sight rails + depth of excavation

W/2

1·000

NB. curve road lengths set out in a similar manner

Junctions and Hammer Heads -

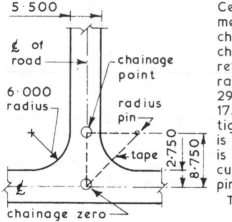

5·500

₵ of road

chainage point

6·000 radius

radius pin

tape

2·750

8·750

₵

chainage zero

Centre lines fixed by traditional methods. Tape hooked over pin at chainage zero and passed around chainage point pin at 8.750 then returned to chainage zero via the radius pin with a tape length of 29.875. Radius pin held tape length 17.500 and tape is moved until tight between all pins. Radius pin is driven and a 6.000 tape length is swung from the pin to trace out curve which is marked with pegs or pins

Tape length = 17.50 + √2 × 8·75
= 29·875

Figure 3.16 Setting out roads

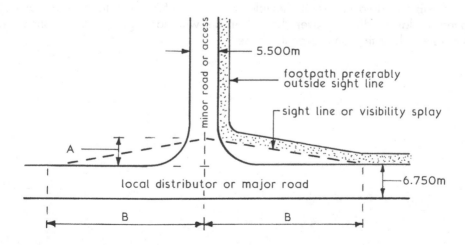

5.500m

footpath preferably outside sight line

sight line or visibility splay

minor road or access

A

local distributor or major road

6.750m

B

B

Figure 3.17 Setting out junction sight lines

83

Road and paving

Table 3.4 Guidance for sight line dimensions A and B

Type of road	Min. dimension A (m)
Access to a single dwelling or a small development of about half a dozen units	2.0
Minor road junctions within developed areas	2.4
Minor road junctions at local distributor roads	4.5
Busy minor road junctions, access roads, district or local distributor roads and other major junctions	9.0
Speed restriction mph (kph)	Min. major road dimension B (m)
20 (32)	45
30 (48)	90
40 (64)	120
50 (80)	160
60 (90)	215
70 (112)	295

approaching junctions. The recommended dimensions vary relative to the category of road classification and the speed restriction.

Guidance for dimensions A and B are shown in Table 3.4.

Planning Policy Guidance (PPG) 13, DCLG.Manual for Streets, DfT and DCLG. Manual for Streets 2, The Chartered Institution of Highways and Transport.

Paving construction

Once the subgrade has been prepared and any drainage or other buried services installed the construction of the paving can be undertaken. Paved surfaces can be either flexible or rigid in format. Flexible or bound surfaces are formed of materials applied in layers directly over the subgrade whereas rigid paving consists of a concrete slab resting on a granular base.

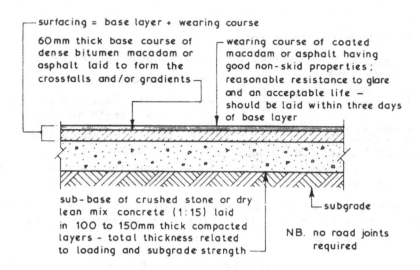

surfacing = base layer + wearing course

60mm thick base course of
dense bitumen macadam or
asphalt laid to form the
crossfalls and/or gradients

wearing course of coated
macadam or asphalt having
good non-skid properties;
reasonable resistance to glare
and an acceptable life –
should be laid within three days
of base layer

sub-base of crushed stone or dry
lean mix concrete (1:15) laid
in 100 to 150mm thick compacted
layers - total thickness related
to loading and subgrade strength

subgrade

NB. no road joints
required

Figure 3.18 Typical flexible paving details

Rigid pavings

These consist of a reinforced or unreinforced in-situ concrete slab laid over a base course of crushed stone, or similar material, which has been blinded to receive a polythene sheet slip membrane. The primary objective of this membrane is to prevent grout loss from the in-situ slab.

Joints in rigid pavings ~ longitudinal and transverse joints are required in rigid pavings to:

1. Limit size of slab.
2. Limit stresses due to subgrade restraint.
3. Provide for expansion and contraction movements.

The main joints used are classified as expansion, contraction or longitudinal, the latter being the same in detail as the contraction joint and differing only in direction.

The spacing of road joints is determined by:

1. Slab thickness.
2. Whether slab is reinforced or unreinforced.
3. Anticipated traffic load and flow rate.
4. Temperature at which concrete is laid.

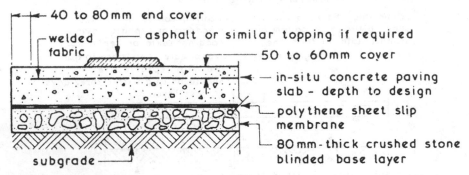

40 to 80 mm end cover

welded fabric

asphalt or similar topping if required

50 to 60 mm cover

in-situ concrete paving slab - depth to design

polythene sheet slip membrane

80 mm-thick crushed stone blinded base layer

subgrade

The paving can be laid between metal road forms or timber edge formwork. Alternatively the kerb stones could be laid first to act as permanent formwork.

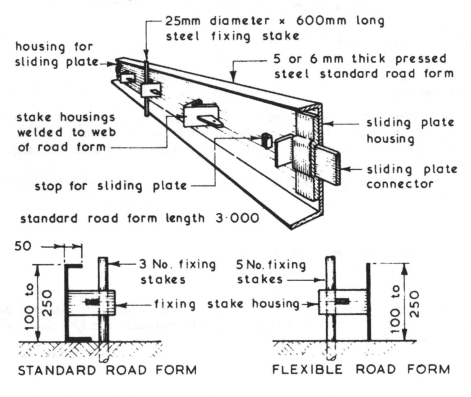

25mm diameter × 600mm long steel fixing stake

housing for sliding plate

5 or 6 mm thick pressed steel standard road form

stake housings welded to web of road form

sliding plate housing

sliding plate connector

stop for sliding plate

standard road form length 3·000

50

3 No. fixing stakes

100 to 250

fixing stake housing

5 No. fixing stakes

100 to 250

STANDARD ROAD FORM

FLEXIBLE ROAD FORM

Figure 3.19 Typical rigid paving details

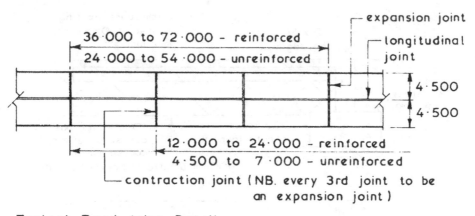

Typical Road Joint Details –

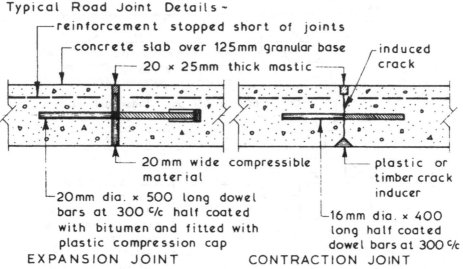

EXPANSION JOINT   CONTRACTION JOINT

Figure 3.20 Typical road joint spacing

Road and paving

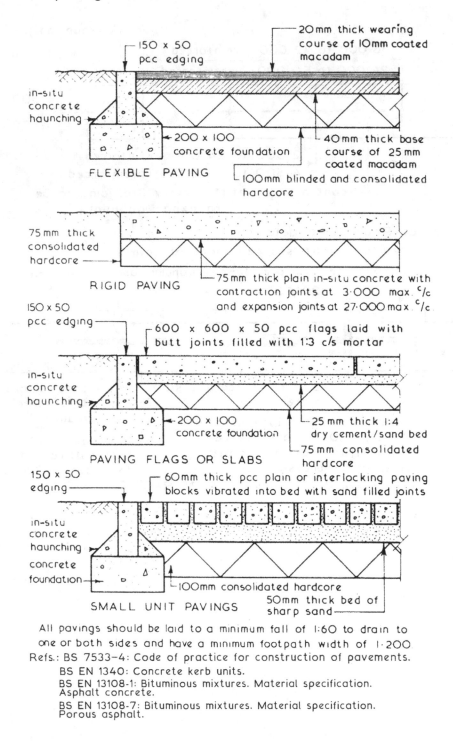

20 mm thick wearing course of 10mm coated macadam

150 x 50 pcc edging

in-situ concrete haunching

200 x 100 concrete foundation

40mm thick base course of 25mm coated macadam

100mm blinded and consolidated hardcore

FLEXIBLE PAVING

75mm thick consolidated hardcore

RIGID PAVING

75mm thick plain in-situ concrete with contraction joints at 3·000 max. $^c/c$ and expansion joints at 27·000 max. $^c/c$

150 x 50 pcc edging

600 x 600 x 50 pcc flags laid with butt joints filled with 1:3 c/s mortar

in-situ concrete haunching

200 x 100 concrete foundation

25 mm thick 1:4 dry cement/sand bed

75 mm consolidated hardcore

PAVING FLAGS OR SLABS

150 x 50 edging

60mm thick pcc plain or interlocking paving blocks vibrated into bed with sand filled joints

in-situ concrete haunching

concrete foundation

100mm consolidated hardcore

50mm thick bed of sharp sand

SMALL UNIT PAVINGS

All pavings should be laid to a minimum fall of 1:60 to drain to one or both sides and have a minimum footpath width of 1·200.
Refs.: BS 7533-4: Code of practice for construction of pavements.
BS EN 1340: Concrete kerb units.
BS EN 13108-1: Bituminous mixtures. Material specification. Asphalt concrete.
BS EN 13108-7: Bituminous mixtures. Material specification. Porous asphalt.

Figure 3.21 Typical examples

Available sections ~ manufactured in 915mm lengths
from silver/grey aggregate concrete.

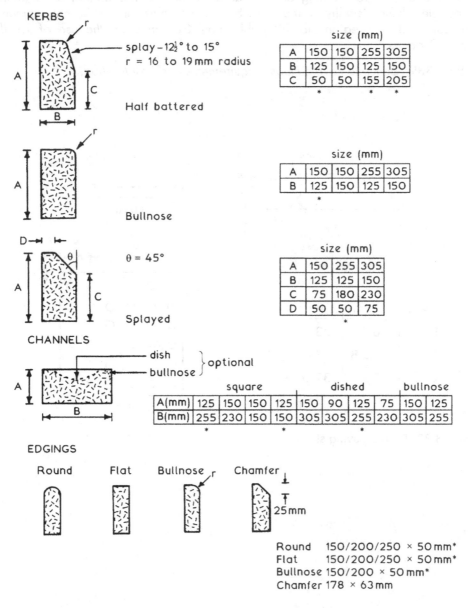

KERBS

splay – 12½° to 15°
r = 16 to 19mm radius

Half battered

size (mm)				
A	150	150	255	305
B	125	150	125	150
C	50	50	155	205

Bullnose

size (mm)				
A	150	150	255	305
B	125	150	125	150

θ = 45°

Splayed

size (mm)			
A	150	255	305
B	125	125	150
C	75	180	230
D	50	50	75

CHANNELS

dish ⎫
bullnose ⎬ optional

	square			dished				bullnose	
A(mm)	125	150	150	125	150	90	125	75	150	125
B(mm)	255	230	150	150	305	305	255	230	305	255

EDGINGS

Round   Flat   Bullnose   Chamfer

25mm

Round   150/200/250 × 50mm*
Flat   150/200/250 × 50mm*
Bullnose 150/200 × 50mm*
Chamfer 178 × 63mm

*denotes BS sections

NB. Further components such as drop/tapered kerbs are available for
vehicle accesses. Quadrants and angles provide for directional change.

Figure 3.22 Precast concrete road kerbs and channels

Road and paving

Tactile flags

Manufactured with a blistered (shown in Figure 3.23) or ribbed surface. Used in walkways to provide warning of hazards or to enable recognition of locations for people whose visibility is impaired. See also Department of the Environment, Transport and the Regions (DETR) publication *Guidance on the use of Tactile Paving Surfaces.*

BS EN 1339: *Concrete Paving Flags. Requirements and Test Methods.*

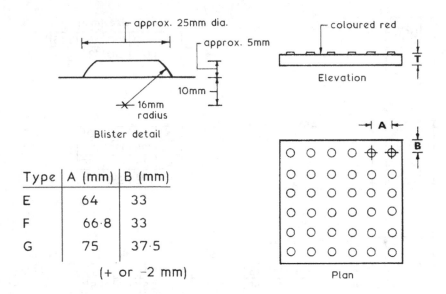

Type	A (mm)	B (mm)
E	64	33
F	66·8	33
G	75	37·5

(+ or −2 mm)

Figure 3.23 Tactile paving slabs

3.5 Landscaping

In the context of building works this would involve reinstatement of the site as a preparation to the landscaping in the form of lawns, paths, pavings, flower and shrub beds and tree planting. The actual planning, lawn laying and planting activities are normally undertaken by a landscape subcontractor. The main contractor's work would involve clearing away all waste and unwanted materials, breaking up and levelling surface areas, removing all unwanted vegetation, preparing the sub-soil for and spreading topsoil to a depth of at least 150mm.

Services ~ the actual position and laying of services is the responsibility of the various service boards and undertakings. The best method is to use the common trench approach; avoid as far as practicable laying services under the highway.

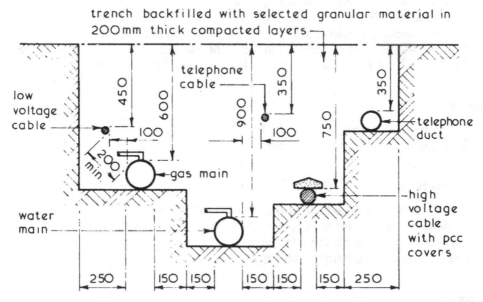

Road Signs ~ these can range from markings painted on roads to define traffic lanes, rights of way and warnings of hazards to signs mounted above the road level to give information, warning or directives, the latter being obligatory

Figure 3.24 Typical depths and arrangement of services within the highway

3.6 Personal protective equipment (PPE)

All equipment (including clothing affording protection against the weather) that is intended to be worn or held by a person at work and which protects them against one or more risks to their health and safety. This is to be provided by the employer at no cost to employees. Self-employed to provide their own, unless in an employer–employee relationship.

Personal Protective Equipment Regulations

These cover the types of PPE generally required.

Face and eyes

- Disposable face masks.
- Respirators/breathing apparatus (special training required).
- Air-fed helmets (special training required).
- Safety glasses, similar to ordinary glasses, may have side shields.
- Shield, one-piece moulded lens worn over other glasses.
- Goggles, flexible plastic frame with elastic headband.
- Face guard, shield or visor to fully enclose the eyes from dust, etc.

Hearing

- Earmuffs – helmet mounted available.
- Earplugs and ear inserts (canal caps).

Legs and feet

- Safety boots/shoes, steel toe cap and slip-resistant sole.
- Wellington boots, steel toe cap and studded soles available.
- Anti-static electricity insulated footwear.
- Knee pads, gaiters, leggings and spats.

Body

- High visibility waistcoats.
- Overalls, coveralls and boiler suits.
- Aprons, chain mail or leather.
- Life jackets.
- Safety harnesses.
- Insulated and waterproof clothing.

Hands

- Gloves, waterproof and/or insulating.
- Gauntlets.
- Armlets and wristcuffs.

Risk assessment

Hazard ~ any situation that has the potential to harm.
Risk ~ possibility of damage, injury or loss caused by a hazard.
Responsibility ~ all persons in and associated with the workplace.

3.7 Demolition

Under the Town and Country Planning Act, demolition is generally not regarded as development, but planning permission will be required if the site is to have a change of use.

Planning (Listed Buildings and Conservation Areas) Act ~ listed buildings and those in conservation areas will require local authority approval for any alterations. Consent for change may be limited to partial demolition, particularly where it is necessary to preserve a building frontage for historic reasons.

Under the Building Act 1984, the intention to demolish a building requires six weeks' written notice of intent. Notice must also be given to utilities providers and adjoining/adjacent building owners, particularly where party walls are involved. Small buildings of volumes less than 50m^3 are generally exempt. Within six weeks of the notice being submitted, the local authority will specify their requirements for shoring, protection of adjacent buildings, debris disposal and general safety requirements under the HSE.

The local authority can issue a demolition enforcement order to a building owner, where a building is considered to be insecure, a danger to the general public and detrimental to amenities under the Public Health Act.

There are concerns for the protection of the general public using a thoroughfare in or near to an area affected by demolition work. The building owner and demolition contractor are required to ensure that debris and other materials are not deposited in the street unless in a suitable receptacle (skip) and the local authority highways department and police are in agreement with its location. Temporary roadworks require protective fencing and site hoardings must be robust and secure. All supplementary provisions, such as hoardings and skips, may also require adequate illumination. Provision must be made for immediate removal of poisonous and hazardous waste.

Demolition works

Hazardous and potentially dangerous work that should only be undertaken by experienced contractors.

Partial demolition

Generally requiring temporary support to the remaining structure. This may involve window strutting, floor props and shoring. The execution of work is likely to be limited to manual handling with minimal use of powered equipment.

Preliminaries

A detailed survey should include:

- An assessment of the condition of the structure and the impact of removing parts on the remainder.
- The effect demolition will have on adjacent properties.

- Photographic records, particularly of any noticeable defects on adjacent buildings.
- Neighbourhood impact, i.e. disruption, disturbance, protection.
- The need for hoardings.
- Potential for salvaging/recycling/reuse of materials.
- Extent of basements and tunnels.
- Services – need to terminate and protect for future reconnections.
- Means for selective removal of hazardous materials.

Insurance

General builders are unlikely to find demolition cover in their standard policies. All risks indemnity should be considered to cover claims from site personnel and others accessing the site. Additional third-party cover will be required for claims for loss or damage to other property, occupied areas, business, utilities, private and public roads.

Salvage

Salvaged materials and components can be valuable: bricks, tiles, slates, steel sections and timber are all marketable. Architectural features such as fireplaces and stairs will command a good price. Reclamation costs will be balanced against the financial gain.

Asbestos

Specialist asbestos removal contractors should be engaged to carry out a full asbestos survey of the building if it is suspected, due to the age of the building, that asbestos products may be present. Asbestos was used for a variety of applications including pipe insulation, fire protection, sheet claddings, linings and roofing. Samples should be taken for laboratory analysis. Following the survey the specialist contractor should be engaged to remove material before any other demolition commences.

Demolition methods

Generally, the reverse order of construction to gradually reduce the height. Where space is not confined, overturning or explosives may be considered.

Piecemeal demolition

This is a progressive, labour intensive taking down of the building piece by piece using handheld equipment such as pneumatic breakers, oxyacetylene cutters, picks and hammers. This method is required for salvaging materials and other reusable components. Chutes should be used to direct debris to a suitable place of collection

Demolition

Machine demolition

Use of tracked vehicles with long reach articulated hydraulic boom fitted with a variety of tools including percussion chisels to demolish concrete and masonry and hydraulic shears to cut through steel sections.

Explosives

Specialised work by licensed operators, charges are set to fire in a sequence that weakens the building to a controlled internal collapse.

Some additional references:

BS 6187: *Code of Practice for Full and Partial Demolition.*
The Construction (Design and Management) Regulations.
The Management of Health and Safety at Work Regulations.

Shoring

This is a form of temporary support which can be given to existing buildings with the primary function of providing the necessary precautions to avoid damage to any person or property from the collapse of the supported structure.

There are three basic systems of shoring which can be used separately or in combination with one another to provide the support(s), namely:

1. Dead shoring – used primarily to carry vertical loadings.
2. Raking shoring – used to support a combination of vertical and horizontal loadings.
3. Flying shoring – an alternative to raking shoring to give a clear working space at ground level.

Dead shores ~ these shores should be placed at approximately 2.000 c/c and positioned under the piers between the windows, with any windows in the vicinity of the shores being strutted to prevent distortion of the openings. A survey should be carried out to establish the location of any underground services so that they can be protected as necessary. The sizes shown in Figure 3.26 are typical; actual sizes should be obtained from tables or calculated from first principles. Any suitable structural material, such as steel, can be substituted for the timber members shown.

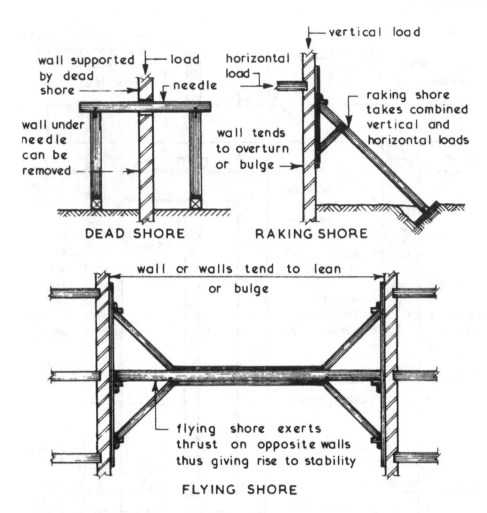

wall supported — load
by dead
shore — needle

wall under
needle
can be
removed —

DEAD SHORE

vertical load

horizontal
load —

wall tends
to overturn
or bulge —

raking shore
takes combined
vertical and
horizontal loads

RAKING SHORE

wall or walls tend to lean
or bulge

flying shore exerts
thrust on opposite walls
thus giving rise to stability

FLYING SHORE

Figure 3.25 Shoring methods

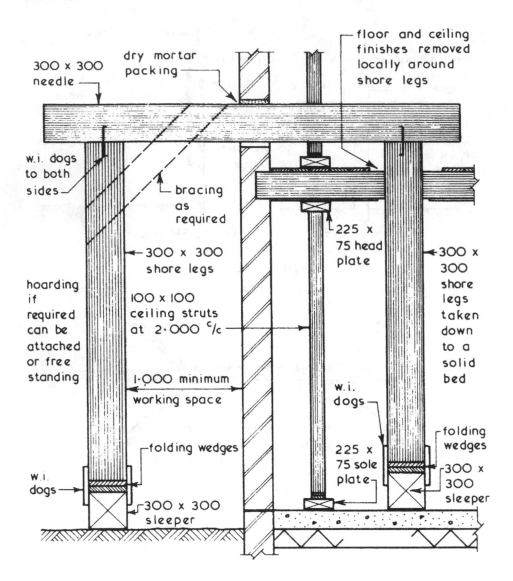

Figure 3.26 Dead or needle shore

4 SOIL IMPROVEMENT AND EXCAVATION

Opening remarks

Soil investigation ~ before a decision is made as to the type of foundation that should be used on any particular site a soil investigation should be carried out to establish existing ground conditions and soil properties. The methods, which can be employed together with other sources of information such as local knowledge, Ordnance Survey and geological maps, mining records and aerial photography, should be familiar to students at this level. If such an investigation reveals a naturally poor subsoil or extensive filling the designer has several options.

4.1 Vibro replacement

The objective of this method is to strengthen the existing soil by rearranging and compacting coarse granular particles to form stone columns with the ground. This is carried out by means of a large poker vibrator, which has an effective compacting radius of 1.5 to 2.7m. On large sites the vibrator is inserted on a regular triangulated grid pattern with centres ranging from 1.5 to 3.0m. In coarse-grained soils extra coarse aggregate is tipped into the insertion positions to make up levels as required, whereas in clay and other fine particle soils the vibrator is surged up and down, enabling the water jetting action to remove the surrounding soft material and thus forming a borehole, which is backfilled with a coarse granular material compacted in situ by the vibrator. The backfill material is usually of 20 to 70mm size of uniform grading within the chosen range. Ground vibration is not a piling system but a means of strengthening ground to increase the bearing capacity within a range of 200 to 500kN/m².

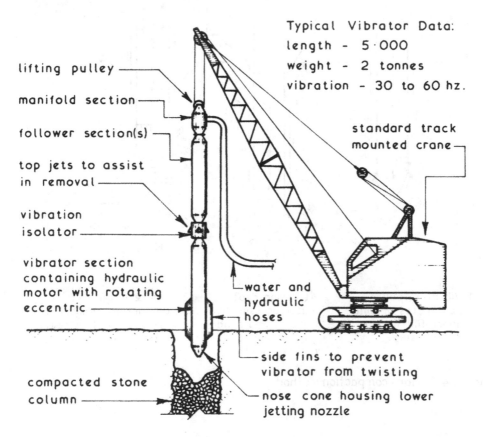

lifting pulley

manifold section

follower section(s)

top jets to assist in removal

vibration isolator

vibrator section containing hydraulic motor with rotating eccentric

compacted stone column

Typical Vibrator Data:
length - 5·000
weight - 2 tonnes
vibration - 30 to 60 hz.

standard track mounted crane

water and hydraulic hoses

side fins to prevent vibrator from twisting

nose cone housing lower jetting nozzle

Figure 4.1 Vibro replacement method

4.2 Vibro compaction

Applied to non-cohesive subsoils where the granular particles are rearranged into a denser condition by poker vibration.

The crane-suspended vibrating poker is water-jetted into the ground using a combination of self-weight and water displacement of the finer soil particles to penetrate the ground. Under this pressure, the soil granules compact to increase in density as the poker descends. At the appropriate depth, which may be determined by building load calculations or the practical limit of plant (generally 30m max.), jetting ceases and fine aggregates or sand are infilled around the poker. The poker is then gradually withdrawn, compacting the granular fill in the process. Compaction continues until sand fill reaches ground level. Spacing of compaction boreholes is relatively close to ensure continuity and an integral ground condition.

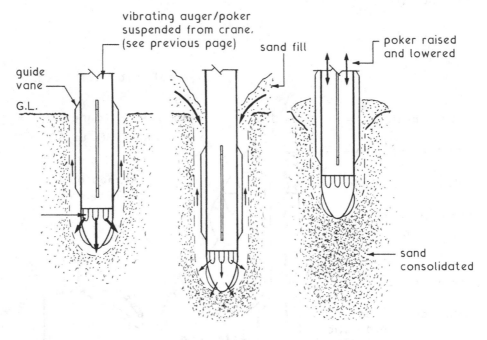

1. Vibrating poker penetrates ground under full water jet pressure.

2. At design depth, water pressure is reduced and sand fill introduced and compacted.

3. With resistance to compaction, poker is raised and lowered to consolidate further sand.

Figure 4.2 Vibro compaction method

4.3 Dynamic compaction

This method of ground improvement consists of dropping a heavy weight from a considerable height and is particularly effective in granular soils. Where water is present in the subsoil, trenches should be excavated to allow the water to escape and not collect in the craters formed by the dropped weight. The drop pattern, size of weight and height of drop are selected to suit each individual site but generally three or four drops are made in each position, forming a crater up to 2.5m deep and 5m in diameter. Vibration through the subsoil can be a problem with dynamic compaction operations; therefore the proximity and condition of nearby buildings must be considered, together with the depth position and condition of existing services on site.

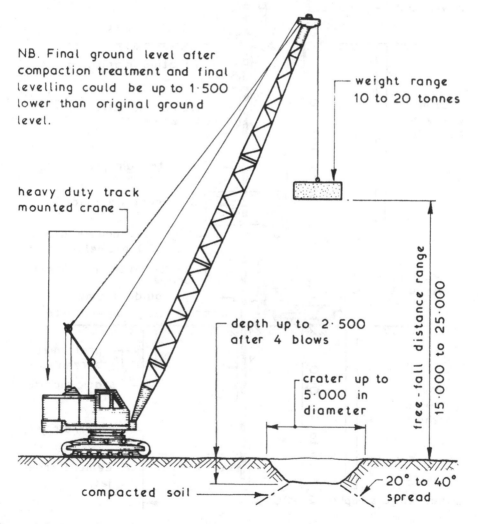

Figure 4.3 Dynamic compaction method

4.4 Jet grouting

This is a means of ground consolidation by lowering a monitor probe into pre-formed boreholes. The probe is rotated and the sides of the borehole are subjected to a jet of pressurised water and air from a single outlet which enlarges and compacts the borehole sides. At the same time a cement grout is introduced under pressure to fill the void being created. The water used by the probe and any combined earth is forced up to the surface in the form of a sludge. The spacing, depth and layout of the boreholes is subject to specialist design.

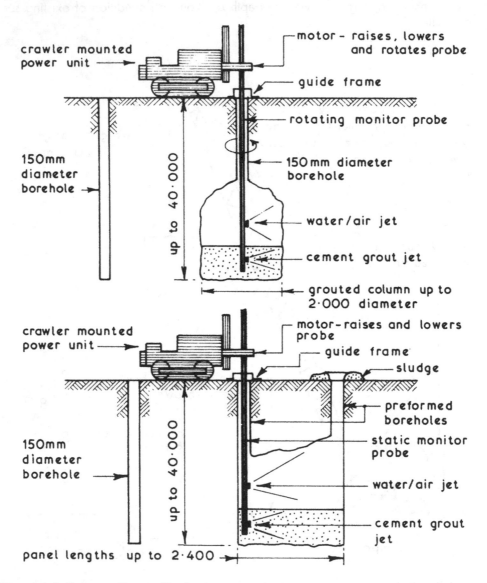

Figure 4.4 Jet grouting methods

4.5 Excavation support

All subsoils have different abilities to remain stable during excavation works. Most will assume a natural angle of repose or rest unless given temporary support. The presence of ground water, apart from creating difficult working conditions, can have an adverse effect on the subsoil's natural angle of repose.

Time factors are relevant, such as period during which excavation will remain open and the time of year when work is carried out.

The need for an assessment of risk with regard to the support of excavations and protection of people within is a legal responsibility under the Health and Safety at Work Act 1974 and the Construction (Design and Management) Regulations 2015.

Trench support methods

A suitable barrier or fence must be provided to the sides of all excavations. Spoil or materials must not be placed near to the edge of any excavation, nor must plant be placed or moved near to any excavation so that persons employed in the excavation are endangered.

If water is present or enters an excavation, a pit or sump should be excavated below the formation level to act as a collection point from which the water can be pumped away.

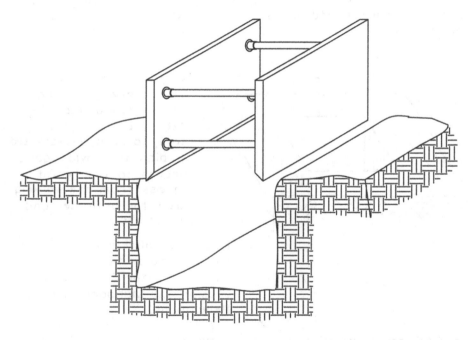

Figure 4.5 Trench box excavation support
Note: Aluminium trench box is lowered into the trench by a sling from a machine.

Excavation support

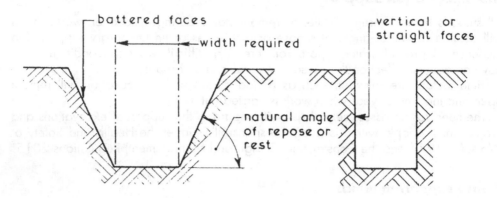

- battered faces
- width required
- natural angle of repose or rest
- vertical or straight faces

Disadvantage ~ extra cost of over-excavating and extra backfilling.

Advantage ~ no temporary support required to sides of excavation.

Disadvantage ~ sides of excavation may require some degree of temporary support.

Advantage ~ minimum amount of soil removed and therefore minimum amount of backfilling.

Pier Holes ~ isolated pits primarily used for foundation pads for columns and piers or for the construction of soakaways.

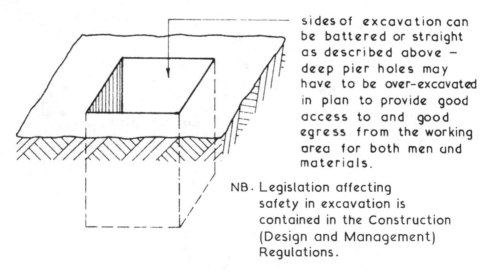

sides of excavation can be battered or straight as described above — deep pier holes may have to be over-excavated in plan to provide good access to and good egress from the working area for both men and materials.

NB. Legislation affecting safety in excavation is contained in the Construction (Design and Management) Regulations.

Figure 4.6 Types of excavation and support requirements

Typical Angles of Repose ~
Excavations cut to a natural angle of repose are called battered.

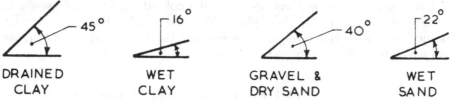

DRAINED
CLAY — 45°

WET
CLAY — 16°

GRAVEL &
DRY SAND — 40°

WET
SAND — 22°

Factors for Temporary Support of Excavations ~

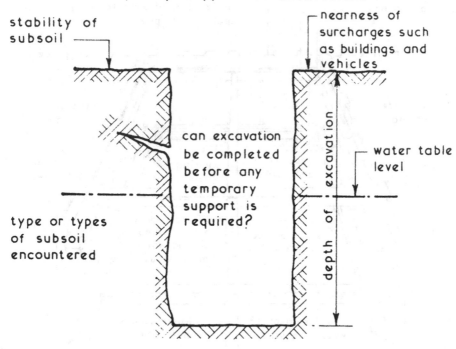

stability of subsoil

nearness of surcharges such as buildings and vehicles

can excavation be completed before any temporary support is required?

type or types of subsoil encountered

depth of excavation

water table level

Figure 4.7 Excavations in different soils

4.6 Cofferdams

These are temporary enclosures installed in soil or water to prevent the ingress of soil and/or water into the working area with the cofferdam. They are usually constructed from interlocking steel sheet piles which are suitably braced or tied back with ground anchors. Alternatively a cofferdam can be installed using any structural material which will fulfil the required function.

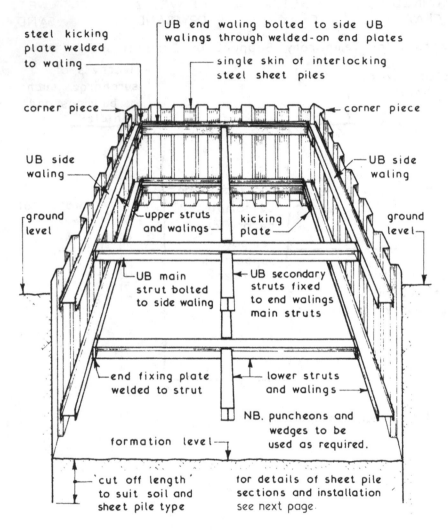

Figure 4.8 Steel sheet pile cofferdam

4.7 Steel sheet piling

Steel sheet piles can be used in excavations and to form permanent retaining walls. Three common formats of steel sheet piles with interlocking joints are available with a range of section sizes and strengths up to a maximum length of 30m.

Installing steel sheet piles

To ensure that the sheet piles are pitched and installed vertically a driving trestle or guide frame is used. These are usually purpose-built to accommodate a panel of 10 to 12 pairs of piles. The piles are lifted into position by a crane and driven by means of a percussion piling hammer or, alternatively, they can be pushed into the ground by hydraulic rams acting against the weight of the power pack, which is positioned over the heads of the pitched piles.

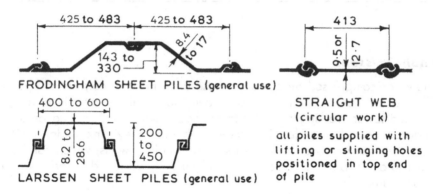

Figure 4.9 Steel sheet pile types

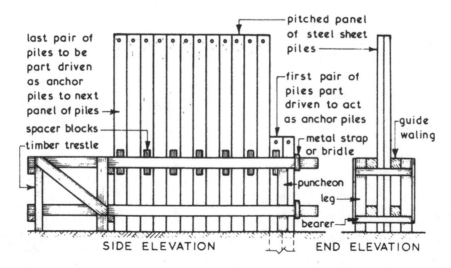

Figure 4.10 Sheet pile installation

109

4.8 Caissons

These are box-like structures that are similar in concept to cofferdams but they usually form an integral part of the finished structure. They can be economically constructed and installed in water or soil where the depth exceeds 18m. There are four basic types of caisson: see Table 4.1.

Table 4.1 Types of caissons

1. Box caissons 2. Open caissons 3. Monolithic caissons	usually of precast concrete and used in water, being towed or floated into position and sunk – land caissons are of the open type and constructed in situ.
4. Pneumatic caissons	used in water – see next section.

Pneumatic caissons

Also known as compressed air caissons, these are similar in concept to open caissons. They can be used in difficult subsoil conditions below water level and have a pressurised lower working chamber to provide a safe, dry working area. Pneumatic caissons can be made of concrete, whereby they sink under their own weight, or they can be constructed from steel with hollow walls, which can be filled with water to act as ballast. These caissons are usually designed to form part of the finished structure.

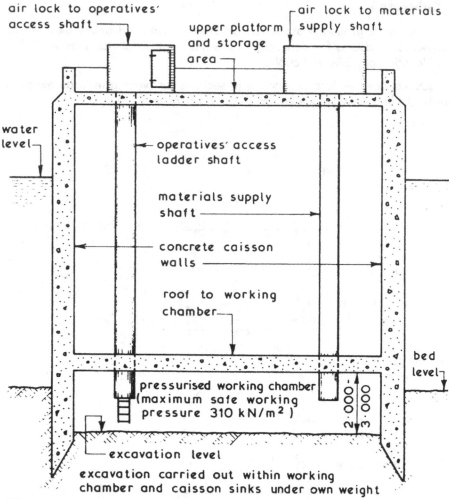

air lock to operatives' access shaft

air lock to materials supply shaft

upper platform and storage area

water level

operatives' access ladder shaft

materials supply shaft

concrete caisson walls

roof to working chamber

bed level

pressurised working chamber (maximum safe working pressure 310 kN/m²)

2.000 - 3.000

excavation level

excavation carried out within working chamber and caisson sinks under own weight

When required depth is reached a concrete slab or plug is cast over the formation level and chamber sealed with mass concrete

Figure 4.11 Typical pneumatic caisson details

4.9 Culverts

Culvert ~ an underground passageway or tunnel, often used below elevated roads and railways as access for pedestrians, sewers, tidal outfalls and as voids for the general location of pipes and cables.

Manufactured from reinforced concrete in sections for placing in large excavations before backfilling; a technique known as *cut and cover*.

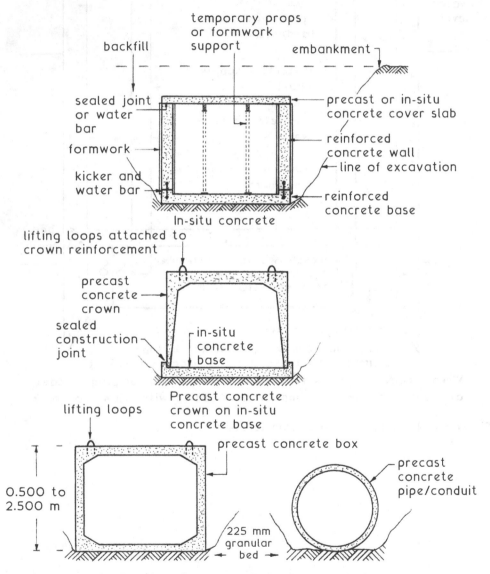

Figure 4.12 Types of pre-cast concrete culvert sections

112

4.10 Dewatering

Dewatering is necessary where the presence of ground water in the soil will have an adverse effect upon the construction process. This can take one of two forms, which are usually referred to as temporary and permanent exclusion.

Permanent exclusion ~ the insertion of an impermeable barrier to stop the flow of water within the ground.
Temporary exclusion ~ the lowering of the water table – within the economic depth range of 1.5m this can be achieved by subsoil drainage methods; for deeper treatment a pump or pumps are usually involved.

Simple sump pumping is suitable for trench work and/or where small volumes of water are involved.

Wellpoint systems

This is a method of lowering the water table to a position below the formation level to give a dry working area. The basic principle is to jet into the subsoil a series of wellpoints, which are connected to a common header pipe, which is then connected to a vacuum pump.

Wellpoint systems are suitable for most subsoils and can encircle an excavation or be laid progressively alongside, as in the case of a trench excavation. If the proposed formation level is below the suction lift capacity of the pump a multi-stage system can be employed.

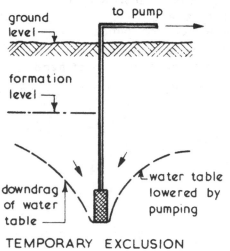

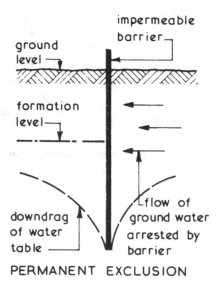

Figure 4.13 Dewatering

Dewatering

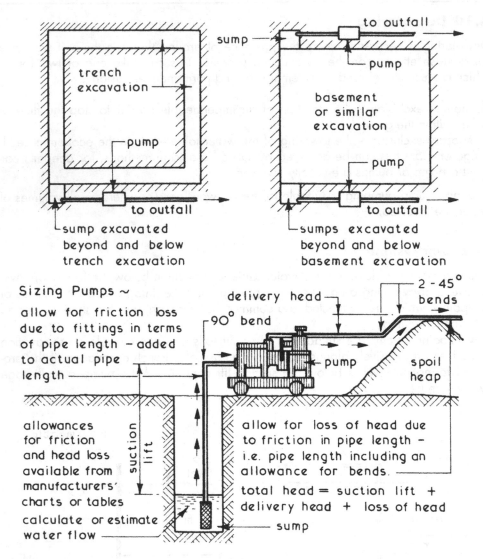

Figure 4.14 Groundwater sump pumping

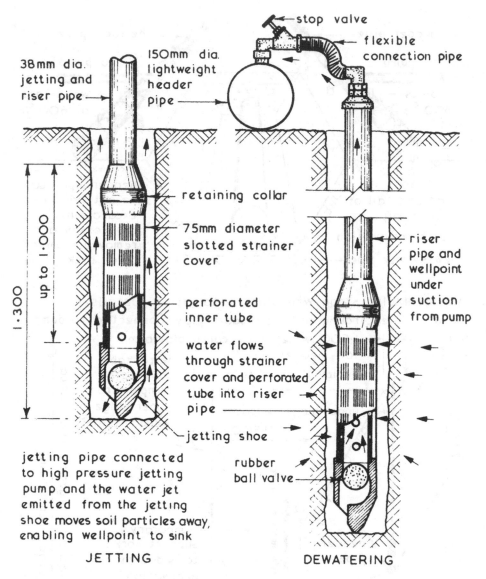

38mm dia. jetting and riser pipe

150mm dia. lightweight header pipe

stop valve

flexible connection pipe

1·300

up to 1·000

retaining collar

75mm diameter slotted strainer cover

perforated inner tube

water flows through strainer cover and perforated tube into riser pipe

jetting shoe

riser pipe and wellpoint under suction from pump

rubber ball valve

jetting pipe connected to high pressure jetting pump and the water jet emitted from the jetting shoe moves soil particles away, enabling wellpoint to sink

JETTING

DEWATERING

Figure 4.15 Well jetting and pumping

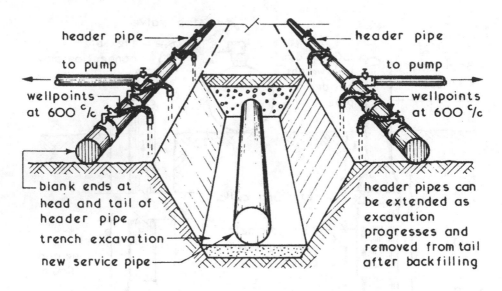

header pipe — to pump

header pipe — to pump

wellpoints at 600 c/c

wellpoints at 600 c/c

blank ends at head and tail of header pipe

header pipes can be extended as excavation progresses and removed from tail after backfilling

trench excavation

new service pipe

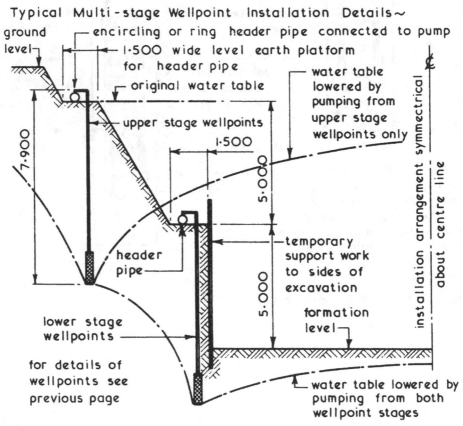

Typical Multi-stage Wellpoint Installation Details~

ground level

encircling or ring header pipe connected to pump

1·500 wide level earth platform for header pipe

original water table

upper stage wellpoints

water table lowered by pumping from upper stage wellpoints only

1·500

7·900

5·000

header pipe

temporary support work to sides of excavation

5·000

installation arrangement symmetrical about centre line

lower stage wellpoints

formation level

for details of wellpoints see previous page

water table lowered by pumping from both wellpoint stages

Figure 4.16 Multi-stage well point example

4.11 Contaminated land

Green-field land has not previously been built upon and is usually part of the 'green-belt' surrounding urban areas, designated inappropriate for development in order to preserve the countryside. Limited development for agricultural purposes only may be permitted on 'green-belt' land. The term brown-field relates to land that is a developed site or has been previously developed and is now derelict. These sites are usually associated with previous construction of industrial buildings and the land may be contaminated with chemicals and metals from the industrial process that took place on the site, along with any contamination from the demolition and clearing of the site. The UK government has an objective to build 60% of new homes on these sites.

Land remediation

Land previously used for industrial buildings could be contaminated with hazardous waste or pollutants, therefore it is essential that a geo-technical survey is undertaken to determine whether contaminants are in the soil and ground water. Of particular concern are acids, salts, heavy metals, cyanides and coal tars, in addition to organic materials that decompose to form the highly explosive gas methane. Analysis of the soil will determine a 'trigger threshold value', above which it will be declared sensitive to the end user. For example, a domestic garden or children's play area will have a low value relative to land designated for a commercial car park.

Site preparation

When building on sites previously infilled with uncontaminated material, a reinforced raft-type foundation may be adequate for light structures. Larger buildings will justify soil consolidation and compaction processes to improve the bearing capacity. Remedial measures for subsoils containing chemicals or other contaminants are varied.

Legislation

The Environment Protection Act of 1990 attempted to enforce responsibility on local authorities to compile a register of all potentially contaminated land. This proved unrealistic and too costly. Since then, requirements under the Environment Act 1995, the Pollution Prevention and Control Act 1999, the Environmental Permitting Regulations 2010 and the subsequent DCLG Planning Policy Statement (PPS 23, 2004): Planning and Pollution Control (Annex 2: Development on land affected by contamination) have made this more of a planning issue. It has become the responsibility of developers to conduct site investigations and to present details of proposed remedial measures as part of their planning application.

Soil removal

The traditional low-technology method for dealing with contaminated sites has been to excavate the soil and remove it to places licensed for depositing. However, with the increase in building work on brown-field sites, suitable dumps are becoming

Contaminated land

scarce. Added to this is the reluctance of ground operators to handle large volumes of this type of waste. Also, where excavations exceed depths of about 5m, it becomes less practical and too expensive. Alternative physical, biological or chemical methods of soil treatment may be considered.

Encapsulation

This process is the in-situ enclosure of the contaminated soil. A perimeter trench is taken down to rock or other sound strata and filled with an impervious agent such as bentonite clay. An impermeable horizontal capping is also required to link with the trenches. A high-specification barrier is necessary where liquid or gas contaminants are present as these can migrate quite easily. A system of monitoring the soil condition is essential as the barrier may decay in time. Suitable for all types of contaminant.

Soil washing

This involves extraction of the soil, sifting to remove large objects and placing it in a scrubbing unit resembling a huge concrete mixer. Within this unit water and detergents are added for a basic wash process, before pressure spraying to dissolve pollutants and to separate clay from silt. This eliminates fuels, metals and chemicals.

Vapour extraction

This is used to remove fuels or industrial solvents and other organic deposits. At variable depths, small diameter boreholes are located at frequent intervals. Attached to these are vacuum pipes to draw air through the contaminated soil. The contaminants are collected at a vapour treatment processing plant on the surface, treated and evaporated into the atmosphere. This is a slow process and it may take several months to cleanse a site.

Electrolysis

This process uses low voltage d.c. electricity in the presence of metals. Electricity flows between an anode and cathode, where metal ions in water accumulate in a sump before pumping to the surface for treatment.

Biological/phytoremediation

This is the removal of contaminants by plants which will absorb harmful chemicals from the ground. The plants are subsequently harvested and destroyed. A variant uses fungal degradation of the contaminants.

Bioremediation

Stimulating the growth of naturally occurring microbes. Microbes consume petrochemicals and oils, converting them to water and carbon dioxide. Conditions must be right, i.e. a temperature of at least 10°C with an adequate supply of nutrients and oxygen. Untreated soil can be excavated and placed over perforated piping, through which air is pumped to enhance the process prior to the soil being replaced.

Chemical

Oxidation ~ subsoil boreholes are used for the pumped distribution of liquid hydrogen peroxide or potassium permanganate. Chemicals and fuel deposits convert to water and carbon dioxide.

Solvent extraction ~ the subsoil is excavated and mixed with a solvent to break down oils, grease and chemicals that do not dissolve in water.

Thermal

Thermal treatment (off-site) ~ an incineration process involving the use of a large heating container/oven. Soil is excavated, dried and crushed prior to heating to 2500°C, where harmful chemicals are removed by evaporation or fusion.

Thermal treatment (in situ) ~ steam, hot water or hot air is pressure-injected through the soil. Variations include electric currents and radio waves to heat water in the ground to become steam. This evaporates chemicals.

Building Regulations, Approved Document, C1: Site preparation and resistance to contaminants. Section 1: Clearance or treatment of unsuitable material. Section 2: Resistance to contaminants.

4.12 Retaining walls

The major function of any retaining wall is to act as an earth-retaining structure for the whole or part of its height on one face, the other being exposed to the elements. Most small height retaining walls are built entirely of brickwork or a combination of brick facing and blockwork or mass concrete backing. To reduce hydrostatic pressure on the wall from ground water, an adequate drainage system in the form of weep holes should be used. Alternatively, subsoil drainage behind the wall could be employed.

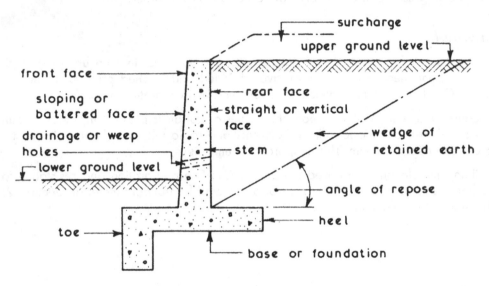

Figure 4.17 Retaining wall terminology

Design of retaining walls

This should allow for the effect of hydrostatics or water pressure behind the wall and the pressure created by the retained earth. Calculations are based on a 1m unit length of wall.

Small height retaining walls

These retaining walls must be stable and the usual rule of thumb for small height brick retaining walls is for the height to lie between two and four times the wall thickness. Stability can be checked by applying the middle third rule (page 123).

Retaining walls up to 6m high

These can be classified as medium height retaining walls and have the primary function of retaining soils at an angle in excess of the soil's natural angle of repose.

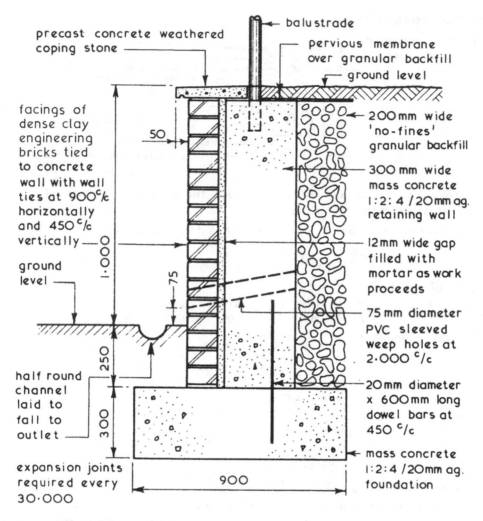

facings of dense clay engineering bricks tied to concrete wall with wall ties at 900 c/c horizontally and 450 c/c vertically

precast concrete weathered coping stone

balustrade

pervious membrane over granular backfill

ground level

50

200 mm wide 'no-fines' granular backfill

300 mm wide mass concrete 1:2:4 /20mm ag. retaining wall

1·000

ground level

75

12mm wide gap filled with mortar as work proceeds

75 mm diameter PVC sleeved weep holes at 2·000 c/c

250

half round channel laid to fall to outlet

20mm diameter x 600mm long dowel bars at 450 c/c

300

mass concrete 1:2:4 /20mm ag. foundation

expansion joints required every 30·000

900

Figure 4.18 Typical example of a combination retaining wall

Walls within this height range are designed to provide the necessary resistance by either their own mass or by the principles of leverage.

Design

The actual design calculations are usually carried out by a structural engineer who endeavours to ensure that:

- Overturning of the wall does not occur.
- Forward sliding of the wall does not occur.
- Materials used are suitable and not overstressed.

121

Retaining walls

1. The resultant thrust
2. The overturning or bending moment

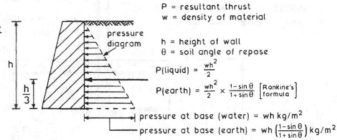

P, the resultant thrust, will act through the centre of gravity of the pressure diagram, i.e. at h/3.

The overturning moment due to water is therefore:

$$\frac{wh^2}{2} \times \frac{h}{3} \text{ or } \frac{wh^3}{6}$$

and for earth:

$$\frac{wh^2}{2} \times \frac{1-\sin\theta}{1+\sin\theta} \times \frac{h}{3} \text{ or } \frac{wh^3}{6} \times \frac{1-\sin\theta}{1+\sin\theta}$$

Typical example ~

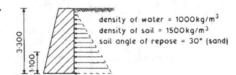

For water:

$$P = \frac{wh^2}{2} = \frac{1000 \times (3.3)^2}{2} = 5445 \text{ kg}$$

NB. kg × gravity = Newtons

Therefore, 5445 kg × 9.81 = 53.42 kN

The overturning or bending moment will be:

$$P \times h/3 = 53.42 \text{ kN} \times 1.1 \text{ m} = 58.8 \text{ kNm}$$

For earth:

$$P = \frac{wh^2}{2} \times \frac{1-\sin\theta}{1+\sin\theta}$$

$$P = \frac{1500 \times (3.3)^2}{2} \times \frac{1-\sin 30°}{1+\sin 30°} = 2723 \text{ kg} \text{ or } 26.7 \text{ kN}$$

The overturning or bending moment will be:
P × h/3 = 26.7 kN × 1.1m = 29.4 kNm

Figure 4.19 Retaining wall thrust factors

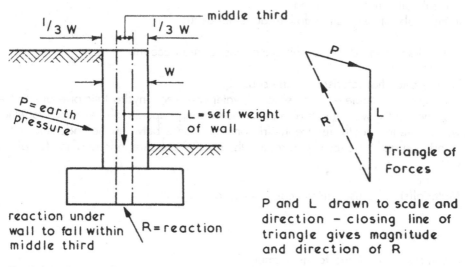

middle third

$^1/3$ W $^1/3$ W

W

P = earth pressure

L = self weight of wall

reaction under wall to fall within middle third

R = reaction

Triangle of Forces

P and L drawn to scale and direction — closing line of triangle gives magnitude and direction of R

Typical Example of Brick Retaining Wall ~

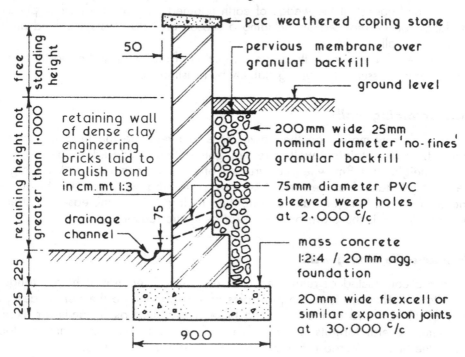

free standing height

retaining height not greater than 1·000

50

retaining wall of dense clay engineering bricks laid to english bond in cm. mt 1:3

drainage channel

75

225   225

900

pcc weathered coping stone

pervious membrane over granular backfill

ground level

200mm wide 25mm nominal diameter 'no-fines' granular backfill

75mm diameter PVC sleeved weep holes at 2·000 $^c/c$

mass concrete 1:2:4 / 20 mm agg. foundation

20mm wide flexcell or similar expansion joints at 30·000 $^c/c$

Figure 4.20 Typical brick retaining wall

Retaining walls

- The subsoil is not overloaded.
- In clay subsoils, slip circle failure does not occur.

The factors that the designer will have to take into account:

- Nature and characteristics of the subsoil(s).
- Height of water table – the presence of water can create hydrostatic pressure on the rear face of the wall, it can also affect the bearing capacity of the subsoil together with its shear strength, reduce the frictional resistance between the underside of the foundation and the subsoil and reduce the passive pressure in front of the toe of the wall.
- Type of wall.
- Material(s) to be used in the construction of the wall.

Earth pressures

These can take one of two forms, namely:

Active earth pressures ~ these are those pressures that tend to move the wall at all times and consist of the wedge of earth retained plus any hydrostatic pressure. The latter can be reduced by including a subsoil drainage system behind and/or through the wall.

Passive earth pressures ~ these are a reaction of an equal and opposite force to any imposed pressure, thus giving stability by resisting movement.

Mass retaining walls

These walls rely mainly on their own mass to overcome the tendency to slide forward. Mass retaining walls are not generally considered to be economic over a height of 1.8m when constructed of brick or concrete and 1m high in the case of natural stonework. Any mass retaining wall can be faced with another material but generally any applied facing will not increase the strength of the wall and is therefore only used for aesthetic reasons.

Cantilever retaining walls

These are constructed of reinforced concrete with an economic height range of 1.2m to 6m. They work on the principles of leverage where the stem is designed as a cantilever fixed at the base and base is designed as a cantilever fixed at the stem. Several formats are possible and in most cases a beam is placed below the base to increase the total passive resistance to sliding.

Crib retaining walls

These are a system of precast concrete or treated timber components comprising headers and stretchers, which interlock to form a three-dimensional framework. During assembly the framework is filled with graded stone to create sufficient mass to withstand ground pressures.

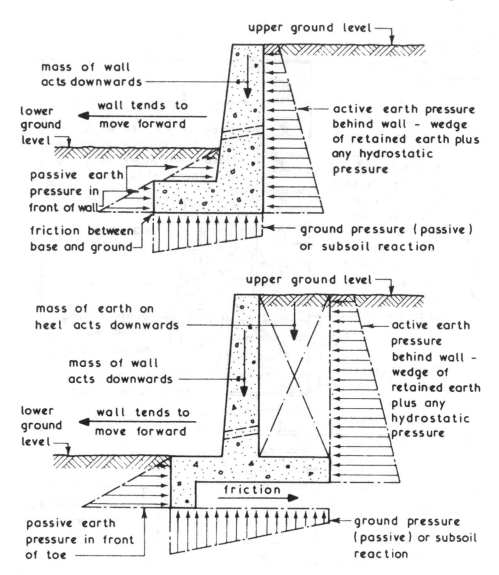

Figure 4.21 Passive and active pressures in retaining wall design

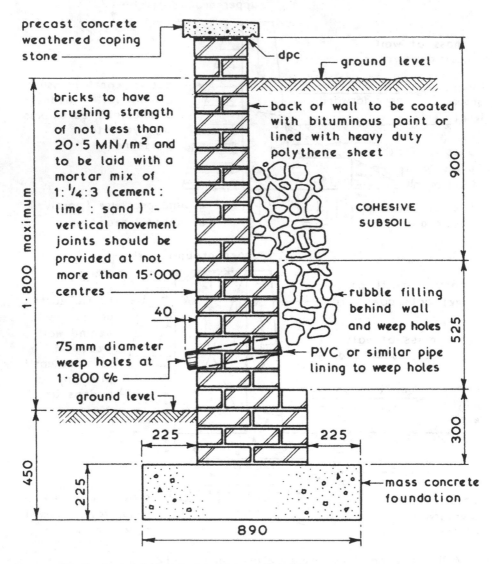

precast concrete weathered coping stone

dpc

ground level

back of wall to be coated with bituminous paint or lined with heavy duty polythene sheet

COHESIVE SUBSOIL

bricks to have a crushing strength of not less than 20·5 MN/m² and to be laid with a mortar mix of 1:¹/₄:3 (cement: lime : sand) – vertical movement joints should be provided at not more than 15·000 centres

1·800 maximum

900

525

rubble filling behind wall and weep holes

40

75 mm diameter weep holes at 1·800 c/c

PVC or similar pipe lining to weep holes

ground level

225

225

300

450

225

mass concrete foundation

890

Figure 4.22 Typical solid brick retaining wall details

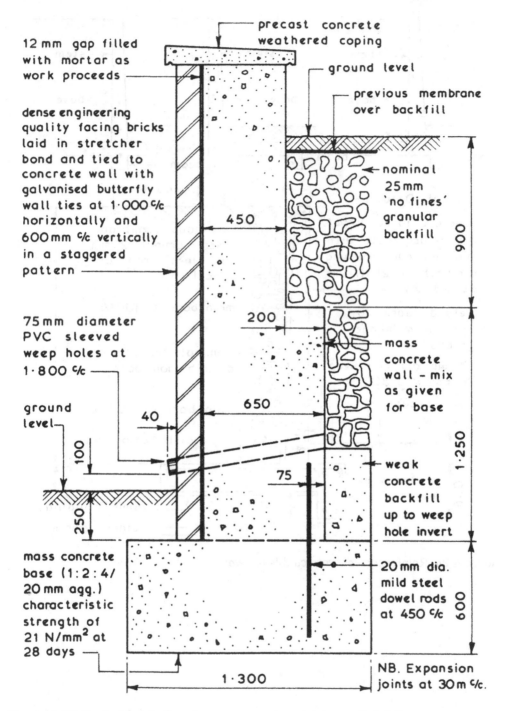

12 mm gap filled with mortar as work proceeds

dense engineering quality facing bricks laid in stretcher bond and tied to concrete wall with galvanised butterfly wall ties at 1·000 c/c horizontally and 600 mm c/c vertically in a staggered pattern

75 mm diameter PVC sleeved weep holes at 1·800 c/c

ground level

mass concrete base (1:2:4/ 20 mm agg.) characteristic strength of 21 N/mm² at 28 days

precast concrete weathered coping

ground level

previous membrane over backfill

nominal 25 mm 'no fines' granular backfill

mass concrete wall – mix as given for base

weak concrete backfill up to weep hole invert

20 mm dia. mild steel dowel rods at 450 c/c

NB. Expansion joints at 30 m c/c.

450

200

650

40

100

250

75

900

1·250

600

1·300

Figure 4.23 Typical brick-faced mass concrete retaining wall detail

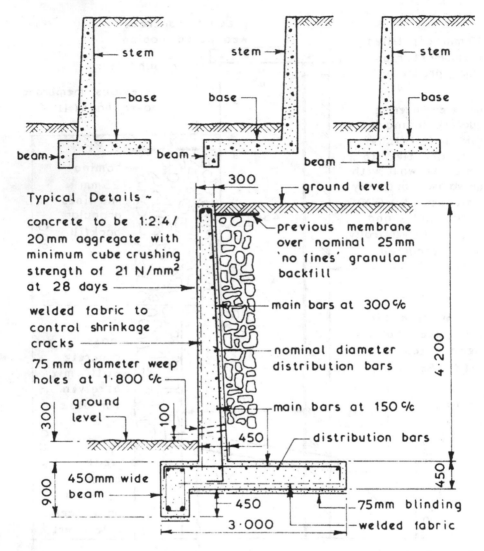

Typical Details ~

concrete to be 1:2:4/ 20mm aggregate with minimum cube crushing strength of 21 N/mm² at 28 days

welded fabric to control shrinkage cracks

75 mm diameter weep holes at 1·800 c/c

ground level

300

100

450

450mm wide beam

900

450

3·000

300 — ground level

previous membrane over nominal 25mm 'no fines' granular backfill

main bars at 300 c/c

nominal diameter distribution bars

main bars at 150 c/c

distribution bars

4·200

450

75mm blinding

welded fabric

Figure 4.24 Reinforced concrete cantilever walls

stem — base — beam

stem — base — beam

stem — base — beam

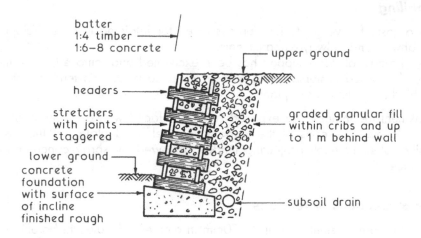

batter
1:4 timber
1:6–8 concrete

upper ground

headers

stretchers
with joints
staggered

graded granular fill
within cribs and up
to 1 m behind wall

lower ground
concrete
foundation
with surface
of incline
finished rough

subsoil drain

NB. height limited to 10 m with timber.

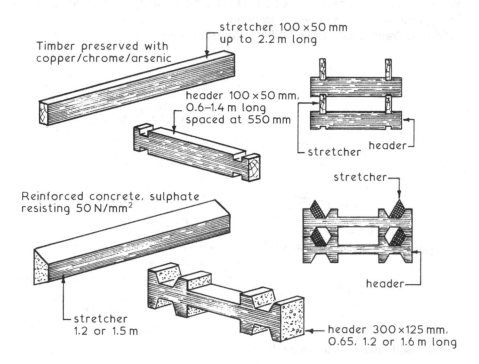

Timber preserved with
copper/chrome/arsenic

stretcher 100 × 50 mm
up to 2.2 m long

header 100 × 50 mm,
0.6–1.4 m long
spaced at 550 mm

stretcher

header

stretcher

header

Reinforced concrete, sulphate
resisting 50 N/mm²

stretcher
1.2 or 1.5 m

header 300 × 125 mm,
0.65, 1.2 or 1.6 m long

Figure 4.25 Crib retaining walls

Retaining walls

Soil nailing

This is a cost-effective geotechnic process used for retaining large soil slopes, notably highway and railway embankments.

After the natural slope support has been excavated and removed, the remaining wedge of exposed unstable soil is pinned or nailed back with tendons into stable soil behind the potential slip plane.

Embankment treatment ~ the exposed surface is faced with a plastic coated wire mesh to fit over the ends of the tendons. A steel head plate is fitted over and centrally bolted to each projecting tendon, followed by spray concreting to the whole face.

Types of soil nails or tendons

- Solid deformed steel rods up to 50mm in diameter, located in boreholes up to 100mm in diameter. Cement grout is pressurised into the void around the rods.
- Hollow steel, typically 100mm diameter tubes with an expendable auger attached. Cement grout is injected into the tube during boring to be ejected through purpose-made holes in the auger.
- Solid glass reinforced plastic (GRP) with resin grouts.

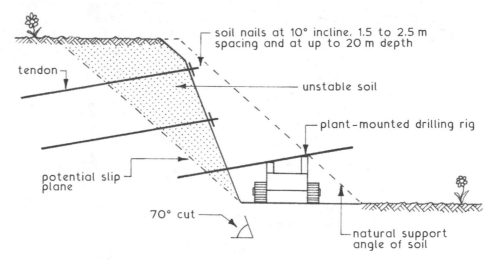

Figure 4.26 Soil nailing to an embankment

Gabions

This is a type of retaining wall produced from individual rectangular boxes made from panels of wire mesh, divided internally and filled with stones. These units are stacked and overlapped (like stretcher bonded masonry) and applied in several layers or courses to retained earth situations. Typical sizes, 1.0m long × 0.5m wide × 0.5m high, up to 4.0m long × 1.0m wide × 1.0m high.

Mattress unit fabrication is similar to a gabion but is thinner and uses a smaller mesh and stone size to provide some flexibility and shaping potential. Application is at a much lower incline. Generally used next to waterways for protection against land erosion where tidal movement and/or water level differentials could scour embankments.

Typical sizes, 3.0m long × 2.0m wide × 0.15m thick, up to 6.0m long × 2.0m wide × 0.3m thick.

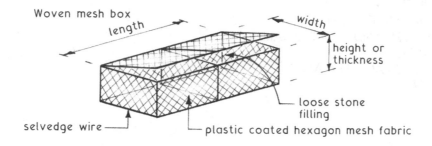

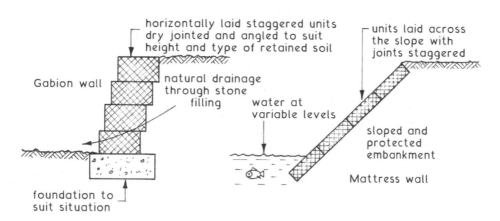

Figure 4.27 Gabion baskets and walls

5 MATERIALS

5.1 Density of building materials

Table 5.1 Density of building materials per m²

Material	Weight (kg/m²)
BRICKS, BLOCKS and PAVING –	
Clay brickwork – 102.5mm	
low density	205
medium density	221
high density	238
Calcium silicate brickwork – 102.5mm	205
Concrete blockwork, aerated	78
.... lightweight aggregate	129
Concrete flagstones (50mm)	115
Glass blocks (100mm thick) 150 × 150	98
200 × 200	83
ROOFING –	
Thatching (300mm thick)	40.00
Tiles – plain clay	63.50
..– plain concrete	93.00
..– single lap, concrete	49.00
Tile battens (50 × 25) and felt underlay	7.70
Bituminous felt underlay	1.00
Bituminous felt, sanded topcoat	2.70
3 layers bituminous felt	4.80
HD/PE breather membrane underlay	0.20
SHEET MATERIALS –	
Aluminium (0.9mm)	2.50
Copper (0.9mm)	4.88
Cork board (standard) per 25mm thickness	4.33
.... (compressed)	9.65
Hardboard (3.2mm)	3.40
Glass (3mm)	7.30
Lead (1.25mm)	14.17
.... (3mm)	34.02
Particle board/chipboard (12mm)	92.6
(22mm)	16.82
Planking, softwoodstrip flooring (ex 25mm)	11.20
.... hardwood..................	16.10
Plasterboard (9.5mm)	8.30
(12.5mm)	11.00
..............(19mm)	17.00
Plywood per 25mm	15.00
PVC floor tiling (2.5mm)	3.90
Strawboard (25mm)	9.80

Density of building materials

Table 5.1 (Cont.)

Material	Weight (kg/m^2)
Weatherboarding (20mm)	7.68
Woodwool (25mm)	14.50
INSULATION	
Glass fibre thermal (100mm)	2.00
.... acoustic.........	4.00
APPLIEDMATERIALS -	
Asphalt (18mm)	42
Plaster, 2 coat work	22
STRUCTURAL TIMBER:	
Rafters and joists (100 × 50 @ 400c/c)	5.87
Floor joists (225×50 @ 400 c/c)	14.93

Densities per m^3

Table 5.2 Density of building materials per m^3

Material	Approx. density (kg/m^3)
Cement	1440
Concrete (aerated)	640
.... (broken brick)	2000
.... (natural aggregates)	2300
.... (no-fines)	1760
.... (reinforced)	2400
Metals:	
Aluminium	2710
Brass	8500
Copper	8930
Lead	11,325
Steel	7860
Tin	7300
Zinc	7140
Timber (softwood/pine)	480 (average)
.... (hardwood, e.g. maple, teak, oak)	720......
Water	1000

5.2 Concrete

Concrete is a mixture of cement + aggregates + water in controlled proportions.

CEMENT

Manufactured from clay and chalk and is the matrix or binder of the concrete mix. Cement powder can be supplied in bags or bulk —

Bags ~

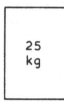

25 kg

airtight sealed bags requiring a dry, damp-free store.

Bulk ~

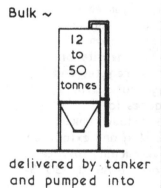

12 to 50 tonnes

delivered by tanker and pumped into storage silo.

AGGREGATES

Coarse aggregate is generally defined as a material which is retained on a 4mm sieve.

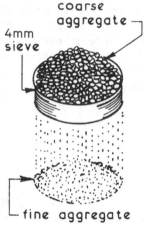

coarse aggregate

4mm sieve

fine aggregate

Fine aggregate is generally defined as a material which passes a 4mm sieve. Aggregates can be either natural rock which has disintegrated or crushed stone or gravel.

WATER

Must be of a quality fit for drinking.

MIXES

These are expressed as a ratio thus:
1:3:6/20mm
which means —
1 part cement.
3 parts of fine aggregate.
6 parts of coarse aggregate.
20mm — maximum size of coarse aggregate for the mix.

Water is added to start the chemical reaction and to give the mix workability ~ the amount used is called the Water/Cement Ratio and is usually about 0·4 to 0·5.

Too much water will produce a weak concrete of low strength whereas too little water will produce a concrete mix of low and inadequate work-ability.

Figure 5.1 Concrete constituents

Concrete

Concrete ~ a mixture of cement + fine aggregate + coarse aggregate + water in controlled proportions and of a suitable quality.

Cement ~ powder produced from clay and chalk or limestone. In general most concrete is made with ordinary or rapid hardening Portland cement, both types being manufactured to the recommendations of BS EN 197-1. Ordinary Portland cement is adequate for most purposes but has a low resistance to attack by acids and sulphates. Rapid hardening Portland cement does not set faster than ordinary Portland cement but it does develop its working strength at a faster rate. For a concrete which must have an acceptable degree of resistance to sulphate attack, the quantity of tricalcium aluminate is reduced during the manufacture of Portland cement (4% instead of 10% in OPC). BS EN 15743 contains reference to this type of sulphate-resisting cement.

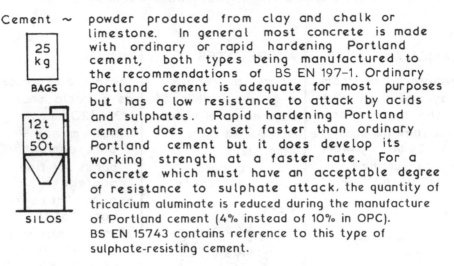

Figure 5.2 Cement storage

Aggregates ~ shape, surface texture and grading (distribution of particle sizes) are factors which influence the workability and strength of a concrete mix. Fine aggregates are generally regarded as those materials which pass through a 4 mm sieve whereas coarse aggregates are retained on a 4 mm sieve. Dense aggregates have a density of more than $1200 \, kg/m^3$ for coarse aggregates and more than $1250 \, kg/m^3$ for fine aggregates. These are detailed in PD 6682-1: Aggregates for concrete. Lightweight aggregates include clinker; foamed or expanded blastfurance slag and exfoliated and expanded materials such as vermiculite, perlite, clay and sintered pulverised fuel ash to BS EN 13055-1.

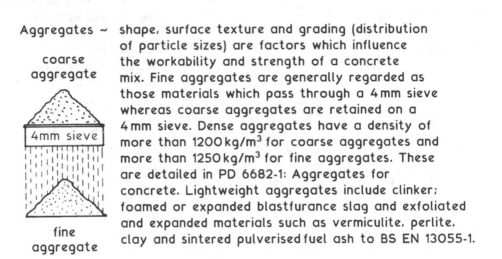

Figure 5.3 Aggregate size

Water ~ must be clean and free from impurities which are likely to affect the quality or strength of the resultant concrete. Pond, river, canal and sea water should not be used and only water which is fit for drinking should be specified.

drinking
water
quality

Figure 5.4 Water

Cement

Whichever type of cement is being used it must be properly stored on site to keep it in good condition. The cement must be kept dry since contact with any moisture, whether direct or airborne, could cause it to set. A rotational use system should be introduced to ensure that the first batch of cement delivered is the first to be used.

Concrete batching

A batch is one mixing of concrete and can be carried out by measuring the quantities of materials required by volume or weight. The main aim of both methods is to ensure that all consecutive batches are of the same standard and quality.

Volume batching

Concrete mixes are often quoted by ratio such as 1:2:4 (cement: fine aggregate or sand: coarse aggregate). Cement weighing 50kg has a volume of 0.033 m³; therefore for the above mix 2 × 0.033 (0.066 m³) of sand and 4 × 0.033 (0.132 m³) of coarse aggregate is required. To ensure accurate amounts of materials are used for each batch a gauge box should be employed, its size being based on convenient handling. Ideally a batch of concrete should be equated to using 50kg of cement per batch. Assuming a gauge box 300mm deep and 300mm wide with a volume of half the required sand the gauge box size would be – volume = length × width × depth = length × 300 × 300.

$$\text{length} = \frac{\text{volume}}{\text{width} \times \text{depth}} = \frac{0.033}{0.3 \times 0.3} = 0.366\,\text{m}$$

For the above given mix fill the gauge box once with cement, twice with sand and four times with coarse aggregate.

An allowance must be made for the bulking of damp sand which can be as much as 33.33%. General rule of thumb unless using dry sand: allow for 25% bulking.

Materials should be well mixed when dry, before water is added.

Concrete

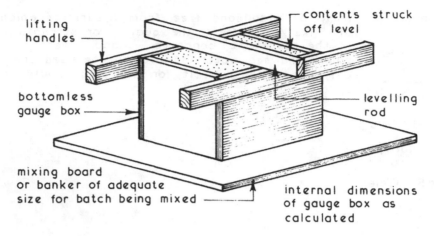

lifting handles

contents struck off level

bottomless gauge box

levelling rod

mixing board or banker of adequate size for batch being mixed

internal dimensions of gauge box as calculated

Figure 5.5 Gauge box

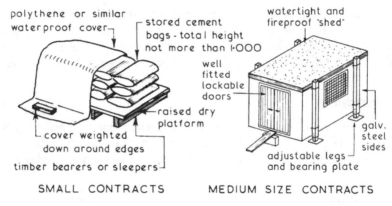

polythene or similar waterproof cover

stored cement bags - total height not more than 1·000

watertight and fireproof "shed"

well fitted lockable doors

cover weighted down around edges

raised dry platform

timber bearers or sleepers

galv. steel sides

adjustable legs and bearing plate

SMALL CONTRACTS   MEDIUM SIZE CONTRACTS

LARGE CONTRACTS — for bagged cement watertight container as above. For bulk delivery loose cement, a cement storage silo.

Aggregates ~ essentials of storage are to keep different aggregate types and/or sizes separate, store on a clean, hard, free draining surface and to keep the stored aggregates clean and free of leaves and rubbish.

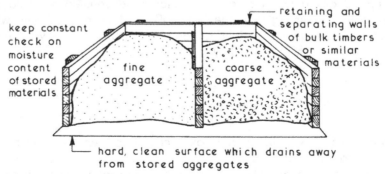

keep constant check on moisture content of stored materials

retaining and separating walls of bulk timbers or similar materials

fine aggregate

coarse aggregate

hard, clean surface which drains away from stored aggregates

Figure 5.6 Typical storage methods

Weight or weigh batching

This is a more accurate method of measuring materials for concrete than volume batching since it reduces considerably the risk of variation between different batches. The weight of sand is affected very little by its dampness, which in turn leads to greater accuracy in proportioning materials.

When loading a weighing hopper the materials should be loaded in a specific order:

1. Coarse aggregates – tend to push other materials out and leave the hopper clean.
2. Cement – this is sandwiched between the other materials since some of the fine cement particles could be blown away if cement is put in last.
3. Sand or fine aggregates – put in last to stabilise the fine lightweight particles of cement powder.

Typical densities ~ cement – 1440 kg/m^3 sand – 1600 kg/m^3 coarse aggregate – 1440 kg/m^3.

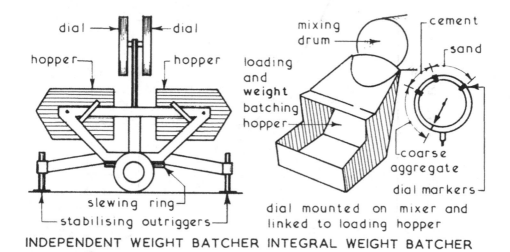

INDEPENDENT WEIGHT BATCHER INTEGRAL WEIGHT BATCHER

Figure 5.7 Weight batching plant

Water/cement ratio

Water in concrete has two functions:

1. Starts the chemical reaction which causes the mixture to set into a solid mass.
2. Gives the mix workability so that it can be placed, tamped or vibrated into the required position.

Concrete

Very little water is required to set concrete (approximately 0.2 w/c ratio) and the surplus evaporates leaving minute voids; therefore, the more water added to the mix to increase its workability, the weaker the resultant concrete. Generally w/c ratios of 0.4 to 0.5 are adequate for most purposes.

Free water ~ water on the surface of aggregates where stored in stock piles plus water added during the mixing process. Does not include any water that is absorbed by the aggregate.

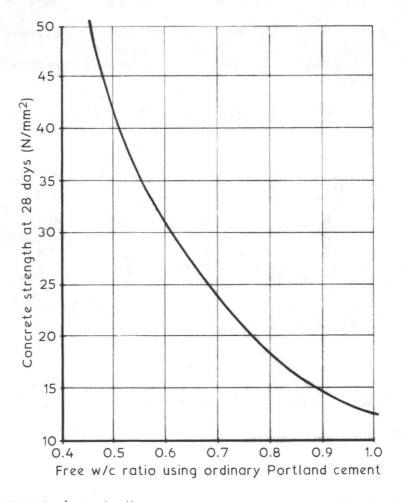

Figure 5.8 Water/cement ratios

Free water/cement ratio

A high w/c ratio increases the workability of concrete. This may be convenient for placing but it should be controlled in order to regulate the final concrete strength.

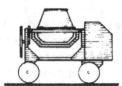

Small Batches ~ small, easily transported mixers with output capacities of up to 100 litres can be used for small and intermittent batches. These mixers are versatile and robust machines which can be used for mixing mortars and plasters as well as concrete.

Figure 5.9 Small concrete mixer

Note: The w/c ratio is found by dividing the weight of water in a batch by the weight of cement. E.g. If a batch contains 50kg of water and 100kg of cement, the w/c ratio is 0.5.

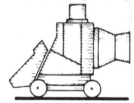

Medium to Large Batches ~ mixers with output capacities from 100 litres to 10m³ with either diesel or electric motors. Many models are available with tilting or reversing drum discharge, integral weigh batching and loading hopper and a controlled water supply.

Figure 5.10 Medium concrete mixer

Concrete

Ready Mixed Concrete ~ used mainly for large concrete batches of up to 6 m³. This method of concrete supply has the advantages of eliminating the need for site space to accommodate storage of materials, mixing plant and the need to employ adequately trained site staff who can constantly produce reliable and consistent concrete mixes. Ready mixed concrete supply depots also have better facilities and arrangements for producing and supplying mixed concrete in winter or inclement weather conditions. In many situations it is possible to place the ready mixed concrete into the required position direct from the delivery lorry via the delivery chute or by feeding it into a concrete pump. The site must be capable of accepting the 20 tonnes laden weight of a typical ready mixed concrete lorry with a turning circle of about 15·000. The supplier will want full details of mix required and the proposed delivery schedule.

Figure 5.11 Ready mix concrete vehicle

BS EN 206–1: *Concrete. Specification, Performance, Production and Conformity.*

Concrete specification

Concrete is a composite with many variables, represented by numerous gradings that indicate components, quality and manufacturing control.

Grade mixes ~ C7.5, C10, C15, C20, C25, C30, C35, C40, C45, C50, C55 and C60; F3, F4 and F5; IT2, IT2.5 and IT3.

$$\left.\begin{array}{l} C = \text{Characteristic compressive} \\ F = \text{Flexural} \\ IT = \text{Indirect tensile} \end{array}\right\} \text{strengths at 28 days } \left(N/mm^2\right)$$

Note: If the grade is followed by a 'P', e.g. C30P, this indicates a prescribed mix.

Table 5.3 Concrete grades

Grades C7.5 and C10	Unreinforced plain concrete.
Grades C15 and C20	Plain concrete or if reinforced containing lightweight aggregate.
Grades C25	Reinforced concrete containing dense aggregate.
Grades C30 and C35	Post tensioned reinforced concrete.
Grades C40 to C60	Pre tensioned reinforced concrete.

Categories of mix

1. Prescribed mix – Components are predetermined (to a recipe) to ensure strength requirements. Variations exist to allow the purchaser to specify particular aggregates, admixtures and colours. All grades permitted.

2. Standard prescribed mix – Applicable to minor works such as house construction, particularly where weight or volume batching is used for small quantities. See next page for standard (ST) mixes with C30 strength class concrete.

3. Designed mix – Concrete is specified to an expected performance. Criteria can include characteristic strength, durability and workability, to which a concrete manufacturer will design and supply an appropriate mix. All grades permitted.

4. Designated mix – Selected for specific applications.

- General (GEN) graded 0–4, 7.5–25N/mm^2 for foundations, floors and external works.
- Foundations (FND) graded 2, 3, 4A and 4B, 35N/mm^2, mainly for sulphate resisting foundations.
- Paving (PAV) graded 1 or 2, 35 or 45N/mm^2 for roads and drives.
- Reinforced (RC) graded 30, 35, 40, 45 and 50N/mm^2, mainly for pre-stressing.

See also BS EN 206:2013+A1:2016 *Concrete, Specification, performance, production and conformity*.

Standard prescribed mixes for site batching are graded for concrete with a characteristic compressive strength of 30N/mm^2 at 28 days by laboratory testing sample cubes.

Consistency class ~ this refers to the amount of water in the mix, affecting workability and final strength.

Concrete supply

This is usually geared to the demand or the rate at which the mixed concrete can be placed. Fresh concrete should always be used or placed within 30 minutes of mixing to prevent any undue drying out. Under no circumstances should more water be added after the initial mixing.

Concrete

Table 5.4 Standard prescribed mix constituents by weight

Weight batched proportions

Standard prescribed mix	Consistency class (slump)	Cement bag (kg)	Fine agg. (kg)	Coarse agg. (kg) max. 20mm	Mix ratio
ST1	S1	25	84	126	1:3.35:5.02
ST2	S2	25	72	107	1:2.87:4.28
ST2	S3	25	65	97	1:2.58:3.88
ST2	S4	25	68	83	1:2.72:3.30
ST3	S2	25	63	95	1:2.52:3.80
ST4	S2	25	56	83	1:2.23:3.33
ST5	S2	25	48	72	1:1.92:2.88

Table 5.5 Standard prescribed mix constituents by volume

Volume batched proportions

Standard prescribed mix	Consistency class (slump)	Cement bag (kg)	Fine aggregate (litres)	Coarse aggregate (litres)
ST1	S1	25	60	85
ST2	S2	25	50	75
ST2	S3	25	45	70
ST2	S4	25	50	60
ST3	S2	25	45	65

Table 5.6 Standard prescribed mix slump classification

Consistency class	Slump (mm)
S1	10–40
S2	50–90
S3	100–150
S4	160–210

Site concrete testing

The quality of concrete should be monitored, especially to ensure uniformity of mix components and consistency of batches. When constituents are known to be constant by design, the water/cement ratio of consecutive batches can be compared by using a slump test. Water for hydration is about one-quarter of the cement weight, but if a w/c ratio of 0.25 were used the mix would be too stiff and unworkable. Therefore, a w/c ratio of 0.5 is more usual (the additional water providing a lubricant for the aggregates and cement). Increasing the w/c ratio above 0.5 will make the concrete easier to work, but it will be detrimental to the final compressive strength unless other measures are incorporated to counter this, e.g. steel or fabric reinforcement.

Slump test equipment

Shown in Figure 5.12 is an open-ended frustum of a cone, tamping rod and a rule. Dimensions in mm.

Procedure

The cone is one-quarter filled with concrete and tamped 25 times. Further filling and tamping is repeated three more times until the cone is full and the sample levelled off. The cone is removed and the concrete slump measured. The measurement should be consistent for all samples of concrete being used in the same situation.

Typical slump specification

Generally between 50 and 100mm depending on application for slump consistency classes applicable to standard prescribed site batched concrete mixes.

BS EN 12350–2: *Testing Fresh Concrete. Slump Test.*

Concrete cube tests

Procedure for testing batched concrete from source:

- Wet concrete samples extracted from the mixer batch are placed into machine faced steel casting moulds in 50mm layers.
- Each layer compacted at least 35 times (150mm cube) or 25 times (100mm cube) with a tamping bar.
- Alternatively, each layer may be consolidated by external vibration from an electric or pneumatic hammer.
- Surplus concrete trowelled off and samples marked with time, date and job reference.
- Samples left for 24 hours ±30 minutes and covered with a damp cloth.
- After 24 hours, samples removed from moulds and submersed in water at 20° C ± 2°C.
- At 7 days, hydraulic compression test applied to sample cubes to determine stress failure strength.

Concrete

- If strength specification not achieved, other samples from the same batch tested at 28 days.
- If 28 day samples fail compressive strength requirement, core samples may be taken from placed concrete for further laboratory analysis.

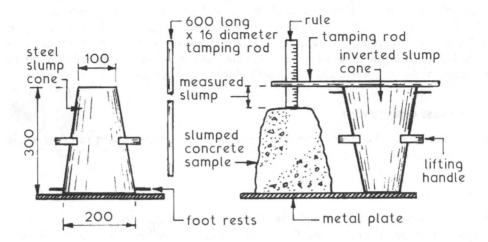

Figure 5.12 Slump test of concrete mix

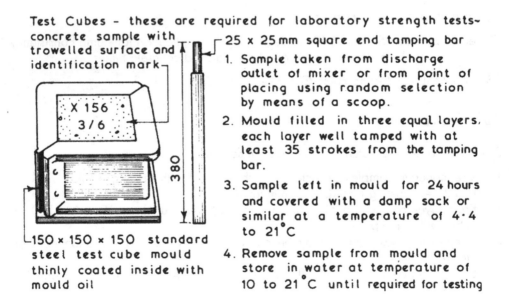

Test Cubes - these are required for laboratory strength tests-

concrete sample with trowelled surface and identification mark⌐

25 x 25 mm square end tamping bar

1. Sample taken from discharge outlet of mixer or from point of placing using random selection by means of a scoop.

2. Mould filled in three equal layers, each layer well tamped with at least 35 strokes from the tamping bar.

3. Sample left in mould for 24 hours and covered with a damp sack or similar at a temperature of 4·4 to 21°C

4. Remove sample from mould and store in water at temperature of 10 to 21°C until required for testing

150 x 150 x 150 standard steel test cube mould thinly coated inside with mould oil

Figure 5.13 Standard concrete cube test

28 day characteristic crushing strength categories, below which not more than 5% of test results are permitted to fail, are: 7.5, 10, 15, 20, 25, 30 and 35N/mm². Categories 40, 45 and 50N/mm² are also specified, mainly for prestressed reinforced concrete.

BS EN 12390–1: *Testing Hardened Concrete. Shape, Dimensions and Other Requirements for Specimens and Moulds.*

Non-destructive testing of concrete (also known as in-place or in-situ tests)

Changes over time and in different exposures can be monitored.
BS EN 13791: 201: *Assessment of in–situ compressive strength in structures and precast con-crete components.*

BS 1881: *Testing concrete.*
BS EN 13791: *Assessment of in-situ compressive strength in structures and precast concrete components.*

This document provides information on strength in situ, voids, flaws, cracks and deterioration.

Rebound hammer test

Attributed to Ernst Schmidt after he devised the impact hammer in 1948, it works on the principle of an elastic mass rebounding off a hard surface. Varying surface densities will affect impact and propagation of stress waves. These can be recorded on a numerical scale known as rebound numbers. It has limited application to smooth surfaces of concrete only. False results may occur where there are local variations in the concrete, such as a large piece of aggregate immediately below the impact surface. Rebound numbers can be graphically plotted to correspond with compressive strength.

BS EN 12504–2: *Testing Concrete in Structures. Non-Destructive Testing. Determination of Rebound Number.*
PD 6682–1: Aggregates for concrete.

Vibration test ~ a number of electronic tests have been devised, which include measurement of ultrasonic pulse velocity through concrete. This applies the principle of recording a pulse at predetermined frequencies over a given distance. The apparatus includes transducers in contact with the concrete, pulse generator, amplifier and time measurement to digital display circuit. For converting the data to concrete compressive strength, see BS EN 12504–4: *Testing Concrete. Determination of Ultrasonic Pulse Velocity.*

A variation, using resonant frequency, measures vibrations produced at one end of a concrete sample against a receiver or pick up at the other. The driving unit or exciter is activated by a variable frequency oscillator to generate vibrations varying in resonance, depending on the concrete quality. The calculation of compressive strength by conversion of amplified vibration data is by formulae found in BS 1881 –209: *Testing Concrete. Recommendations for the Measurement of Dynamic Modulus of Elasticity.*

Concrete

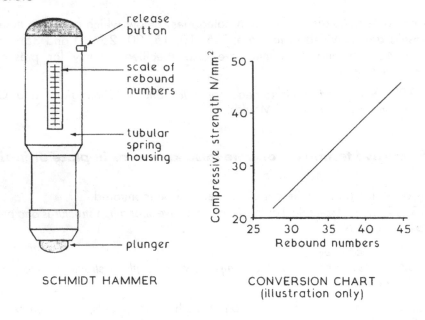

Figure 5.14 Rebound test for concrete

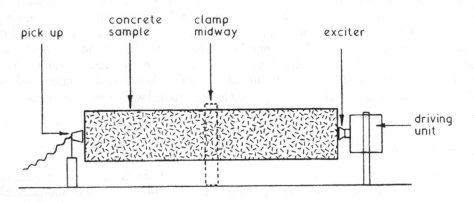

Figure 5.15 Resonant frequency test

Other relevant standards

BS 1881–122: *Testing Concrete. Method for Determination of Water Absorption.*
BS 1881–124: *Testing Concrete. Methods for Analysis of Hardened Concrete.*
BS EN 12390–7: *Testing Hardened Concrete. Density of Hardened Concrete.*

5.3 Timber

The quality of softwood timber for structural use depends very much on the environment in which it is grown and the species selected. Timber can be visually strength graded, but this is unlikely to occur at the construction site except for a general examination for obvious handling defects and damage during transit. Site inspection will be to determine that the grading authority's markings on the timber comply with that specified for the application.

Format of strength grade markings on softwood timber for structural uses are shown in Figure 5.16.

BS 4978: *Visual Strength Grading of Softwood. Specification.*

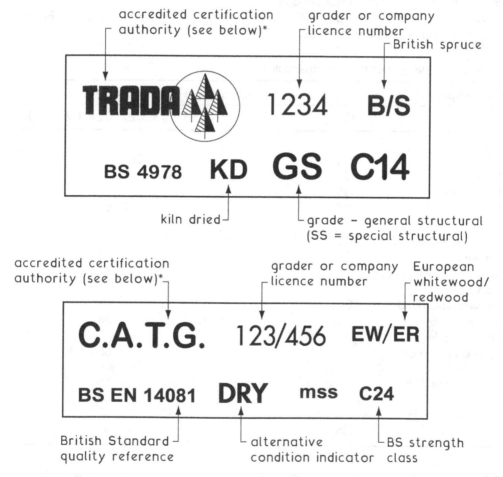

*Accredited certification authorities include
BM TRADA Certification Ltd. and Certification And Timber Grading Ltd.

Figure 5.16 Timber grading stamps

Timber

BS EN 14081: *Timber Structures. Strength Graded Structural Timber with Rectangular Cross Section* (three parts).

Grading is done either visually or by computerised machine. Individual rectangular timber sections are assessed against permissible defect limitations and grade marked accordingly.

- UK grading standard: BS 4978.
- European grading standard: BS EN 14081 (three parts).

The two principal grades apart from rejects are GS (general structural) and SS (special structural), preceded with an M if graded by machine.

Additional specification is to BS EN 338: *Structural Timber. Strength Classes.* This standard provides softwood strength classifications from C14 to C40 as well as a separate classification of hardwoods.

A guide to softwood grades with strength classes for timber from the UK, Europe and North America is given in Table 5.7.

Table 5.7 Timber species strength classifications

Source/species	Strength class (BS EN 338)						
	C14	C16	C18	C22	C24	C27	C30
UK:							
British pine	GS				SS		
British spruce	GS			SS			
Douglas fir	GS			SS			
Larch	GS					SS	
Ireland:							
Sitka and Norway spruce	GS			SS			
Europe:							
Redwood or white–wood			GS			SS	
USA:							
Western whitewood	GS			SS			
Southern pine			GS			SS	
USA/Canada:							
Spruce/pine/fir or hemlock			GS			SS	
Douglas fir and larch			GS			SS	
Canada:							
Western red cedar	GS			SS			
Sitka spruce	GS			SS			

BS EN 338: *Structural Softwood Classifications and Typical Strength Properties* is illustrated in Table 5.8.

Table 5.8 Structural softwood strength properties

BSEN 338 strength class	Bending parallel to grain (N/mm²)	Tension parallel to grain (N/mm²)	Compression parallel to grain (N/mm²)	Compression perpendicular to grain (N/mm²)	Shear parallel to grain (N/mm²)	Modulus of elasticity		Characteristic density (kg/m³)	Average density (kg/m³)
						Mean (N/mm²)	Minimum (N/mm²)		
C14	4.1	2.5	5.2	2.1	0.60	6800	4600	290	350
C16	5.3	3.2	6.8	2.2	0.67	8800	5800	310	370
C18	5.8	3.5	7.1	2.2	0.67	9100	6000	320	380
C22	6.8	4.1	7.5	2.3	0.71	9700	6500	340	410
C24	7.5	4.5	7.9	2.4	0.71	10,800	7200	350	420
TR26	10.0	6.0	8.2	2.5	1.10	11,000	7400	370	450
C27	10.0	6.0	8.2	2.5	1.10	12,300	8200	370	450
C30	11.0	6.6	8.6	2.7	1.20	12,300	8200	380	460
C35	12.0	7.2	8.7	2.9	1.30	13,400	9000	400	480
C40	13.0	7.8	8.7	3.0	1.40	14,500	10,000	420	500

1. Strength class TR26 is specifically for the manufacture of trussed rafters.
2. Characteristic density values are given specifically for the design of joints. Average density is appropriate for calculation of dead load.

Timber

Visual strength grading ~ 'process by which a piece of timber can be sorted, by means of visual inspection, into a grade to which characteristic values of strength, stiffness and density may be allocated.' Definition from BS EN 14081–1.

Characteristics

Knots

Branch growth from or through the main section of timber weakens the overall structural strength. Measured by comparing the sum of the projected cross-sectional knot area with the cross-sectional area of the piece of timber. This is known as the knot area ratio (KAR). Knots close to the edge of the section have greater structural significance; therefore this area is represented as a margin condition at the top and bottom quarter of a section. A margin condition exists when more than half the top or bottom quarter of a section is occupied by knots.

MKAR = Margin knot area ratio.
TKAR = Total knot area ratio.

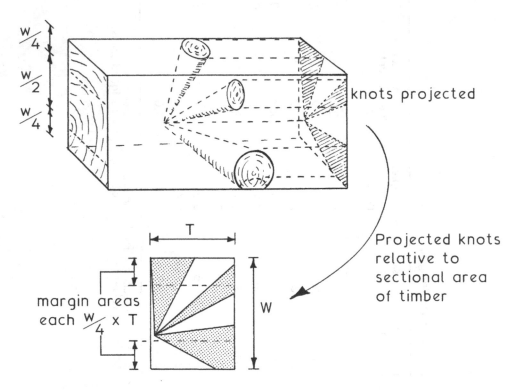

Figure 5.17 Growth of knots in timber

Fissures and resin pockets

Defects in growth, fissures, also known as shakes, are usually caused by separation of annual growth rings. Fissures and resin pockets must be limited in structural timber as they reduce resistance to shear and bending parallel to the grain.

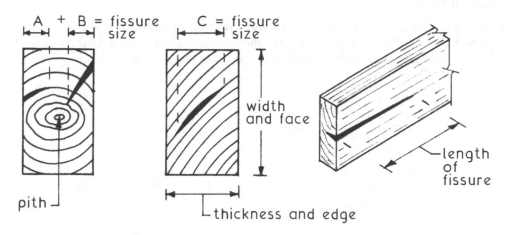

Figure 5.18 Defects in growth

Slope of grain

This is an irregularity in growth or where the log is not cut parallel to the grain. If excessive this will produce a weakness in shear. Measurement is taken by scoring a line along the grain of the timber surface and comparing this with the parallel sides of the section.

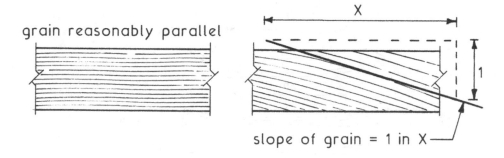

Figure 5.19 Slope of grain

Timber

Wane or waney edge

This occurs on timber cut close to the outer surface of the log, producing incomplete corners. Measurement is parallel to the edge or face of the section and it is expressed as a fraction of the surface dimension.

Growth rate

Measurement is applied to the annual growth ring separation averaged over a line 75mm long. If pith is present the line should commence 25mm beyond and if 75mm is impractical to achieve, the longest possible line is taken.

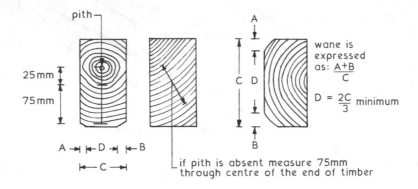

Figure 5.20 Wane/waney edge in timber

Distortion

Measurement is made over the length and width of section to determine the amount of bow, spring and twist.

Characteristics and tolerances of GS and SS graded timber are shown in Table 5.9.

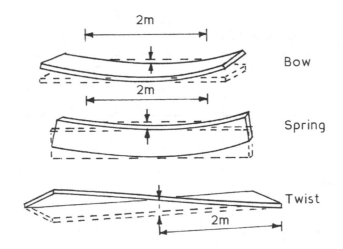

Figure 5.21 Common types of deformation in timber

154

Table 5.9 Timber distortion characteristics and tolerances for GS and SS graded timber

Criteria	GS	SS
KAR: No margin condition	MKAR≤1/2 TKAR≤1/2	MKAR≤1/2 TKAR≤1/3
	Or:	
KAR: Margin condition	MKAR > 1/2 TKAR≤1/3	MKAR > 1/2 TKAR≤1/5
Fissures and resin pockets:	Defects≤1/2 timber thickness	Defects≤1/2 timber thickness
Not through Thickness	<1.5m or 1/2 timber length take lesser	<1.0m or 1/4 timber length take lesser
Through Thickness	<1.0m or 1/4 timber length take lesser	<0.5m or 1/4 timber length take lesser
	If at ends fissure length maximum 2 × timber width	If at ends fissure length<width of timber section
Slope of grain:	Maximum 1 in 6	Maximum 1 in 10
Wane:	Maximum 1/3 of the full edge and face of the section – length not limited	
Resin pockets:		
Not through thickness	Unlimited if shorter than width of section, otherwise as for fissures	
Through thickness	Unlimited if shorter than 1/2 width of section, otherwise as for fissures	
Growth rate of annual rings:	Average width or growth <10mm	Average width or growth <6mm
Distortion:		
–bow	<20mm over 2m	<10mm over 2m
–spring	<12mm over 2m	<8mm over 2m
–twist	<2mm per 25mm width over 2m	<1mm per 25mm width over 2m

Timber

Timber dimensioning

Structural softwood cross-sectional size has established terminology such as sawn, basic and unwrought, as produced by conversion of the log into commercial dimensions, e.g. 100 × 50mm and 225 × 75mm (4″ × 2″ and 9″ × 3″ respectively, as the nearest imperial sizes).

Timber is converted in imperial and metric sizes depending on its source in the world. Thereafter, standardisation can be undertaken by machine planing the surfaces to produce uniformly compatible and practically convenient dimensions, i.e. 225mm is not the same as 9″. Planed timber has been variously described as nominal, regularised and wrought, e.g. 100 × 50mm sawn becomes 97 × 47mm when planed and is otherwise known as ex 100 × 50mm, where ex means out of.

Guidance in BS EN 336 requires the sizes of timber from a supplier to be redefined as 'target sizes' within the following tolerances:

T1 ~ Thickness and width <100mm, −1 to +3mm. Thickness and width > 100mm, −2 to +4mm.
T2 ~ Thickness and width<100mm, −1 to +1mm.
Thickness and width>100mm, −1.5 to +1.5mm.

T1 applies to sawn timber, e.g. 100 × 75mm. T2 applies to planed timber, e.g. 97 × 72mm. Further example ~ a section of timber required to be 195mm planed × 50mm sawn is specified as: 195 (T2) × 50 (T1).

Target sizes for sawn softwood (T1) ~
50, 63, 75, 100, 125, 150, 175, 200, 225, 250 and 300mm.
Target sizes for planed/machined softwood (T2) ~
47, 60, 72, 97, 120, 145, 170, 195, 220, 245 and 295mm.

BS EN 336: *Structural Timber. Sizes, Permitted Deviations.*

Conversion ~ after timber is sawn into commercially useable sections, it is seasoned (oven dried) to a moisture content of between 10% and 15% depending on species.

$$\text{Moisture content} = \frac{\text{Mass when wet} - \text{Mass when dried}}{\text{Mass when dried}} \times 100\%$$

Protection ~ the moisture content of seasoned timber is much less than timber in its natural state, therefore timber will readily absorb water if it is unprotected.

The effects of water absorption may be:

- Deformities and distortion.
- Rot and decay.
- Swelling and shrinkage.

Swelling and shrinkage will be most noticeable after fixing, unless the seasoned moisture content is maintained by correct storage at the suppliers, on site and by adequate protection in use.

Movement ~ shrinkage occurs as wood dries below its fibre saturation point. It will also expand if water is allowed to penetrate the open fibres. Volume change is not the same, or even proportionally the same, as potential for movement varies directionally:

- Longitudinally – minimal, i.e. 0.1% to 0.3%.
- Transversely/radially – 2% to 6%.
- Tangentially – 5% to 10%.

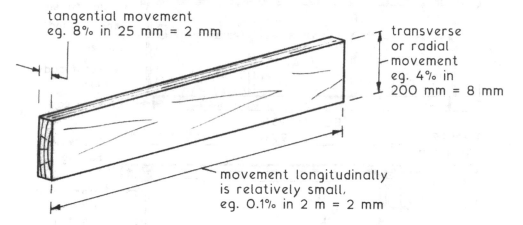

Figure 5.22 Shrinkage and movement of timber

5.4 Joinery production

Purpose-made joinery items in the form of doors, windows, stairs and cupboard fitments can be purchased as stock items from manufacturers. There is also a need for purpose-made joinery to fulfil client/designer/user requirements to suit a specific need, to fit into a non-standard space, as a specific decor requirement or to complement a particular internal environment. These purpose-made joinery items can range from the simple to the complex, requiring high degrees of workshop and site skills.

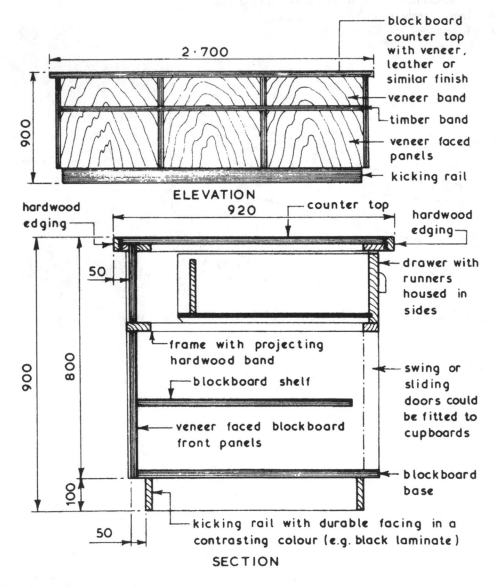

Figure 5.23 Timber counter details

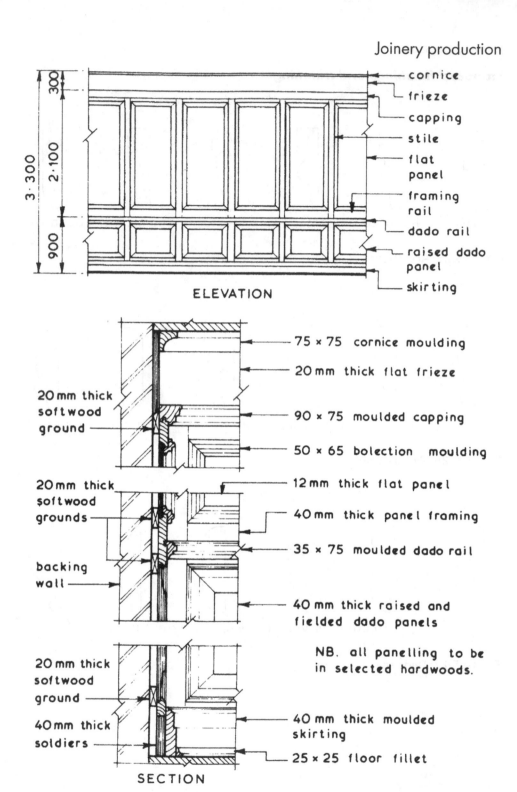

ELEVATION

SECTION

Figure 5.24 Timber wall panelling features

Joinery production

Standard finishings and trimmings

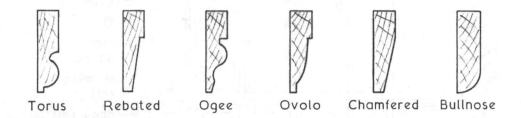

Torus   Rebated   Ogee   Ovolo   Chamfered   Bullnose

Figure 5.25 Architraves

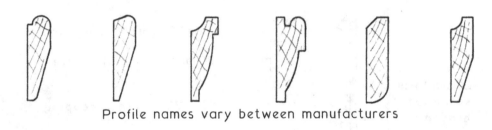

Profile names vary between manufacturers

Figure 5.26 Picture rails

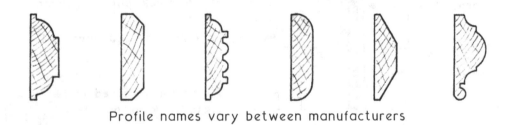

Profile names vary between manufacturers

Figure 5.27 Dado rails

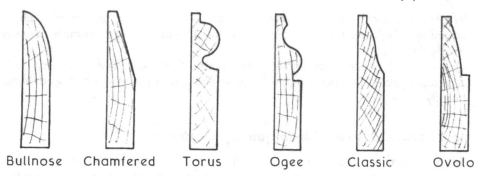

Figure 5.28 Skirtings

Joinery timbers

Both hardwoods and softwoods can be used for joinery works. Softwoods can be selected for their stability, durability and/or workability if the finish is to be paint, but if it is left in its natural colour with a sealing coat, the grain texture and appearance should be taken into consideration. Hardwoods are usually left in their natural colour and treated with a protective clear sealer or polish; therefore texture, colour and grain pattern are important when selecting hardwoods for high-class joinery work.

Typical softwoods suitable for joinery work

1. Douglas fir – sometimes referred to as Columbian pine or Oregon pine. It is available in long lengths and has a straight grain. Colour is reddish-brown to pink. Suitable for general and high-class joinery. Approximate density 530kg/m^3.
2. Redwood – also known as Scots pine, Red pine, Red deal and Yellow deal. It is a widely used softwood for general joinery work having good durability, a straight grain and is reddish-brown to straw in colour. Approximate density 430kg/m^3.
3. European spruce – similar to redwood but with a lower durability. It is pale yellow to pinkish-white in colour and is used mainly for basic framing work and simple internal joinery. Approximate density 650kg/m^3.
4. Sitka spruce – originates from Alaska, Western Canada and Northwest USA. The long, white strong fibres provide a timber quality for use in board or plywood panels. Approximate density 450kg/m^3.
5. Pitch pine – durable softwood suitable for general joinery work. It is light red to reddish-yellow in colour and tends to have large knots, which in some cases can be used as a decorative effect. Approximate density 650kg/m^3.
6. Parana pine – moderately durable, straight grained timber available in a good range of sizes. Suitable for general joinery work especially timber stairs. Light to dark brown in colour with the occasional pink stripe. Approximate density 560kg/m^3.

7. Western hemlock – durable softwood suitable for interior joinery work such as panelling. Light yellow to reddish-brown in colour. Approximate density 500kg/m^3.
8. Western red cedar – originates from British Columbia and Western USA. A straight grained timber suitable for flush doors and panel work. Approximate density 380kg/m^3.

Typical hardwoods suitable for joinery works

1. Beech – hard, close-grained timber with some silver grain in the predominantly reddish-yellow to light brown colour. Suitable for all internal joinery. Approximately density 700kg/m^3.
2. Iroko – hard durable hardwood with a figured grain and is usually golden brown in colour. Suitable for all forms of high-class joinery. Approximate density 660kg/m^3.
3. Mahogany (African) – interlocking grained hardwood with good durability. It has an attractive light brown to deep red colour and is suitable for panelling and all high-class joinery work. Approximate density 560kg/m^3.
4. Mahogany (Honduras) – durable hardwood usually straight grained but can have a mottled or swirl pattern. It is light red to pale reddish-brown in colour and is suitable for all high-class joinery work. Approximate density 530kg/m^3.
5. Mahogany (South American) – a well-figured, stable and durable hardwood with a deep-red or brown colour that is suitable for all high-class joinery, particularly where a high polish is required. Approximate density 550kg/m^3.
6. Oak (English) – very durable hardwood with a wide variety of grain patterns. It is usually a light yellow brown to a warm brown in colour and is suitable for all forms of joinery but should not be used in conjunction with ferrous metals due to the risk of staining caused by an interaction of the two materials. (The gallic acid in oak causes corrosion in ferrous metals.) Approximate density 720kg/m^3.
7. Sapele – close texture timber of good durability, dark reddish-brown in colour with a varied grain pattern. It is suitable for most internal joinery work especially where a polished finish is required. Approximate density 640kg/m^3.
8. Teak – very strong and durable timber but hard to work. It is light golden brown to dark golden yellow in colour, which darkens with age, and is suitable for high-class joinery work and laboratory fittings. Approximate density 650kg/m^3.
9. Jarrah (Western Australia) – hard, dense, straight grained timber. Dull red colour, suited to floor and stair construction subjected to heavy wear. Approximate density 820kg/m^3.

5.5 Composite boards

Factory-manufactured, preformed sheets with a wide range of properties and applications. The most common size is 2440 × 1220mm or 2400 × 1200mm in thicknesses from 3 to 50mm.

- Plywood (BS EN 636) – produced in a range of laminated thicknesses from 3 to 25mm, with the grain of each layer normally at right angles to that adjacent. 3,7,9 or 11 plies make up the overall thickness and inner layers may have lower strength and different dimensions to those in the outer layers. Adhesives vary considerably from natural vegetable and animal glues to synthetics such as urea, melamine, phenol and resorcinol formaldehydes. Quality of laminates and type of adhesive determine application. Surface finishes include plastics, decorative hardwood veneers, metals, rubber and mineral aggregates.
- Block and stripboards (BS EN 12871) – range from 12 to 43mm thickness, made up from a solid core of glued softwood strips with a surface-enhancing veneer. Appropriate for dense panelling and doors.

Battenboard – strips over 30mm wide (unsuitable for joinery).
Blockboard – strips up to 25mm wide.
Laminboard – strips up to 7mm wide.

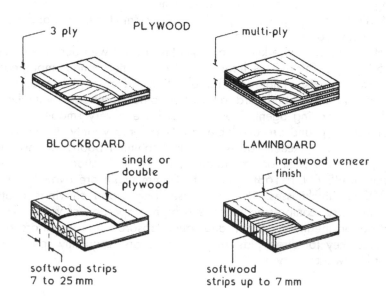

Figure 5.29 Plywood board constructions

- Compressed strawboard (BS 4046) – produced by compacting straw under heat and pressure and edge binding with paper. Used as panels with direct decoration or as partitioning with framed support. Also for insulated roof decking with 58mm slabs spanning 600mm joist spacing.

Composite boards

Particle board

- Chipboard (BS EN 319) – bonded waste wood or chip particles in thicknesses from 6 to 50mm, popularly used for floors in 18 and 22mm at 450 and 600mm maximum joist spacing, respectively. Sheets are produced by heat pressing the particles in thermosetting resins.
- Wood cement board – approximately 25% wood particles mixed with water and cement, to produce a heavy and dense board often preferred to plasterboard and fibre cement for fire cladding. Often three-layer boards, from 6 to 40mm in thickness.
- Oriented strand board (BS EN 300) – composed of wafer-thin strands of wood, approximately 80mm long × 25mm wide, resin bonded and directionally oriented before superimposed by further layers. Each layer is at right angles to adjacent layers, similar to the structure of plywood. A popular alternative for wall panels, floors and other chipboard and plywood applications, they are produced in a range of thicknesses from 6 to 25mm.
- Fibreboards (BS EN 622–4) – basically wood in composition, reduced to a pulp and pressed to achieve three categories – hardboard, mediumboard and softboard:
    - Hardboard density at least 800kg/m^3 in thicknesses from 3.2 to 8mm. Provides an excellent base for coatings and laminated finishes.
    - Mediumboard (low density) 350 to 560kg/m^3 for pinboards and wall linings in thicknesses of 6.4,9, and 12.7mm.
    - Mediumboard (high density) 560 to 800kg/m^3 for linings and partitions in thicknesses of 9 and 12mm.
    - Softboard, otherwise known as insulating board with density usually below 250kg/m^3. Thicknesses from 9 to 25mm, often found impregnated with bitumen in existing flat roofing applications. Ideal as pinboard.
    - Medium density fibreboard, differs from other fibreboards with the addition of resin bonding agent. These boards have a very smooth surface, ideal for painting, and are available moulded for a variety of joinery applications. Density exceeds 600kg/m^3 and common board thicknesses are 9, 12, 18 and 25mm for internal and external applications.
- Woodwool (BS EN 13168) – units of 600mm width are available in 50, 75 and 100mm thicknesses. They comprise long wood shavings coated with a cement slurry, compressed to leave a high proportion of voids. These voids provide good thermal insulation and sound absorption. The perforated surface is an ideal key for direct plastering and they are frequently specified as permanent formwork.

5.6 Plasters and plasterboards

Lime based plasters with a variety of binding materials have been used to provide a flat, even finish to internal walls from ancient times up to the mid-20th century, when gypsum based plasters became the standard material.

Gypsum plasters

Gypsum is a crystalline combination of calcium sulphate and water. The raw material is crushed, screened and heated to dehydrate the gypsum and this process, together with various additives, defines its type as set out in BS EN 13279–1: *Gypsum Binders and Gypsum Plasters*.

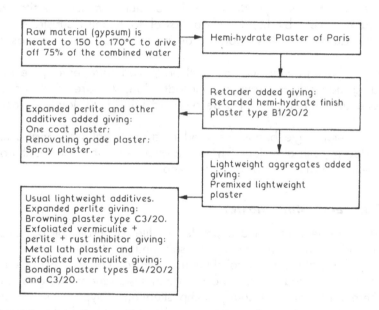

Figure 5.30 Gypsum plaster manufacture process

Gypsum plasters are supplied as a dry, loose material in strong paper bags of max. 20kg for mixing on site with water. All plasters should be stored in dry conditions since any absorption of moisture before mixing may accelerate the setting time, reducing workability and working time. It can also reduce the strength of the set plaster. A wide variety of gypsum plasters are available for standard and special applications.

Gypsum plasters fall into three categories:

- Undercoat plasters.
- Finishing plasters.
- Universal (a single plaster that works as an undercoat and finishing coat).

The undercoat plasters' function is to bond to the substrate, fill all gaps and irregularities and produce a flat, uniform surface to receive the finishing plaster. The thickness of the coat depends upon the quality of the substrate but is typically 10–12mm thick. Before the undercoat hardens it is scraped with a tool to produce a mechanical key for the finishing plaster.

Finishing plasters are applied to the undercoat or backing plaster, after it has dried and set sufficiently, typically in two coats to a thickness of 3–5mm to prove a flat, smooth finish to receive decoration.

Universal plasters are typically used in renovation of small areas of defective plastering. Variants of this type of plaster can be applied mechanically by spraying machine and ruled to a straight level finish.

Plastering work falls into two distinct methods: solid plastering and dry-lining.

Solid plastering

This is the traditional wet system where dry plaster is mixed on site with water and applied in layers by hand with a trowel to a solid substrate to achieve a smooth and durable finish suitable for decorative treatments, such as paint and wallpaper.

Non porous backgrounds such as steel or glazed surfaces require application of a bonding agent to improve plaster adhesion. A wire mesh or expanded metal surface attachment may also be required with metal lathing plaster as the undercoat.

Expanded metal angle beads are used at all external wall angles to provide a straight edge in both directions. These are attached with plaster dabs or galvanised nails before finishing just below the nosing.

Lightweight renovating plaster

This consists of cement, lime and expanded perlite, generally proportioned 1:1:6. Additives include a waterproofing agent, salt inhibitor to prevent surface efflorescence and synthetic fibres to control shrinkage and to improve flexural strength. LRP is an alternative undercoat plaster to cement and sand render and gypsum plasters, for use where renovating previously damp situations. Gypsum is less suitable in these situations as it is naturally hygroscopic. LRP can be applied directly to sound, dry substrates such as bricks and blocks (see Figure 5.31).

Dry lining

Dry lining can be applied to solid walls or stud partition walls, the former instead of using wet undercoat plaster to form a straight even surface to receive the finishing coat of plaster. For solid walls 2.4m high × 1.2m wide × 9.5mm, thick plasterboards are bonded to the wall with an adhesive, a process called 'dot and dab'. The boards are lined and levelled to create a straight flat surface. There are two types of board that can be used: square edge and taper edge. If the boards are to be finished with wet finishing plaster then square edge boards are used with a glass-fibre scrim tape applied over the joint before being plastered. Alternatively, taper edge boards can be used to provide a flush seamless surface without a coat of finishing plaster, this is obtained by filling the tapered joint with a special filling plaster, applying a joint tape over the filling and finishing with a thin layer of joint filling plaster, the edge of which is feathered out using a jointing tool. When the joint is dry it is sanded flat.

The main advantage of dry lining walls is that the drying out period required with wet finishes is eliminated. By careful selection and fixing of some dry lining materials it is possible to improve the thermal insulation properties of a wall. Dry linings can be fixed direct to the backing by means of a recommended adhesive or they can be fixed to a suitable arrangement of wall battens.

For plasterboard types see:

BS EN 520: *Gypsum Plasterboards. Definitions, Requirements and Test Methods.*

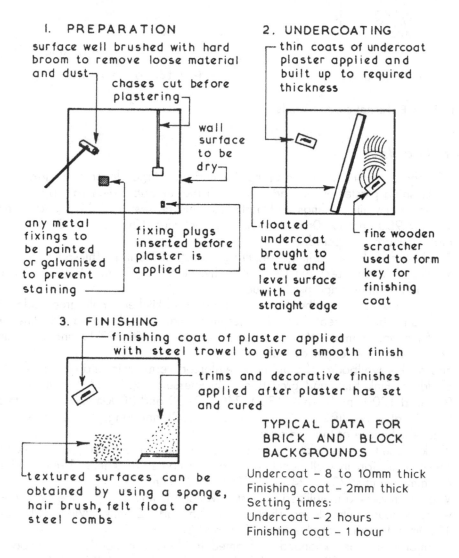

1. PREPARATION
surface well brushed with hard broom to remove loose material and dust
chases cut before plastering
wall surface to be dry
any metal fixings to be painted or galvanised to prevent staining
fixing plugs inserted before plaster is applied

2. UNDERCOATING
thin coats of undercoat plaster applied and built up to required thickness
floated undercoat brought to a true and level surface with a straight edge
fine wooden scratcher used to form key for finishing coat

3. FINISHING
finishing coat of plaster applied with steel trowel to give a smooth finish
trims and decorative finishes applied after plaster has set and cured
textured surfaces can be obtained by using a sponge, hair brush, felt float or steel combs

TYPICAL DATA FOR BRICK AND BLOCK BACKGROUNDS
Undercoat – 8 to 10mm thick
Finishing coat – 2mm thick
Setting times:
Undercoat – 2 hours
Finishing coat – 1 hour

Figure 5.31 Plaster application

Plasters and plasterboards

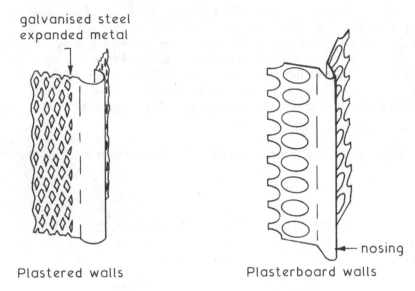

Figure 5.32 External corner beads

Plasterboards

- Wallboard – ivory faced for taping, jointing and direct decoration; grey faced for finishing plaster or wall adhesion with plaster dabs. General applications, i.e. internal walls, ceilings and partitions. Thicknesses: 9.5, 12.5 and 15mm. Widths: 900 and 1200mm. Lengths: vary between 1800 and 3000mm. Edge profile square or tapered.
- Baseboard – lining ceilings requiring direct plastering. Thickness: 9.5mm. Width: 900mm. Length: 1220mm. Thickness: 12.5mm. Width: 600mm. Length: 1220mm. Edge profile square.
- Moisture resistant – wallboard for bathrooms and kitchens. Pale green colour to face and back. Ideal base for ceramic tiling or plastering. Thicknesses: 12.5mm and 15mm. Width: 1200mm. Lengths: 2400, 2700 and 3000mm. Square and taper edges available.
- Firecheck – wallboard of glass fibre reinforced vermiculite and gypsum for fire cladding. Pink face and grey back. Thicknesses: 12.5 and 15mm. Widths: 900 and 1200mm. Lengths: 1800, 2400, 2700 and 3000mm. A 25mm thickness is also produced, 600mm wide × 3000mm long. Plaster finished if required. Square or tapered edges.
- Plank – used as fire protection for structural steel and timber, in addition to sound insulation in wall panels and floating floors. Thickness: 19mm. Width: 600mm. Lengths: 2350, 2400, 2700 and 3000mm.
- Vapour check – a metallized polyester wallboard lining to provide an integral water vapour control layer. Thicknesses: 9.5 and 12.5mm. Widths: 900 and 1200mm. Lengths: vary between 1800 and 3000mm.
- Thermal – various expanded or foamed insulants are bonded to wallboard. Approximately 25–50mm overall thickness in board sizes 1200 × 2400mm.

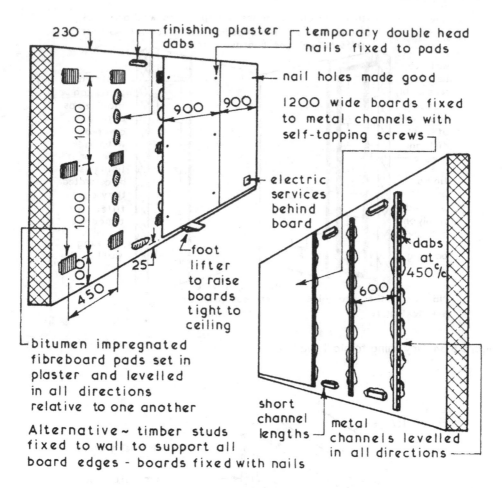

230 — finishing plaster dabs — temporary double head nails fixed to pads

— nail holes made good

1200 wide boards fixed to metal channels with self-tapping screws

900   900

1000

1000

1000

100

25   450

— electric services behind board

— foot lifter to raise boards tight to ceiling

dabs at 450 c/c

600

bitumen impregnated fibreboard pads set in plaster and levelled in all directions relative to one another

short channel lengths — metal channels levelled in all directions —

Alternative ~ timber studs fixed to wall to support all board edges - boards fixed with nails

Figure 5.33 Dry lining to block walls 1

Plasters and plasterboards

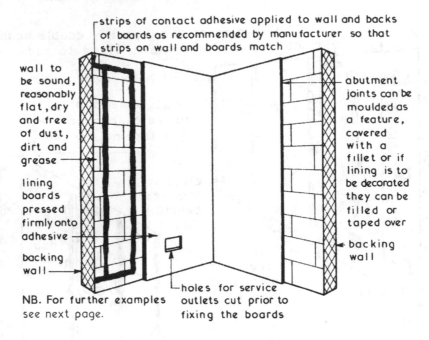

strips of contact adhesive applied to wall and backs of boards as recommended by manufacturer so that strips on wall and boards match

wall to be sound, reasonably flat, dry and free of dust, dirt and grease

lining boards pressed firmly onto adhesive

backing wall

abutment joints can be moulded as a feature, covered with a fillet or if lining is to be decorated they can be filled or taped over

backing wall

holes for service outlets cut prior to fixing the boards

NB. For further examples see next page.

Figure 5.34 Dry lining to block walls 2

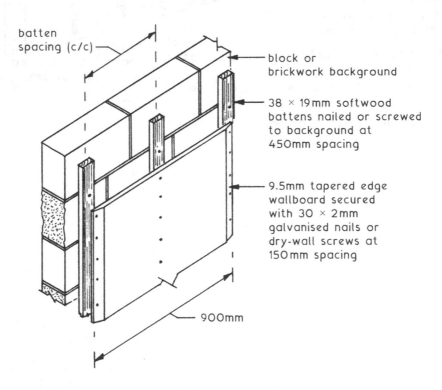

batten spacing (c/c)

block or brickwork background

38 × 19mm softwood battens nailed or screwed to background at 450mm spacing

9.5mm tapered edge wallboard secured with 30 × 2mm galvanised nails or dry-wall screws at 150mm spacing

900mm

Figure 5.35 Taper edge boards on battens

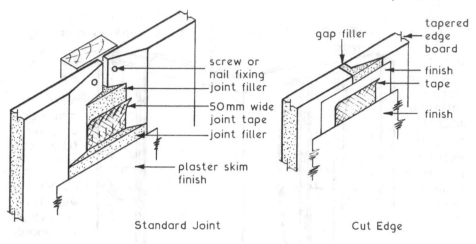

screw or
nail fixing
joint filler

50 mm wide
joint tape

joint filler

plaster skim
finish

Standard Joint

gap filler

tapered
edge
board

finish
tape

finish

Cut Edge

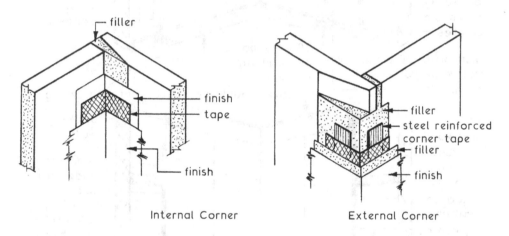

filler

finish
tape

finish

Internal Corner

filler

steel reinforced
corner tape

filler

finish

External Corner

Figure 5.36 Tape and fill jointing of taper edge boards

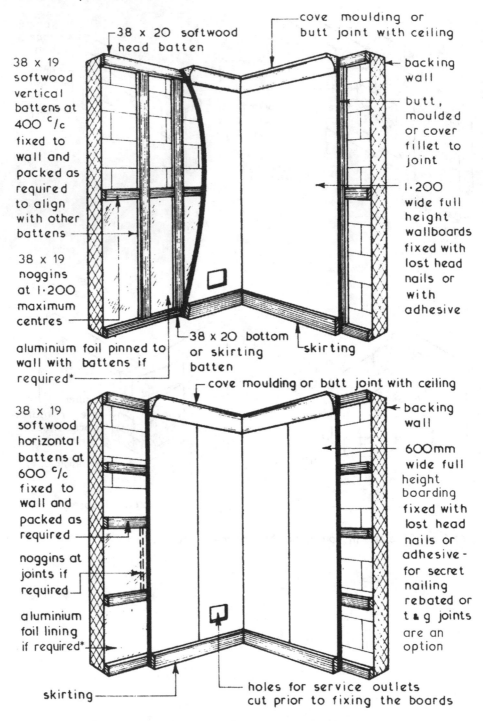

38 x 20 softwood head batten

cove moulding or butt joint with ceiling

38 x 19 softwood vertical battens at 400 ᶜ/c fixed to wall and packed as required to align with other battens

38 x 19 noggins at 1·200 maximum centres

backing wall

butt, moulded or cover fillet to joint

1·200 wide full height wallboards fixed with lost head nails or with adhesive

aluminium foil pinned to wall with battens if required*

38 x 20 bottom or skirting batten

skirting

cove moulding or butt joint with ceiling

38 x 19 softwood horizontal battens at 600 ᶜ/c fixed to wall and packed as required

noggins at joints if required

aluminium foil lining if required*

backing wall

600mm wide full height boarding fixed with lost head nails or adhesive - for secret nailing rebated or t & g joints are an option

skirting

holes for service outlets cut prior to fixing the boards

*alternatively use polythene sheet as a vapour check.

Figure 5.37 Plywood grounds

172

5.7 Plastics

The term plastic can be applied to any group of substances based on synthetic or modified natural polymers, which during manufacture are moulded by heat and/or pressure into the required form. Plastics can be classified by their overall grouping such as polyvinyl chloride (PVC) or they can be classified as thermoplastic or thermosetting. The former soften on heating whereas the latter are formed into permanent, non-softening materials. The range of plastics available give the designer and builder a group of materials that are strong, reasonably durable, easy to fit and maintain and, since most are mass produced, are of relatively low cost.

Typical applications of plastics in buildings are shown in Table 5.10.

Table 5.10 Plastic types and applications

Application	Plastics used
Rainwater goods	Unplasticised PVC (uPVC or PVC-U).
Soil, waste, water and gas pipes and fittings	uPVC; polyethylene (PE); acrylonitrile butadiene styrene (ABS), polypropylene (PP).
Hot and cold water pipes	Chlorinated PVC; ABS; polypropylene; polyethylene; PVC (not for hot water).
Bathroom and kitchen fittings	Glass-fibre reinforced polyester (GRP); acrylic resins.
Cold water cisterns	Polypropylene; polystyrene; polyethylene.
Rooflights and sheets	GRP; acrylic resins; uPVC.
DPCs and membranes, vapour control layers	Low-density polyethylene (LDPE); PVC film; polypropylene.
Doors and windows	GRP; uPVC.
Electrical conduit and fittings	Plasticised PVC; uPVC; phenolic resins.
Thermal insulation	Generally cellular plastics such as expanded polystyrene bead and boards; expanded PVC; foamed polyurethane; foamed phenol formaldehyde; foamed urea formaldehyde.
Floor finishes	Plasticised PVC tiles and sheets; resin-based floor paints; uPVC.
Wall claddings and internal linings	Unplasticised PVC; polyvinyl fluoride film laminate; melamine resins; expanded polystyrene tiles and sheets.

5.8 Mastics and sealants

Used to weather- and leak-proof junctions and abutments between separate elements and components that may be subject to differential movement. Also to fill gaps where irregularities occur.

Properties:

- Thermal movement to facilitate expansion and contraction.
- Strength to resist wind and other non-structural loading.
- Ability to accommodate tolerance variations.
- Stability without loss of shape.
- Colour fast and non-staining to adjacent finishes.
- Weather resistant.

Maintenance: of limited life, perhaps 10 to 25 years depending on composition, application and use. Future accessibility is important for ease of removal and replacement.

Mastics

Generally regarded as non-setting gap fillers applied in a plastic state. Characterised by a hard surface skin over a plastic core that remains pliable for several years. Based on a viscous material such as bitumen, polyisobutylene or butyl rubber. Applications include bitumen treatment to rigid road construction joints and linseed oil putty glazing. In older construction, a putty-based joint may also be found between WC pan spigot outlet and cast iron socket. In this situation the putty was mixed with red lead pigments (oxides of lead), a material now considered a hazardous poison, therefore protective care must be taken when handling an old installation of this type. Modern pushfit plastic joints are much simpler, safer to use and easier to apply.

Sealants

Applied in a plastic state by hand, knife, disposable cartridge gun, pouring or tape strip to convert by chemical reaction with the atmosphere (one part) or with a vulcanising additive (two part) into an elastomer or synthetic rubber. An elastomer is generally defined as a natural or synthetic material with a high strain capacity or elastic recovery, i.e. it can be stretched to twice its length before returning to its original length. Formed of polysulphide rubber, polyurethane, silicone or some butyl rubbers.

Applications:

- Polysulphide: façades, glazing, fire protection, roads and paving joints. High modulus or hardness but not completely elastic.
- Polyurethane: general uses, façades and civil engineering. Highly elastic and resilient to abrasion and indentation, moderate resistance to ultraviolet light and chemicals.

- Silicone: general uses, façades, glazing, sanitary, fire protection and civil engineering. Mainly one part but set quickly relative to others in this category. Highly elastic and available as high (hard) or low (soft) modulus.

Two part, polysulphide- and polyurethane-based sealants are often used with a curing or vulcanising additive to form a synthetic rubber on setting. After the two parts are mixed the resulting sealant remains workable for up to about four hours. It remains plastic for a few days and during this time cannot take any significant loading. Thereafter it has exceptional resistance to compression and shear.

One part, otherwise known as room temperature vulcanising (RTV) types are usually of a polysulphide, polyurethane or silicone base. Polysulphide and polyurethane cure slowly and convert to a synthetic rubber or elastomer sealant by chemical reaction with moisture in the atmosphere. Generally of less movement and loading resistance than two part sealants, but are frequently used in non-structural situations such as sealing around door and window frames, bathroom and kitchen fitments.

Other sealants

- One part acrylic (water-based) RTV. Flexible but with limited elasticity. Internal uses such as sealing around door and window frames, fire protection and internal glazing.
- Silane modified polymer in one part RTV or two parts. Highly elastic and can be used for general applications as well as for façades and civil engineering situations.

Prior to 2003, several separate British Standards existed to provide use and application guidance for a range of sealant products. As independent publications these are now largely superseded, their content rationalised and incorporated into the current standard, BS EN ISO 11600: *Building Construction. Jointing Products. Classification and Requirements for Sealants.*

This International Standard covers materials application to jointing, classification of materials, quality grading and performance testing. This enables specific definition of a sealant's requirements in terms of end use without having to understand the chemical properties of the various sealant types. Typical criteria are movement potential, elasticity and hardness when related to particular substrate surfaces such as aluminium, glass or masonry.

Grading summary:

BS EN ISO 11600 G, for use in glazing.
BS EN ISO 11600 F, for façade and similar applications such as movement joints.

Other suffixes or sub-classes:

E = elastic sealant, i.e. high elastic recovery or elastomeric.
P = plastic sealant, i.e. low elastic recovery.
HM = high modulus, indicates hardness.
LM = low modulus, indicates softness.

Mastics and sealants

By definition, HM and LM are high movement (20–25%) types of elastic sealants, therefore the suffix E is not shown with these.

Associated standards:

BS EN ISO 6927: *Buildings and Civil Engineering Works. Sealants. Vocabulary.*
BS 6213: *Selection of Construction Sealants. Guide.*
BS 6093: *Design of Joints and Jointing in Building Construction. Guide.*

5.9 Metals

Table 5.11 Metal types and applications

Aluminium	Used extensively in commercial construction for window and door frames, curtain walling profiles, cladding and roofing sheets, rainwater goods.
Steel	Universal steel frame sections are constructed from hot rolled steel. Cold rolled steel is used for cladding rails and lightweight steel framing which is galvanised. Stainless steel is used for cavity wall ties and many other components and fixings that must be resistant to corrosion and rusting.
Copper	Main uses are plumbing pipework and electric cables, also used for sheet roofing and cladding for aesthetic purposes.
Lead	Primary use is sheet form for roof gutters and flashings and also sheet roofing on conservation projects.
Brass Alloys	Primary use is for plumbing components, including valves.
Zinc	Used in the galvanising process to protect steel from rusting.

6 PLANT

EARTHMOVING
MATERIAL HANDLING
SCAFFOLDS

6.1 Earthmoving

Bulldozers

This is a track or wheel mounted power unit with a bucket or mould blade at the front which is controlled by hydraulic rams.

The main functions of a bulldozer are:

1. Shallow excavations up to 300 m deep.
2. Grading of soil.
3. Pushing earth into piles.
4. Clearance of shrubs and small trees.

Scrapers

These machines consist of a scraper bowl which is lowered to cut and collect soil where site stripping and levelling operations are required involving large volume of earth.

Graders

These machines are similar in concept to bulldozers in that they have a long, slender, adjustable mould blade, which is usually slung under the centre of the machine.

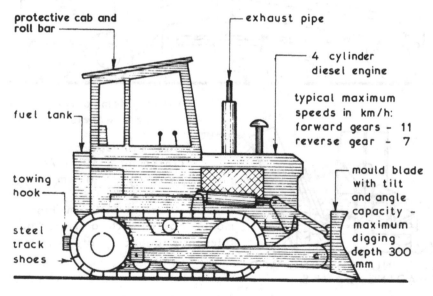

Figure 6.1 Bulldozer

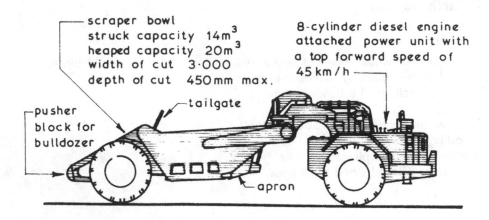

scraper bowl
struck capacity 14m³
heaped capacity 20m³
width of cut 3·000
depth of cut 450mm max.

8-cylinder diesel engine
attached power unit with
a top forward speed of
45 km/h

pusher
block for
bulldozer

tailgate

apron

Figure 6.2 Scraper

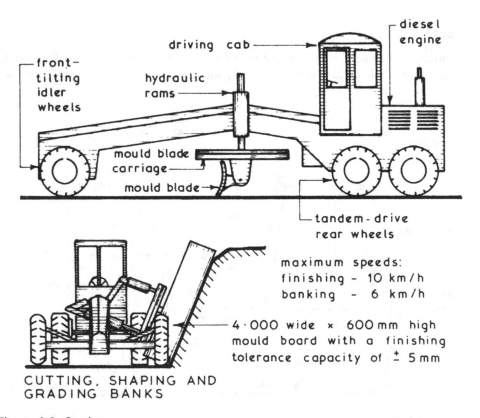

driving cab

diesel
engine

front-
tilting
idler
wheels

hydraulic
rams

mould blade
carriage

mould blade

tandem-drive
rear wheels

maximum speeds:
finishing – 10 km/h
banking – 6 km/h

4·000 wide × 600mm high
mould board with a finishing
tolerance capacity of ± 5mm

CUTTING, SHAPING AND
GRADING BANKS

Figure 6.3 Grader

A grader's main function is to finish or grade the upper surface of a large area, usually as a follow-up operation to scraping or bulldozing. They can produce a fine and accurate finish but do not have the power of a bulldozer; therefore they are not suitable for over-site excavation work.

Tractor shovels

Track or wheel mounted power unit with an articulated front bucket, their primary function is to scoop up loose materials and deposit them into an attendant transport vehicle. To increase their versatility tractor shovels can be fitted with a four-in-one bucket, enabling them to carry out bulldozing, excavating, clam lifting and loading activities.

Backacters

These machines are suitable for trench, foundation and basement excavations and are available as a universal power unit-based machine or as a purpose-designed hydraulic unit. A range of bucket types and widths are available for trenching operations.

Multi-purpose excavators

These versatile machines are fitted with a loading/excavating front bucket and a rear backacter bucket, both being hydraulically controlled. When in operation using the backacter bucket the machine is raised off its axles by rear-mounted hydraulic outriggers or jacks and, in some models, by placing the front bucket on the ground. A variety of bucket widths and various attachments such as breakers and post hole auger borers can be fitted.

Dumpers

These are used for the transportation of materials by means of an integral tipping skip. A wide range of dumpers are available of various carrying capacities and options for gravity or hydraulic discharge control with front tipping, side tipping or elevated tipping facilities.

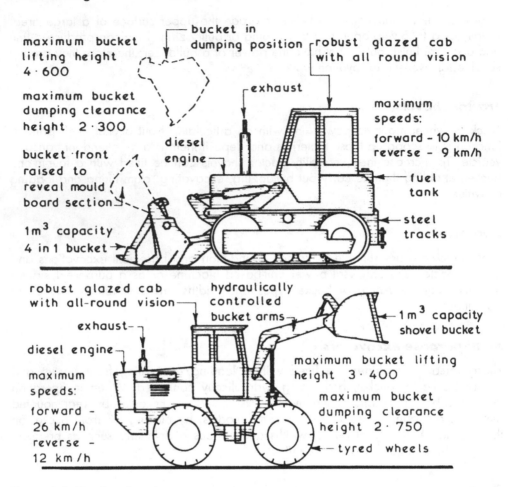

maximum bucket
lifting height
4·600

bucket in
dumping position

robust glazed cab
with all round vision

exhaust

maximum
speeds:
forward – 10 km/h
reverse – 9 km/h

maximum bucket
dumping clearance
height 2·300

diesel
engine

bucket front
raised to
reveal mould
board section

fuel
tank

steel
tracks

1m³ capacity
4 in 1 bucket

robust glazed cab
with all-round vision

hydraulically
controlled
bucket arms

1 m³ capacity
shovel bucket

exhaust

diesel engine

maximum
speeds:
forward –
26 km/h
reverse –
12 km/h

maximum bucket lifting
height 3·400

maximum bucket
dumping clearance
height 2·750

tyred wheels

Figure 6.4 Tractor shovel

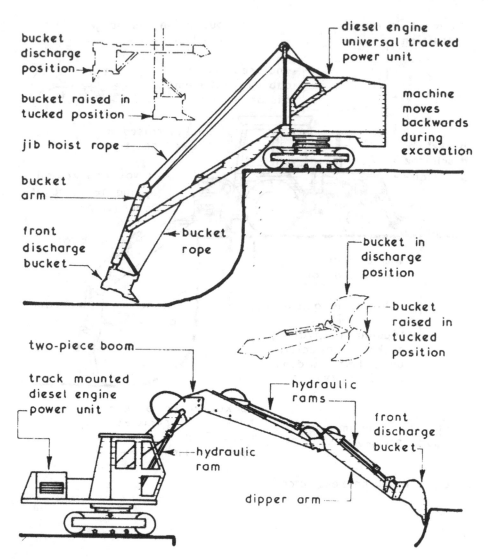

Figure 6.5 Backactor

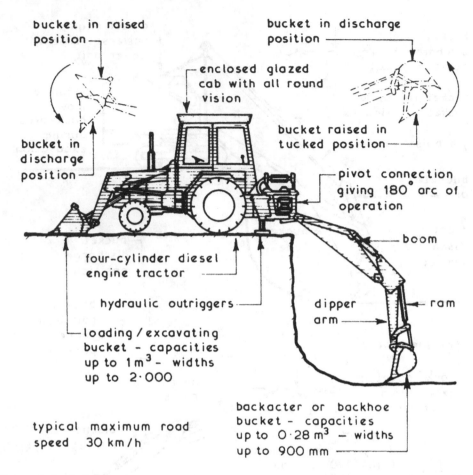

bucket in raised position

bucket in discharge position

enclosed glazed cab with all round vision

bucket in discharge position

bucket raised in tucked position

pivot connection giving 180° arc of operation

boom

four-cylinder diesel engine tractor

hydraulic outriggers

dipper arm

ram

loading / excavating bucket - capacities up to 1 m³ - widths up to 2·000

backacter or backhoe bucket - capacities up to 0·28 m³ — widths up to 900 mm

typical maximum road speed 30 km/h

Figure 6.6 Multi-purpose excavator

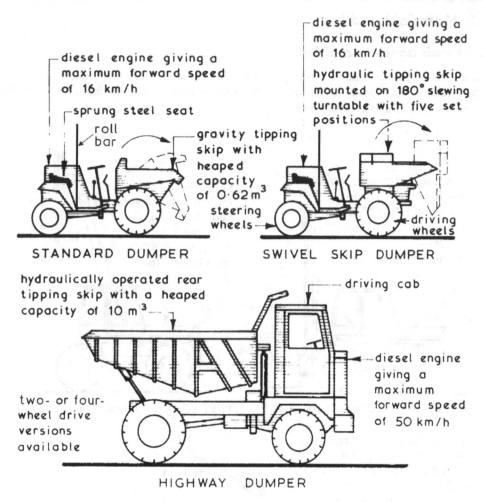

diesel engine giving a
maximum forward speed
of 16 km/h

sprung steel seat

roll
bar

gravity tipping
skip with
heaped
capacity
of 0·62 m³

steering
wheels

STANDARD DUMPER

diesel engine giving a
maximum forward speed
of 16 km/h

hydraulic tipping skip
mounted on 180° slewing
turntable with five set
positions

driving
wheels

SWIVEL SKIP DUMPER

hydraulically operated rear
tipping skip with a heaped
capacity of 10 m³

driving cab

diesel engine
giving a
maximum
forward speed
of 50 km/h

two- or four-
wheel drive
versions
available

HIGHWAY DUMPER

Figure 6.7 Dumpers

6.2 Material handling

Telescopic handlers and fork lifts

These are wheeled vehicles utilizing a twin fork lifting platform for pallets. The telescopic boom can move in the horizontal and vertical plane to transport and lift materials on site up to three storeys high.

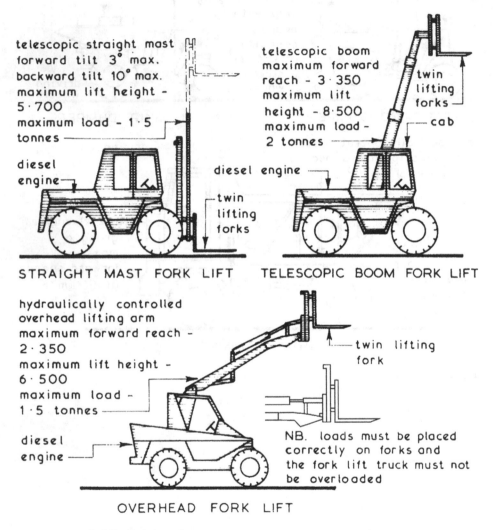

telescopic straight mast forward tilt 3° max. backward tilt 10° max. maximum lift height – 5·700 maximum load – 1·5 tonnes

diesel engine

twin lifting forks

STRAIGHT MAST FORK LIFT

telescopic boom maximum forward reach – 3·350 maximum lift height – 8·500 maximum load – 2 tonnes

twin lifting forks

cab

diesel engine

TELESCOPIC BOOM FORK LIFT

hydraulically controlled overhead lifting arm maximum forward reach – 2·350 maximum lift height – 6·500 maximum load – 1·5 tonnes

diesel engine

twin lifting fork

NB. loads must be placed correctly on forks and the fork lift truck must not be overloaded

OVERHEAD FORK LIFT

Figure 6.8 Fork lifts/telehandlers

Material handling-hoists

These are designed for the vertical transportation of materials and/or passengers. Most material hoists are of a mobile format which can be dismantled, folded onto

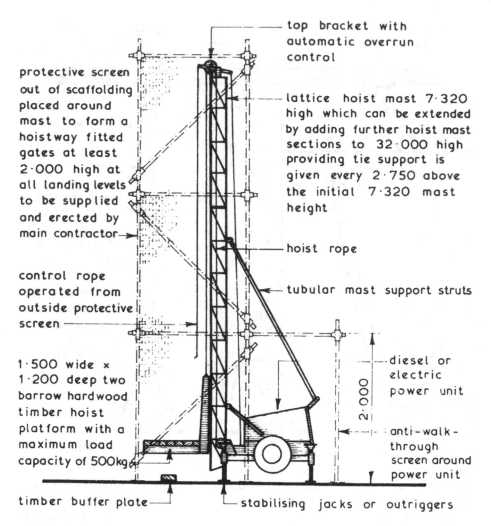

top bracket with automatic overrun control

protective screen out of scaffolding placed around mast to form a hoistway fitted gates at least 2·000 high at all landing levels to be supplied and erected by main contractor

lattice hoist mast 7·320 high which can be extended by adding further hoist mast sections to 32·000 high providing tie support is given every 2·750 above the initial 7·320 mast height

hoist rope

control rope operated from outside protective screen

tubular mast support struts

1·500 wide × 1·200 deep two barrow hardwood timber hoist platform with a maximum load capacity of 500kg

diesel or electric power unit

2·000

anti-walk-through screen around power unit

timber buffer plate

stabilising jacks or outriggers

Figure 6.9 Mobile hoist

the chassis and moved to another position or site under their own power or towed by a haulage vehicle. When in use material hoists need to be stabilised and/or tied to the structure and enclosed with a protective screen.

Cranes

The range of cranes available is very wide and therefore choice must be based on the loads to be lifted, height and horizontal distance to be covered, time period(s) of lifting operations, utilization factors and degree of mobility required. Crane types can range from mobile to static and tower cranes.

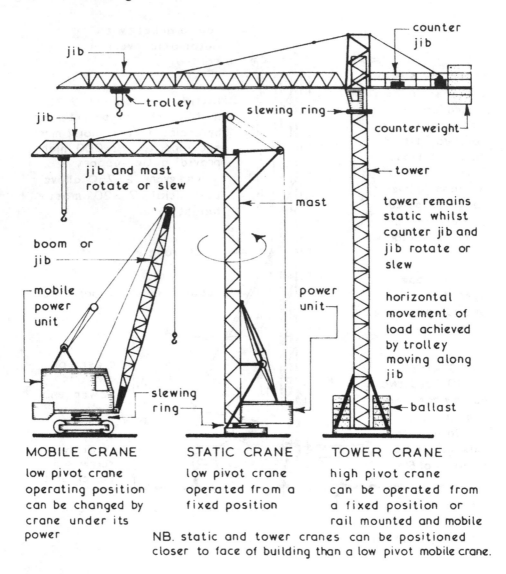

Figure 6.10 Types of crane

Mobile cranes

Tracked and wheeled versions are available, using either a fixed lattice boom or extendable hydraulic telescopic boom. The lifting capacity of these cranes can be increased by using outrigger stabilizing jacks and the approach distance to the face of the building decreased by using a fly jib.

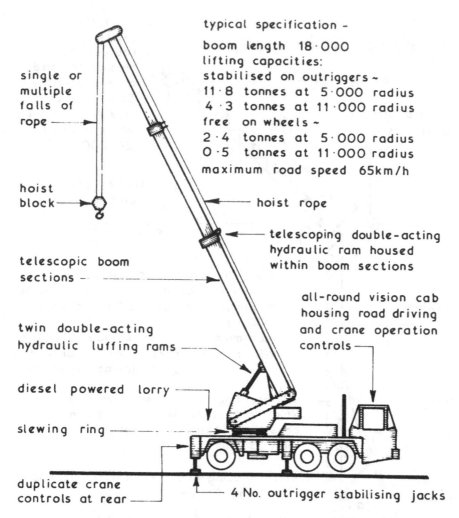

single or
multiple
falls of
rope

hoist
block

telescopic boom
sections –

twin double-acting
hydraulic luffing rams

diesel powered lorry

slewing ring

duplicate crane
controls at rear

typical specification –

boom length 18·000
lifting capacities:
stabilised on outriggers ~
11·8 tonnes at 5·000 radius
4·3 tonnes at 11·000 radius
free on wheels ~
2·4 tonnes at 5·000 radius
0·5 tonnes at 11·000 radius
maximum road speed 65km/h

hoist rope

telescoping double-acting
hydraulic ram housed
within boom sections

all-round vision cab
housing road driving
and crane operation
controls

4 No. outrigger stabilising jacks

Figure 6.11 Mobile hydraulic crane

Mast cranes

These are similar in appearance to tower cranes but they have one major difference in that the mast or tower is mounted on the slewing ring and thus rotates, whereas a tower crane has the slewing ring at the top of the tower and, therefore, only the jib portion rotates. Mast cranes are often mobile, self-erecting, of relatively low lifting capacity and are usually fitted with a luffing jib.

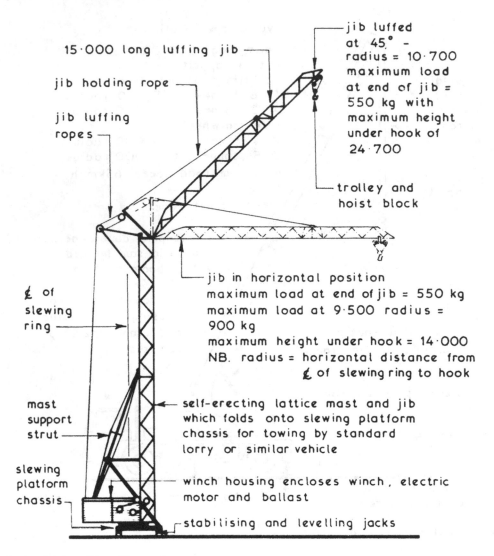

15·000 long luffing jib

jib holding rope

jib luffing ropes

jib luffed at 45° - radius = 10·700 maximum load at end of jib = 550 kg with maximum height under hook of 24·700

trolley and hoist block

℄ of slewing ring

jib in horizontal position
maximum load at end of jib = 550 kg
maximum load at 9·500 radius = 900 kg
maximum height under hook = 14·000
NB. radius = horizontal distance from ℄ of slewing ring to hook

mast support strut

self-erecting lattice mast and jib which folds onto slewing platform chassis for towing by standard lorry or similar vehicle

slewing platform chassis

winch housing encloses winch, electric motor and ballast

stabilising and levelling jacks

Figure 6.12 Luffing jib mast crane

Tower cranes

Most tower cranes have to be assembled and erected on site prior to use and can be equipped with a horizontal or luffing jib. The wide range of models available often makes it difficult to choose a crane suitable for any particular site, but most tower cranes can be classified into one of four basic groups thus:

*   Self-supporting static tower cranes – high lifting capacity with the mast or tower fixed to a foundation base. They are suitable for confined and open sites.

- Supported static tower cranes – similar in concept to self-supporting cranes and are used where high lifts are required, the mast or tower being tied at suitable intervals to the structure to give extra stability.
- Climbing cranes – these are used in conjunction with tall buildings and structures. The climbing mast or tower is housed within the structure and raised as the height of the structure is increased. Upon completion the crane is dismantled into small sections and lowered down the face of the building
- Travelling (rail mounted) tower crane – these cranes have a limited range of linear movement on rail tracks by means of powered bogies.

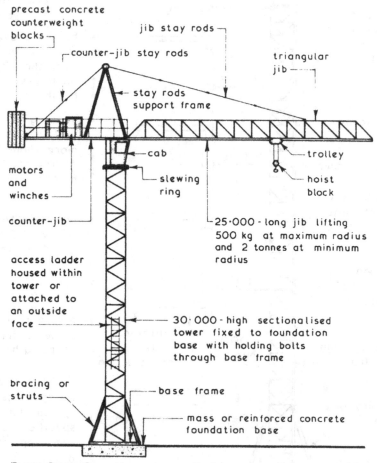

Tower Crane Operation – two methods are in general use:

1. Cab Control – the crane operator has a good view of most of the lifting operations from the cab mounted at the top of the tower but a second person or banksman is required to give clear signals to the crane operator and to load the crane.
2. Remote Control – the crane operator carries a control box linked by a wandering lead to the crane controls.

Figure 6.13 Free standing tower crane

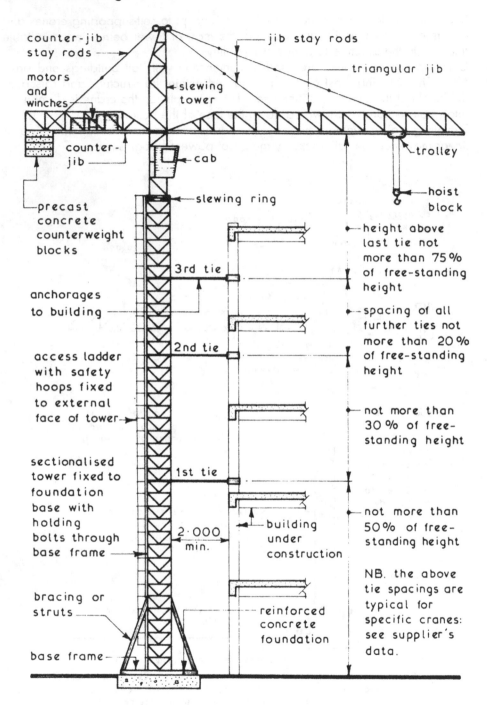

Figure 6.14 Tower crane tied to building

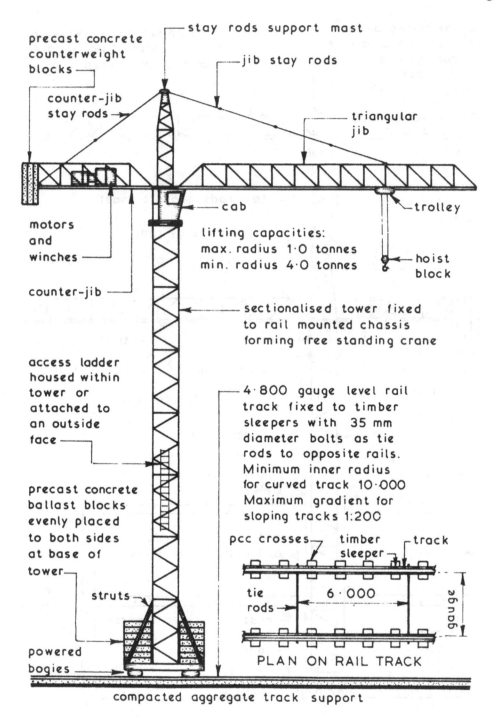

precast concrete
counterweight
blocks

stay rods support mast

jib stay rods

counter-jib
stay rods

triangular
jib

cab

trolley

motors
and
winches

lifting capacities:
max. radius 1·0 tonnes
min. radius 4·0 tonnes

hoist
block

counter-jib

sectionalised tower fixed
to rail mounted chassis
forming free standing crane

access ladder
housed within
tower or
attached to
an outside
face

4·800 gauge level rail
track fixed to timber
sleepers with 35 mm
diameter bolts as tie
rods to opposite rails.
Minimum inner radius
for curved track 10·000
Maximum gradient for
sloping tracks 1:200

precast concrete
ballast blocks
evenly placed
to both sides
at base of
tower

pcc crosses

timber
sleeper

track

struts

tie
rods

6·000

gauge

powered
bogies

PLAN ON RAIL TRACK

compacted aggregate track support

Figure 6.15 Tracked tower crane

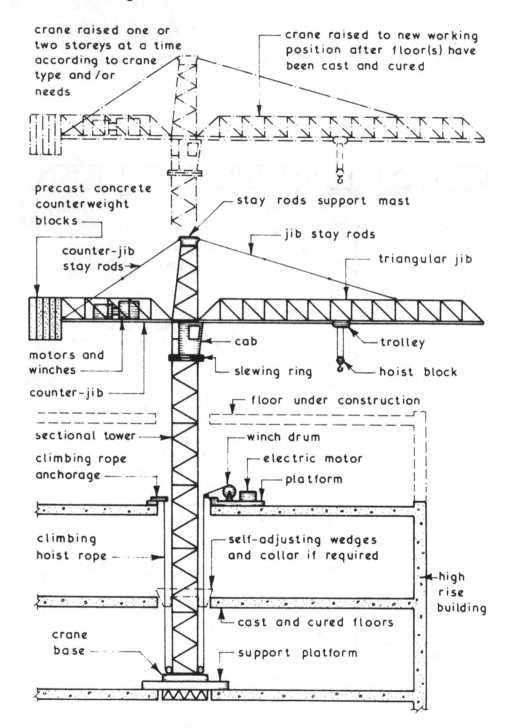

crane raised one or two storeys at a time according to crane type and/or needs

crane raised to new working position after floor(s) have been cast and cured

precast concrete counterweight blocks

stay rods support mast

jib stay rods

counter-jib stay rods→

triangular jib

cab

trolley

motors and winches

slewing ring

hoist block

counter-jib

floor under construction

sectional tower

winch drum

climbing rope anchorage

electric motor

platform

climbing hoist rope

self-adjusting wedges and collar if required

high rise building

crane base

cast and cured floors

support platform

Figure 6.16 Climbing crane

Concrete pumps

These are used to transport large volumes of concrete in a short time period (up to 100m³ per hour), in both the vertical and horizontal directions, from the pump position to the point of placing. Concrete pumps can be trailer or lorry mounted and are usually of a twin cylinder, hydraulically driven format with a small bore pipeline (100mm diameter). The pump is supplied with pumpable concrete by means of a constant flow of ready mixed concrete lorries throughout the pumping period, after which the pipeline is cleared and cleaned. Usually a concrete pump and its operator(s) are hired for the period required.

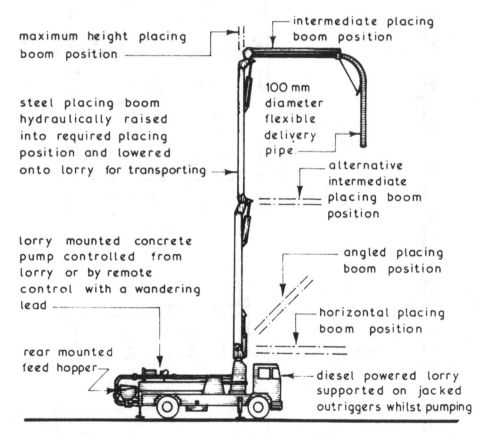

Figure 6.17 Concrete pump

6.3 Scaffolds

These are temporary platforms to provide a safe working place at a convenient height. Scaffolds are usually required when the working height is 1.5m or more above ground level.

BS EN 39: *Loose Steel Tubes for Tube and Coupler Scaffolds.*
BS 1139–1.2: *Metal Scaffolding. Tubes. Specification for Aluminium Tube.*

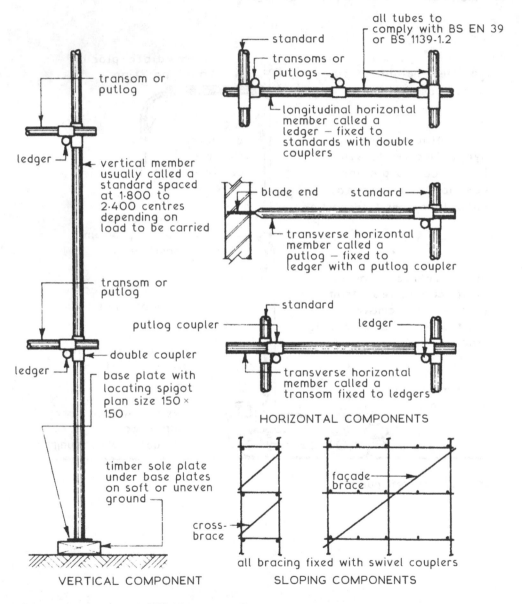

Figure 6.18 Tubular scaffold components

BS EN 12,811–1: *Temporary Equipment. Scaffolds. Performance Requirements and General Design.*
Work at Height Regulations.
Occupational Health and Safety Act. Scaffold regulations.

Putlog scaffolds

These are scaffolds that have an outer row of standards joined together by ledgers, which in turn support the transverse putlogs that are built into the bed joints or perpends as the work proceeds; they are therefore only suitable for new work in bricks or blocks.

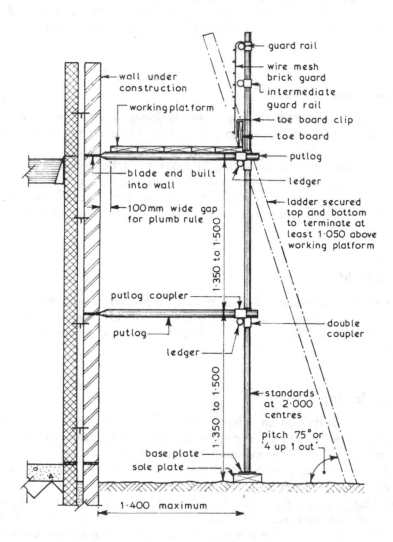

Figure 6.19 Putlog scaffold

Scaffolds

Independent scaffolds

These are scaffolds that have two rows of standards, each row joined together with ledgers, which in turn support the transverse transoms. The scaffold is erected clear of the existing or proposed building but is tied to the building or structure at suitable intervals.

Scaffold working platforms

These are close-boarded or plated level surfaces at a height at which work is being carried out and they must provide a safe working place of sufficient strength to support the imposed loads of operatives and/or materials. All working platforms above ground level must be fitted with a toe board and a guard rail.

Tying-in scaffold

All scaffolds should be tied securely to the building or structure at alternate lift heights vertically and at not more than 6.0m centres horizontally. Putlogs should not be classified as ties.

Tying-in methods include connecting to tubes fitted between sides of window openings or to internal tubes fitted across window openings; the former method should not be used for more than 50% of the total number of ties. If there is an insufficient number of window openings for the required number of ties external rakers should be used.

Mobile scaffolds

Otherwise known as mobile tower scaffolds, these can be assembled from preformed framing components or from standard scaffold tube and fittings. Used mainly for property maintenance. Must not be moved whilst occupied by persons or equipment.

Patent scaffolding

These are systems based on an independent scaffold format in which the members are connected together using an integral locking device, instead of conventional clips and couplers used with traditional tubular scaffolding. They have the advantages of being easy to assemble and take down using semi-skilled labour and should automatically comply with the requirements set out in the Work at Height Regulations 2005. Generally cross bracing is not required with these systems but facade bracing can be fitted if necessary. Although simple in concept, patent systems of scaffolding can lack the flexibility of traditional tubular scaffolds in complex layout situations.

Principal considerations

- Structural stability and integrity.
- Safety of personnel using a scaffold and for those in the vicinity.

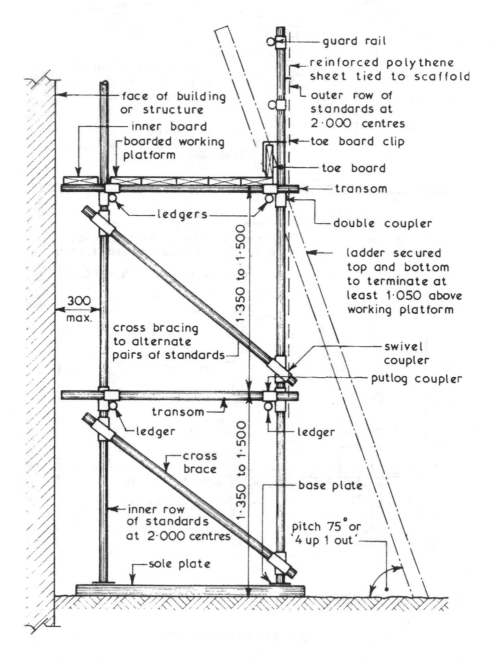

Figure 6.20 Independent scaffold

Scaffolds

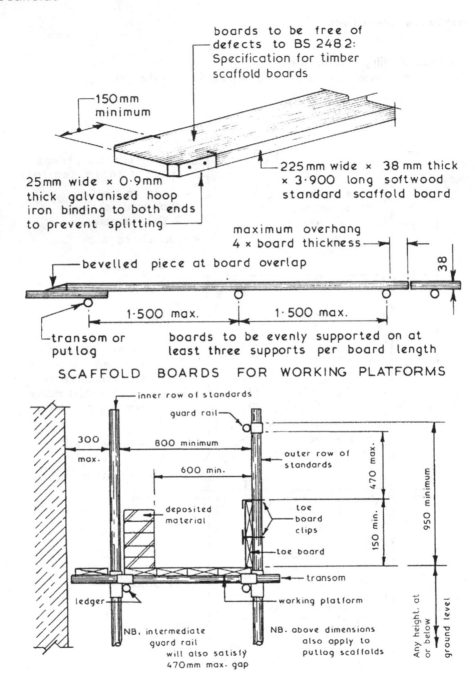

boards to be free of
defects to BS 2482:
Specification for timber
scaffold boards

150 mm
minimum

25mm wide × 0·9mm
thick galvanised hoop
iron binding to both ends
to prevent splitting

225 mm wide × 38 mm thick
× 3·900 long softwood
standard scaffold board

maximum overhang
4 × board thickness

38

bevelled piece at board overlap

1·500 max.   1·500 max.

transom or
putlog

boards to be evenly supported on at
least three supports per board length

SCAFFOLD BOARDS FOR WORKING PLATFORMS

inner row of standards

guard rail

300
max.

800 minimum

600 min.

outer row of
standards

470 max.

950 minimum

deposited
material

toe
board
clips

150 min.

toe board

transom

ledger

working platform

NB. intermediate
guard rail
will also satisfy
470mm max. gap

NB. above dimensions
also apply to
putlog scaffolds

Any height. at
or below

ground level

Figure 6.21 Independent scaffold

200

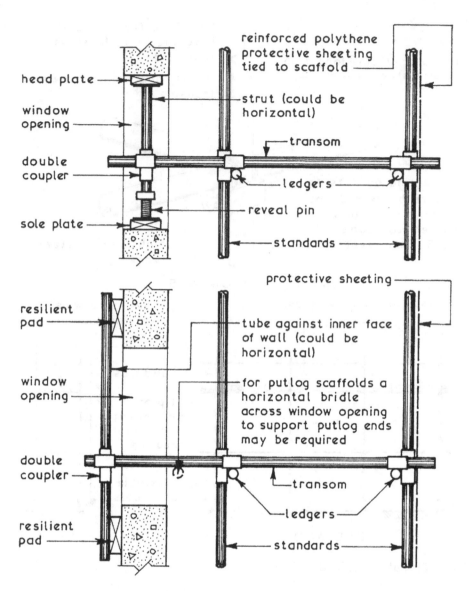

Figure 6.22 Tying-in scaffold

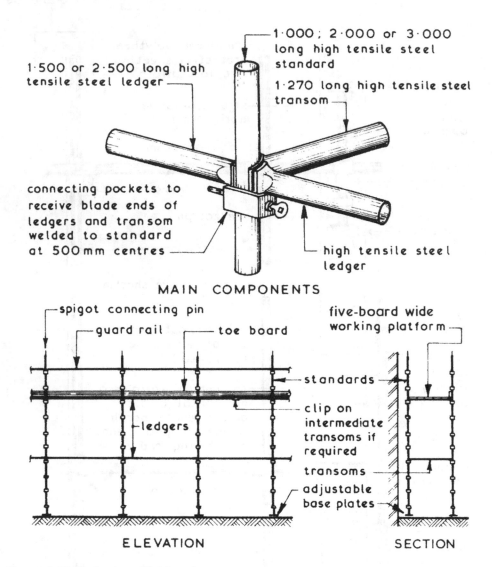

1.000 ; 2.000 or 3.000 long high tensile steel standard

1.270 long high tensile steel transom

1.500 or 2.500 long high tensile steel ledger

connecting pockets to receive blade ends of ledgers and transom welded to standard at 500 mm centres

high tensile steel ledger

MAIN COMPONENTS

spigot connecting pin

guard rail

toe board

ledgers

five-board wide working platform

standards

clip on intermediate transoms if required

transoms

adjustable base plates

ELEVATION

SECTION

Figure 6.23 Patent scaffold system

202

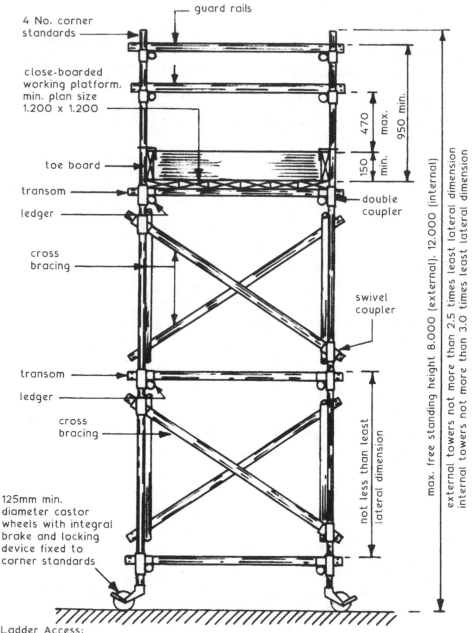

guard rails

4 No. corner standards

close-boarded working platform, min. plan size 1.200 x 1.200

toe board

transom

ledger

cross bracing

transom

ledger

cross bracing

125mm min. diameter castor wheels with integral brake and locking device fixed to corner standards

double coupler

swivel coupler

470 max.

150 min.

950 min.

max. free standing height 8.000 (external), 12.000 (internal)

external towers not more than 2.5 times least lateral dimension
internal towers not more than 3.0 times least lateral dimension

not less than least lateral dimension

Ladder Access:
Inclined within the tower to hinged access door in platform.
Secured to tower so as not to foul the ground.
Lowest rung max. 400mm above the ground.

Figure 6.24 Mobile tower scaffold

203

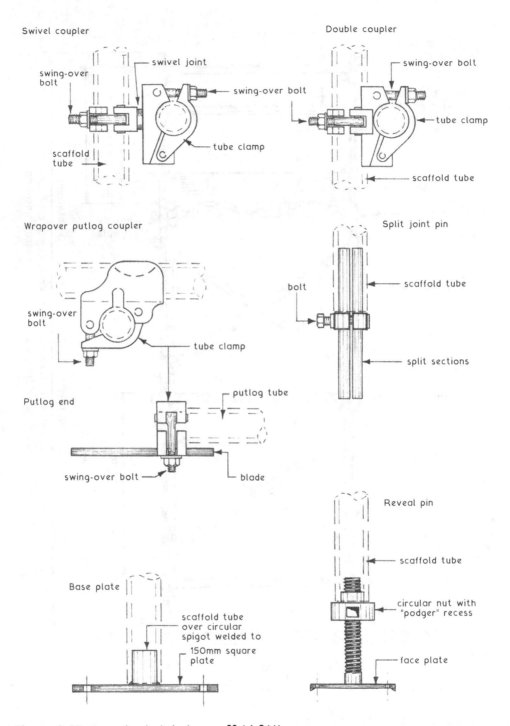

Swivel coupler

swing-over bolt

swivel joint

swing-over bolt

tube clamp

scaffold tube

Double coupler

swing-over bolt

tube clamp

scaffold tube

Wrapover putlog coupler

swing-over bolt

tube clamp

Putlog end

swing-over bolt

putlog tube

blade

Split joint pin

bolt

scaffold tube

split sections

Base plate

scaffold tube over circular spigot welded to 150mm square plate

Reveal pin

scaffold tube

circular nut with "podger" recess

face plate

Figure 6.25 Some basic tubular scaffold fittings

Relevant factors

- Simple configurations of putlog and independent scaffold, otherwise known as *basic systems* (see National Access and Scaffolding Confederation [NASC] technical guidance) do not require specific design calculations. Structural calculations are to be prepared by an appropriately qualified person for other applications.
- Assembly, alterations and dismantling to be undertaken in accordance with NASC technical guidance or system scaffolding manufacturer's instructions.
- Scaffold erectors to be competence qualified.
- On completion, the scaffold supplier should provide a handover certificate endorsed with references to design drawings and calculations. Any limitations of use, with particular mention of platform safe working loads, to be documented.
- Completed scaffolds to be inspected by a competent person, i.e. suitably qualified by experience and training. Inspection reports to be undertaken daily before work commences, after adverse weather and when alterations, modifications or additions are made. Any defects noted and corrective action taken.
- Inspection records to be documented and filed. These to include location and description of the scaffold, time and date of inspection, result of inspection and any actions taken. The report to be authorised by signature and endorsed with the inspector's job title.

7 FOUNDATIONS

Opening remarks

The function of any foundation is to safely sustain and transmit to the ground on which it rests the combined dead, imposed and wind loads in such a manner as not to cause any settlement or other movement which would impair the stability or cause damage to any part of the building.

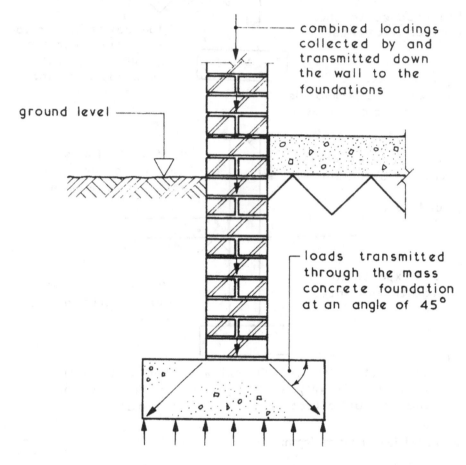

Figure 7.1 Strip foundation

Subsoil beneath foundation is compressed and reacts by exerting an upward pressure to resist foundation loading. If foundation load exceeds maximum passive pressure of ground (i.e. bearing capacity) a downward movement of the foundation could occur. The remedy is to increase plan size of foundation to reduce the load per unit area or, alternatively, reduce the loadings being carried by the foundations.

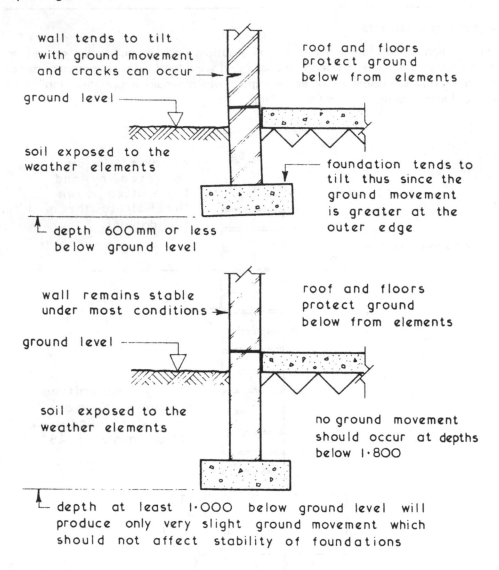

Figure 7.2 Strip foundation depths

Foundations can be affected by internal forces due to changes in volume if the water content of the subsoil changes around or below the foundations.

Compact granular soils, such as gravel, suffer very little movement whereas cohesive soils, such as clay, do suffer volume changes near the upper surface. Similar volume changes can occur due to water held in the subsoil freezing and expanding — this is called frost heave.

7.1 Trees

Trees draw up large quantities of water from the surrounding subsoil, which can cause changes in the water content and volume of the soil. The type of ground and the size, species and proximity of the tree to foundations are all variable factors. Clay subsoils are susceptible to moisture shrinkage or heave.

Subsoil shrinkage from trees is most evident in long periods of dry weather from oak, elm, poplar and willow species.

Heave is the opposite. It occurs during wet weather and is compounded by previous removal of moisture-dependent trees that would otherwise effect some drainage and balance to subsoil conditions.

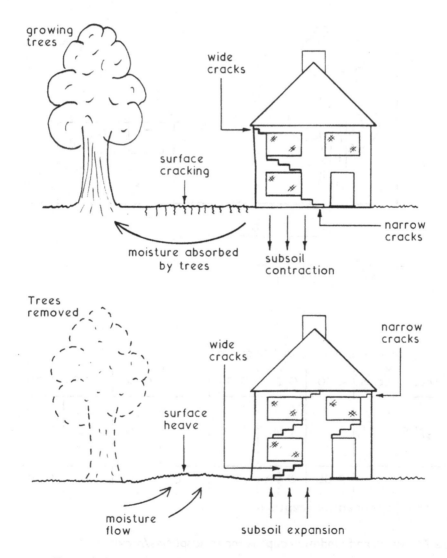

Figure 7.3 Effect on foundation of nearby trees

Trees

Proximity of trees to foundations

Trees up to 30m away may have an effect on foundations, therefore reference to local authority building control policy should be undertaken before specifying construction techniques.

Traditional strip foundations are practically unsuited, but at excavation depths up to 2.5 or 3.0m, deep strip or trench fill (preferably reinforced) may be appropriate. Short bored pile foundations are likely to be more economical and particularly suited to depths exceeding 3.0m.

For guidance only, the illustration and table in Figure 7.4 provide an indication of foundation depths in shrinkable subsoils.

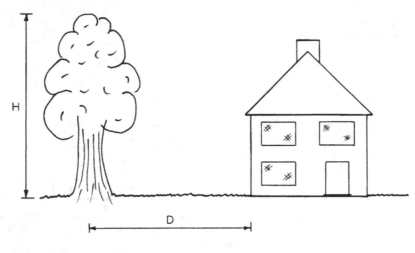

H = Mature height of tree
D = Distance to centre of tree

D/H – Distance from tree/Height of tree

Tree species	0·10	0·25	0·33	0·50	0·66	0·75	1·00
Oak, elm, poplar and willow	3·00	2·80	2·60	2·30	2·10	1·90	1·50
All others	2·80	2·40	2·10	1·80	1·50	1·20	1·00

Minimum foundation depth (m)

Figure 7.4 Minimum foundation depths for tree species/proximities

7.2 Foundation design

Foundations may be considered as either *shallow* or *deep* and are usually classified by their type: strips, pads, rafts and piles. It is also possible to combine foundation types, such as strip foundations connected by beams to and working in conjunction with pad foundations.

Apart from simple domestic strip foundations most foundation types are constructed in reinforced concrete.

Most shallow types of foundation are constructed within 2m of the ground level. Generally, foundations that need to be taken deeper are cheaper when designed and constructed as piled foundations and such foundations are classified as deep foundations.

Design principles

The main objectives of foundation design are to ensure that the structural loads are transmitted to the subsoil(s) safely, economically and without any unacceptable movement during the construction period and throughout the anticipated life of the building or structure.

Basic design procedure

1. Assessment of site conditions in the context of the site and soil investigation report.
2. Calculation of anticipated structural loading(s).
3. Choosing the foundation type, taking into consideration:
   - Soil conditions.
   - Proximity of trees.
   - Type of structure.
   - Structural loading(s).
   - Economic factors.
   - Time factors relative to the proposed contract period.
   - Construction problems.

4. Sizing the chosen foundation in the context of loading(s), ground bearing capacity and any likely future movements of the building or structure.

Foundation load calculation

The first stage is to calculate the loads of the building. The example in Figure 7.5 and Table 7.1 shows the process for calculating the linear load per metre run for a simple strip foundation.

Foundation design

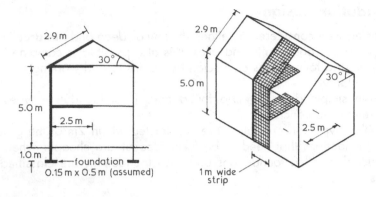

Figure 7.5 Calculating foundation loads
Note: Loads for half the span of the floor and roof are taken.

Table 7.1 Foundation loading calculation

Dead load per m run		
Substructure brickwork, 1m × 1m × 476kg/m²	=	476kg
Cavity conc. (50mm), 1m × 1m × 2300kg/m³	=	115kg
Foundation concrete, 0.15m × 1m × 0.5m × 2300kg/m³	=	173kg
Superstructure brickwork, 5m × 1m × 221kg/m²	=	1105kg
Blockwork & ins., 5m × 1m × 79kg/m²	=	395kg
2 coat plasterwork, 5m × 1m × 22kg/m²	=	110kg
Floor joists/boards/plstrbrd, 2.5m × 1m × 42.75kg/m²	=	107kg
Ceiling joists/plstrbrd/ins., 2.5m × 1m × 19.87kg/m²	=	50kg
Rafters, battens & felt, 2.9m × 1m × 12.10kg/m²	=	35kg
Single lap tiling, 2.9m × 1m × 49kg/m²	=	142kg
		2708kg

Note: Kg × 9.81* = force in Newtons [*9.81 is the acceleration due to the force of gravity in m/s²]
Therefore: 2708kg × 9.81 = 26,565N or 26.56kN
Imposed load per m run (see BS EN 1991-1-1: *Densities, Self-Weight, Imposed Loads for Buildings*)
Floor, 2.5m × 1m × 1.5kN/m² =   3.75kN
Roof, 2.9m × 1m × 1.5kN/m² (snow) = 4.05kN
                              7.80kN
Note: For roof pitch >30°, snow load = 0.75kN/m²
Dead + imposed load is 26.56kN + 7.80kN = 34.36kN
Given that the subsoil has a safe bearing capacity of 75kN/m²,
W = load ÷ bearing capacity = 34.36 ÷ 75 = 0.458m or 458mm
Therefore a foundation width of 500mm is adequate.
Note: This example assumes the site is sheltered. If it is necessary to make allowance for wind loading,
reference should be made to BS EN 1991-1-4: *Wind Actions.*

7.3 Strip foundations

This is the most common foundation type for low rise/domestic buildings and is formed from a mass concrete strip (min. 150mm thick) located centrally under the load-bearing walls in a trench excavated to a depth where the ground is suitable to support the loads without movement of the ground. Substructure masonry walls are then built up to DPC level.

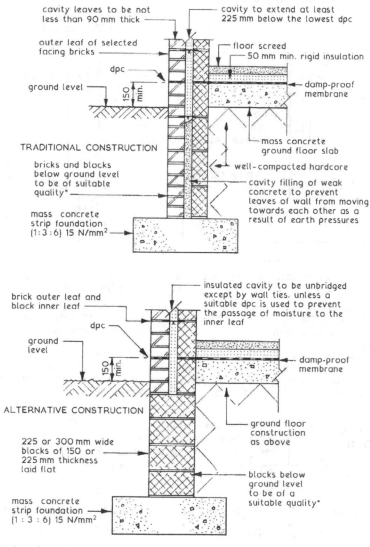

cavity leaves to be not less than 90 mm thick

cavity to extend at least 225 mm below the lowest dpc

outer leaf of selected facing bricks

floor screed

50 mm min. rigid insulation

dpc

ground level

150 min.

damp-proof membrane

TRADITIONAL CONSTRUCTION

bricks and blocks below ground level to be of suitable quality*

mass concrete ground floor slab

well-compacted hardcore

cavity filling of weak concrete to prevent leaves of wall from moving towards each other as a result of earth pressures

mass concrete strip foundation (1 : 3 : 6) 15 N/mm²

brick outer leaf and block inner leaf

insulated cavity to be unbridged except by wall ties, unless a suitable dpc is used to prevent the passage of moisture to the inner leaf

dpc

ground level

150 min.

damp-proof membrane

ALTERNATIVE CONSTRUCTION

225 or 300 mm wide blocks of 150 or 225 mm thickness laid flat

ground floor construction as above

blocks below ground level to be of a suitable quality*

mass concrete strip foundation (1 : 3 : 6) 15 N/mm²

*Min. compressive strength depends on building height and loading. See Building Regulations A.D. A: Section 2C

Figure 7.6 Strip foundations for cavity walls

Strip foundations

Calculation (typical basic procedure for guidance only)

The size of a foundation is basically dependent on two factors:

1. Load being transmitted (max. 70kN/m (dwellings up to three storeys)).
2. Bearing capacity of subsoil under proposed foundation.

For guidance on bearing capacities for different types of subsoil see BS EN 1997-1: *Geotechnical Design. General Rules* and BS 8103–1: *Structural Design of Low Rise Buildings*. Also, directly from soil investigation results.

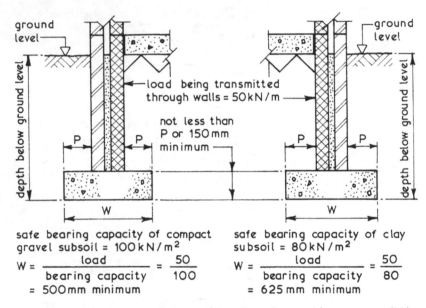

safe bearing capacity of compact gravel subsoil = 100 kN/m²

$$W = \frac{\text{load}}{\text{bearing capacity}} = \frac{50}{100}$$
$$= 500 \text{mm minimum}$$

safe bearing capacity of clay subsoil = 80 kN/m²

$$W = \frac{\text{load}}{\text{bearing capacity}} = \frac{50}{80}$$
$$= 625 \text{mm minimum}$$

The above widths may not provide adequate working space within the excavation and can be increased to give required space. Guidance on the minimum width for a limited range of applications can be taken from the table on the next page.

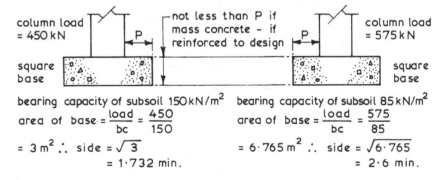

bearing capacity of subsoil 150 kN/m²

$$\text{area of base} = \frac{\text{load}}{\text{bc}} = \frac{450}{150}$$
$$= 3 \text{ m}^2 \therefore \text{ side} = \sqrt{3}$$
$$= 1\cdot732 \text{ min.}$$

bearing capacity of subsoil 85 kN/m²

$$\text{area of base} = \frac{\text{load}}{\text{bc}} = \frac{575}{85}$$
$$= 6\cdot765 \text{ m}^2 \therefore \text{ side} = \sqrt{6\cdot765}$$
$$= 2\cdot6 \text{ min.}$$

Figure 7.7 Calculating strip foundation widths

Table 7.2 Foundation widths

Ground type	Ground condition	Field test	Max. total load on load-bearing wall (kN/m)					
			20	30	40	50	60	70
			Minimum width (mm)					
Rock	Not infer-ior to sand-stone, lime-stone or firm chalk	Requires a mechanical device to excavate.	At least equal to the width of the wall					
Gravel Sand	Medium density Compact	Pick required to excavate. 50mm square peg hard to drive beyond 150mm.	250	300	400	500	600	650
Clay Sandy clay	Stiff Stiff	Requires pick or mechanical device to aid removal. Can be indented slightly with thumb.	250	300	400	500	600	650
Clay Sandy clay	Firm Firm	Can be moulded under substantial pressure by fingers.	300	350	450	600	750	850
Sand Silty sand	Loose Loose	Can be excavated by	400	600	Conventional strip foundations unsuitable for a total load exceeding 30 kN/m			

(Continued)

215

Table 7.2 Foundation widths (Cont.)

			Max. total load on load-bearing wall (kN/m)					
			20	30	40	50	60	70
Clayey sand	Loose	spade. 50mm square peg easily driven.						
Silt	Soft	Finger pushed in up to 10mm.	450	650				
Clay	Soft	Easily moulded with fingers.						
Sandy clay	Soft							
Silty clay	Soft							
Silt	Very soft	Finger easily pushed in up to 25mm. Wet sample exudes between fingers when squeezed.	Conventional strip inappropriate. Steel-reinforced wide strip, deep strip or piled foundation selected subject to specialist advice					
Clay	Very soft							
Sandy clay	Very soft							
Silty clay	Very soft							

Adapted from Table 10 in the Bldg. Regs., A.D: A – Structure.

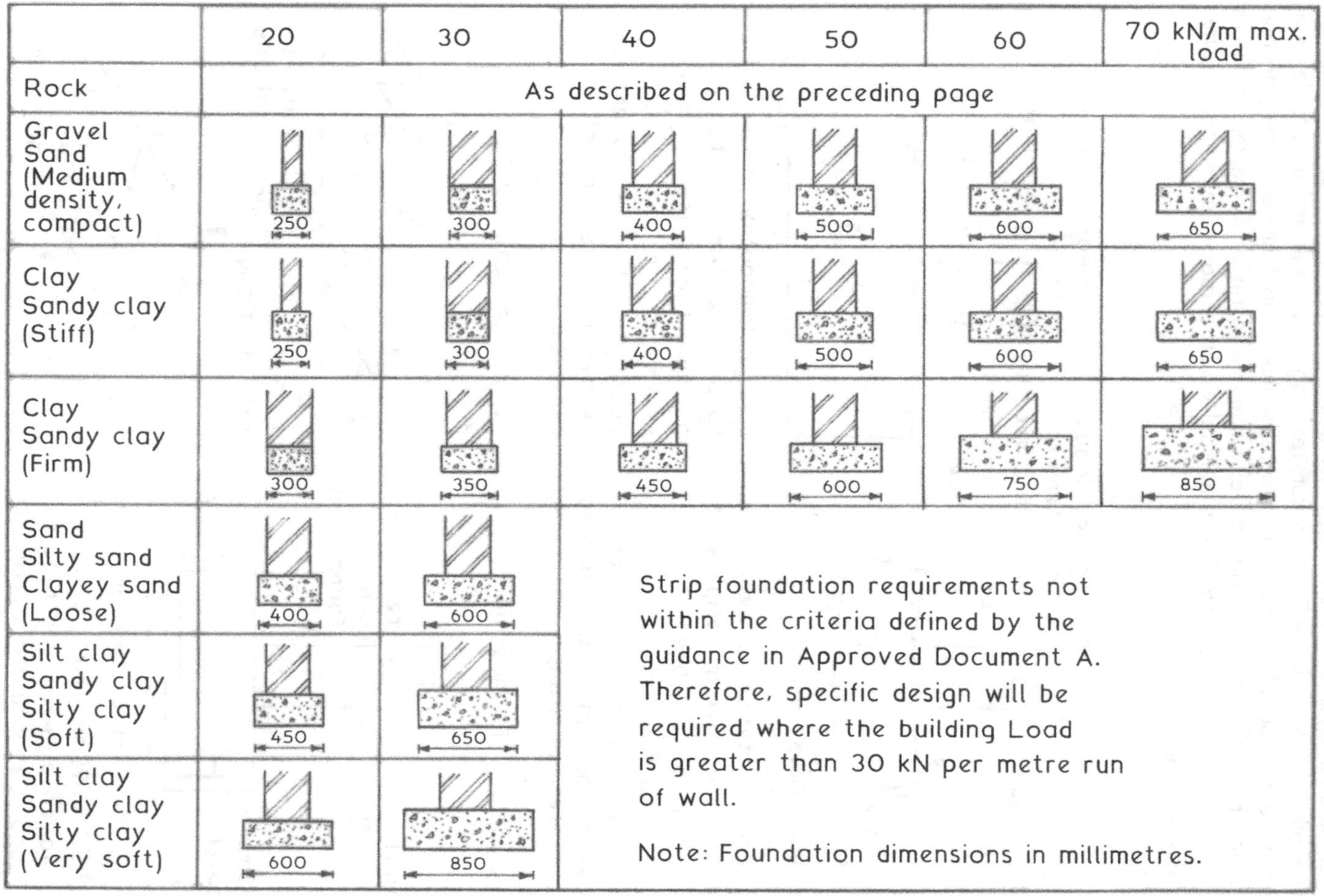

	20	30	40	50	60	70 kN/m max. load
Rock	As described on the preceding page					
Gravel Sand (Medium density, compact)	250	300	400	500	600	650
Clay Sandy clay (Stiff)	250	300	400	500	600	650
Clay Sandy clay (Firm)	300	350	450	600	750	850
Sand Silty sand Clayey sand (Loose)	400	600				
Silt clay Sandy clay Silty clay (Soft)	450	650				
Silt clay Sandy clay Silty clay (Very soft)	600	850				

Strip foundation requirements not within the criteria defined by the guidance in Approved Document A. Therefore, specific design will be required where the building Load is greater than 30 kN per metre run of wall.

Note: Foundation dimensions in millimetres.

Figure 7.8 Strip foundations from Table 7.2

Strip foundations

Steel reinforced concrete strip foundations

Concrete is a material which is strong in compression but weak in tension. If its tensile strength is exceeded cracks will occur, resulting in a weak and unsuitable foundation. One method of providing tensile resistance is to include in the concrete foundation bars of steel as a form of reinforcement to resist all the tensile forces induced into the foundation. Steel is a material which is readily available and has high tensile strength.

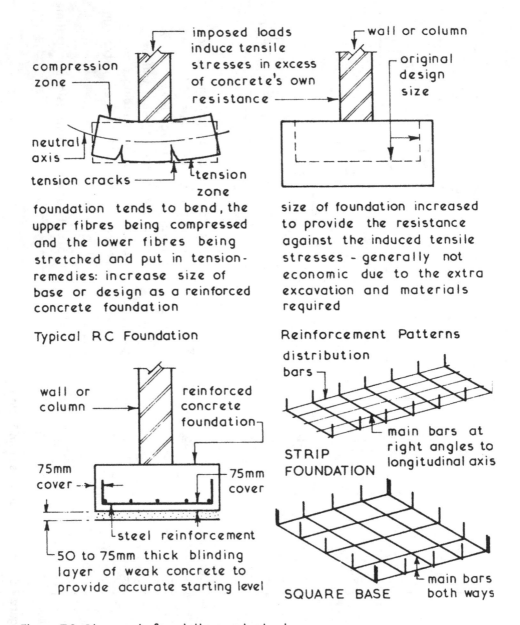

foundation tends to bend, the upper fibres being compressed and the lower fibres being stretched and put in tension - remedies: increase size of base or design as a reinforced concrete foundation

size of foundation increased to provide the resistance against the induced tensile stresses – generally not economic due to the extra excavation and materials required

Figure 7.9 Stresses in foundations under load

Stepped foundations

On sloping sites, to reduce the amount of excavation and materials required to produce an adequate foundation, strip foundations may be 'stepped'.

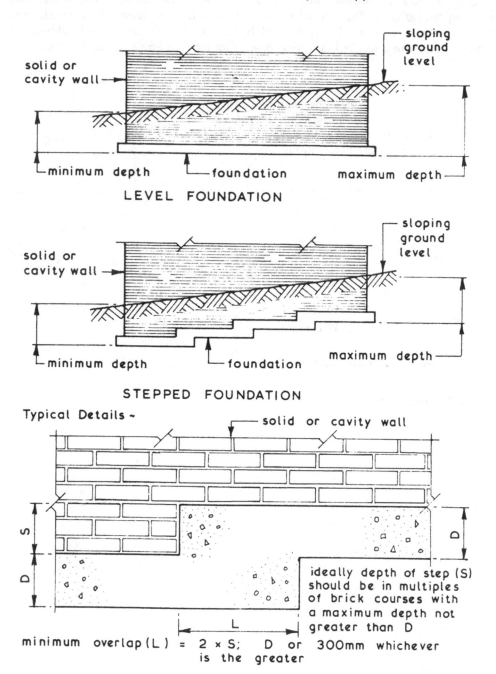

Figure 7.10 Stepped foundations

7.4 Raft foundations

A raft foundation spreads the loads of the structure over the entire footprint of the building instead of concentrating it in a strip under the load-bearing walls, thus reducing the building load per m^2 to be supported by the ground. Raft foundations are therefore suitable for lightly loaded buildings on soils with low load-bearing capacity; they are also used in mining areas that may be liable to subsidence.

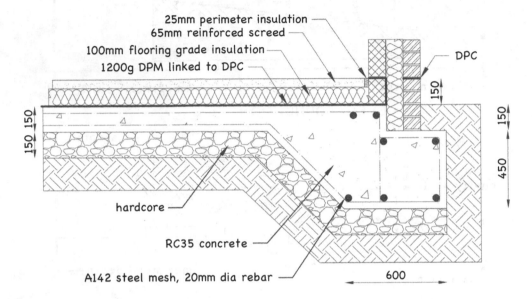

25mm perimeter insulation
65mm reinforced screed
100mm flooring grade insulation
1200g DPM linked to DPC
DPC
150
150 150
150
450
hardcore
RC35 concrete
A142 steel mesh, 20mm dia rebar
600

Figure 7.11 Raft edge detail

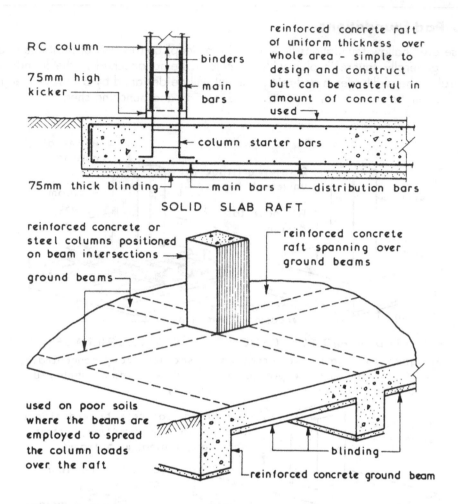

RC column

binders

75mm high kicker

main bars

reinforced concrete raft of uniform thickness over whole area - simple to design and construct but can be wasteful in amount of concrete used

column starter bars

75mm thick blinding

main bars

distribution bars

SOLID SLAB RAFT

reinforced concrete or steel columns positioned on beam intersections

reinforced concrete raft spanning over ground beams

ground beams

used on poor soils where the beams are employed to spread the column loads over the raft

blinding

reinforced concrete ground beam

NB. Ground beams can be designed as upstand beams with a precast concrete suspended floor at ground level thus creating a void space between raft and ground floor.

BEAM AND SLAB RAFT

Figure 7.12 Pad foundation for in-situ concrete frame

7.5 Pad foundations

These are shallow foundations (less than 2.0m deep) constructed of reinforced concrete, generally square in plan with the column being attached to the foundation centrally. The width, depth and reinforcement will be designed by a structural engineer, as will the column connection to avoid rotation and punching shear.

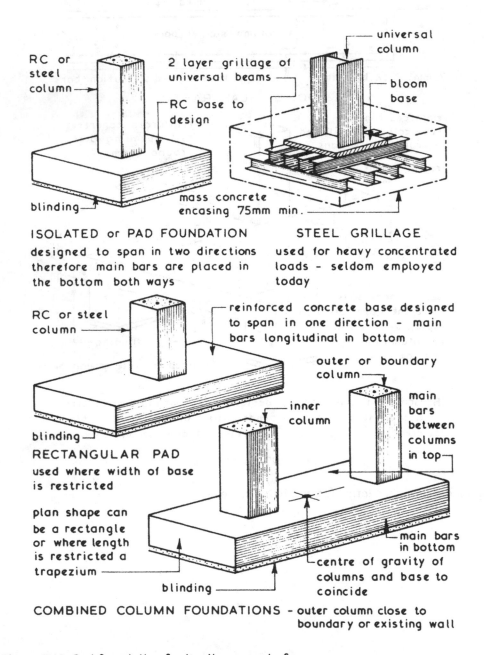

RC or steel column

2 layer grillage of universal beams

RC base to design

universal column

bloom base

blinding

mass concrete encasing 75mm min.

ISOLATED or PAD FOUNDATION
designed to span in two directions therefore main bars are placed in the bottom both ways

STEEL GRILLAGE
used for heavy concentrated loads - seldom employed today

RC or steel column

reinforced concrete base designed to span in one direction - main bars longitudinal in bottom

outer or boundary column

inner column

main bars between columns in top

blinding

RECTANGULAR PAD
used where width of base is restricted

plan shape can be a rectangle or where length is restricted a trapezium

main bars in bottom

centre of gravity of columns and base to coincide

blinding

COMBINED COLUMN FOUNDATIONS - outer column close to boundary or existing wall

Figure 7.13 Pad foundation for in-situ concrete frame

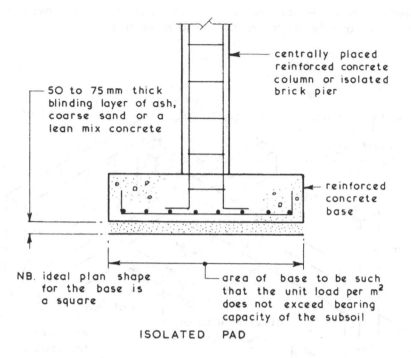

50 to 75 mm thick
blinding layer of ash,
coarse sand or a
lean mix concrete

centrally placed
reinforced concrete
column or isolated
brick pier

reinforced
concrete
base

NB. ideal plan shape
for the base is
a square

area of base to be such
that the unit load per m^2
does not exceed bearing
capacity of the subsoil

ISOLATED PAD

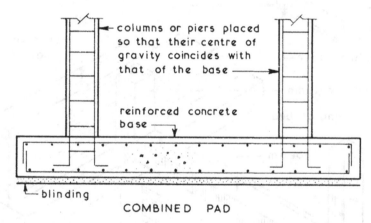

columns or piers placed
so that their centre of
gravity coincides with
that of the base

reinforced concrete
base

blinding

COMBINED PAD

Figure 7.14 Pad foundation for in-situ concrete frame

Pad foundations

Cantilever foundations ~ these can be used where it is necessary to avoid imposing any pressure on an adjacent foundation or underground service.

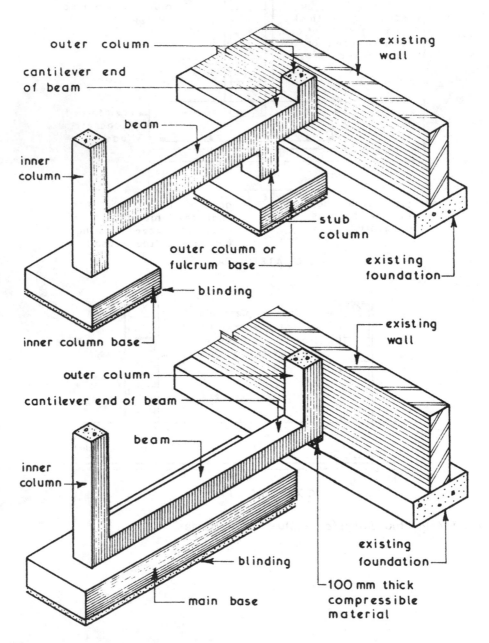

Cantilever foundations designed and constructed in reinforced concrete

Figure 7.15 Typical cantilever foundation types

7.6 Small diameter piles and ground beams

Small diameter piles, aka mini/micro piles, are necessary where the required depth of the foundation is uneconomic for strip foundations. The system consists of two parts.

1. Small diameter pile.
2. Reinforced concrete ground beams.

Small diameter pile types

- Mini augered piles, 150mm to 300mm diameter.
- Steel cased bottom driven mini displacement piles: 75mm to 323mm.
- CFA piles: 300mm to 600mm.
- Pre-cast concrete displacement piles.

Reinforced concrete ground beams

These are formed at a reduced level below ground by forming trenches to contain and form the ground beam with its reinforcement and connection with the piles.

Small diameter piles and ground beams

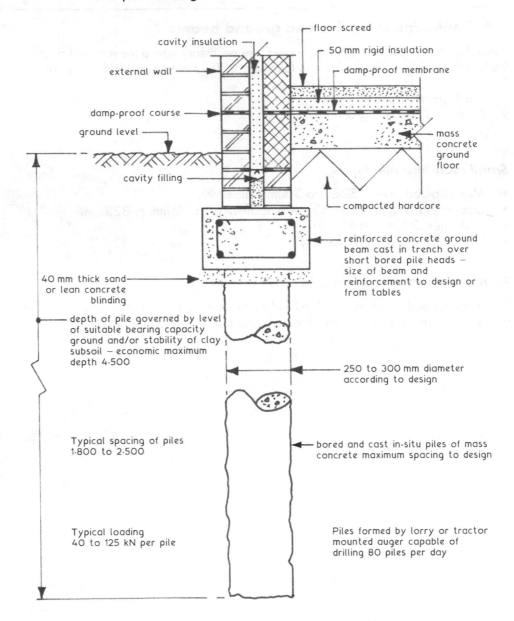

cavity insulation

floor screed

50 mm rigid insulation

external wall

damp-proof membrane

damp-proof course

ground level

mass concrete ground floor

cavity filling

compacted hardcore

reinforced concrete ground beam cast in trench over short bored pile heads – size of beam and reinforcement to design or from tables

40 mm thick sand or lean concrete blinding

depth of pile governed by level of suitable bearing capacity ground and/or stability of clay subsoil – economic maximum depth 4·500

250 to 300 mm diameter according to design

Typical spacing of piles 1·800 to 2·500

bored and cast in-situ piles of mass concrete maximum spacing to design

Typical loading 40 to 125 kN per pile

Piles formed by lorry or tractor mounted auger capable of drilling 80 piles per day

Figure 7.16 Pile foundation and ground beam

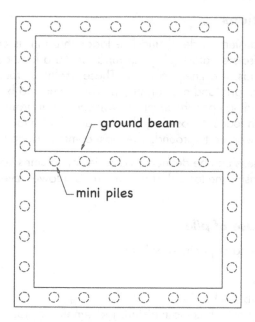

Figure 7.17 Ground beams under load bearing walls

Precast concrete foundations

These are proprietary fully precast reinforced concrete systems using driven precast piles supporting pile caps with wall support beams spanning between the pile caps. The advantages of this system are that it is a minimal dig system and is fast to install, but it is not suitable for all sites.

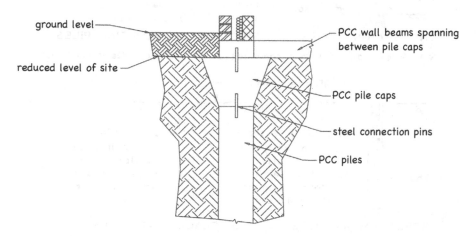

Figure 7.18 Elements of a pre-cast concrete foundation system

7.7 Pile foundations

A framed building transfers its dead and live loads through its columns into concentrated point loads in isolated areas of the ground, unlike a load-bearing wall structure which transfers its loads in a linear manner. These 'isolated' foundations are set out centrally on the column grid and may or may not be connected by reinforced concrete ground beams, depending upon the external wall chosen at ground floor level.

Because of the high loads imposed by the foundations, suitable load-bearing strata are often only found deep underground. The most common deep foundations are:

Piled foundations ~ these can be defined as a series of columns constructed or inserted into the ground to transmit the load(s) of a structure to a lower level of subsoil.

Other factors for use of pile

- Natural low bearing capacity of subsoil.
- Filled ground.
- High water table.
- Highly compressible subsoils.
- Subsoils that may be subject to moisture movement or plastic failure.

Piles may be classified by their basic design function or by their method of construction:

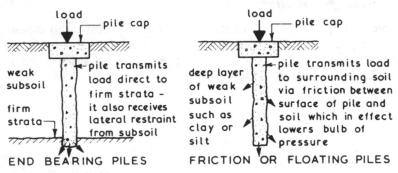

NB. Piles can work in a combination of the above design functions

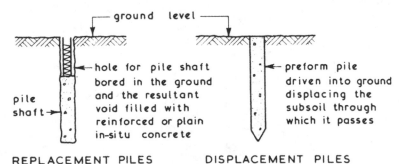

Figure 7.19 Pile types

Replacement piles

These are often called bored piles since the removal of the spoil to form the hole for the pile is always carried out by a boring technique. They are used primarily in cohesive subsoils for the formation of friction piles and when forming pile foundations close to existing buildings where the allowable amount of noise and/or vibration is limited.

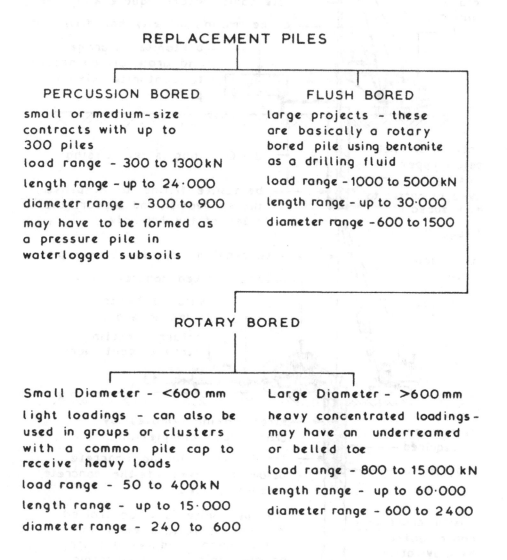

REPLACEMENT PILES

PERCUSSION BORED

small or medium-size contracts with up to 300 piles

load range - 300 to 1300 kN

length range - up to 24·000

diameter range - 300 to 900

may have to be formed as a pressure pile in waterlogged subsoils

FLUSH BORED

large projects - these are basically a rotary bored pile using bentonite as a drilling fluid

load range - 1000 to 5000 kN

length range - up to 30·000

diameter range - 600 to 1500

ROTARY BORED

Small Diameter - <600 mm

light loadings - can also be used in groups or clusters with a common pile cap to receive heavy loads

load range - 50 to 400 kN

length range - up to 15·000

diameter range - 240 to 600

Large Diameter - >600 mm

heavy concentrated loadings - may have an underreamed or belled toe

load range - 800 to 15000 kN

length range - up to 60·000

diameter range - 600 to 2400

NB. The above given data depicts typical economic ranges. More than one pile type can be used on a single contract.

Figure 7.20 Replacement pile types

Pile foundations

Flush bored pile process

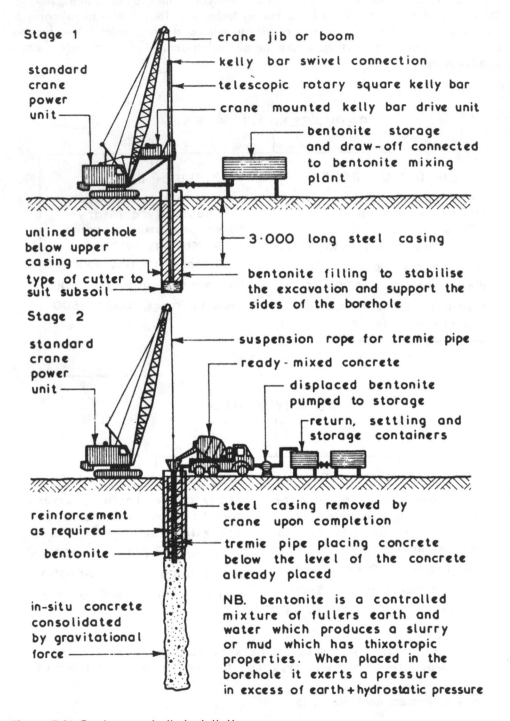

Stage 1

standard crane power unit

crane jib or boom

kelly bar swivel connection

telescopic rotary square kelly bar

crane mounted kelly bar drive unit

bentonite storage and draw-off connected to bentonite mixing plant

unlined borehole below upper casing

3·000 long steel casing

type of cutter to suit subsoil

bentonite filling to stabilise the excavation and support the sides of the borehole

Stage 2

standard crane power unit

suspension rope for tremie pipe

ready-mixed concrete

displaced bentonite pumped to storage

return, settling and storage containers

reinforcement as required

bentonite

in-situ concrete consolidated by gravitational force

steel casing removed by crane upon completion

tremie pipe placing concrete below the level of the concrete already placed

NB. bentonite is a controlled mixture of fullers earth and water which produces a slurry or mud which has thixotropic properties. When placed in the borehole it exerts a pressure in excess of earth+hydrostatic pressure

Figure 7.21 Replacement pile installation

Large diameter rotary bored piles

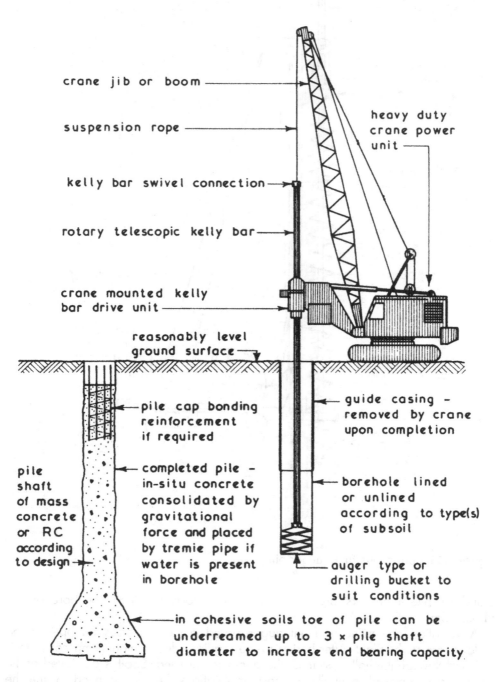

crane jib or boom

suspension rope

kelly bar swivel connection

rotary telescopic kelly bar

crane mounted kelly bar drive unit

heavy duty crane power unit

reasonably level ground surface

pile cap bonding reinforcement if required

guide casing – removed by crane upon completion

completed pile – in-situ concrete consolidated by gravitational force and placed by tremie pipe if water is present in borehole

borehole lined or unlined according to type(s) of subsoil

pile shaft of mass concrete or RC according to design

auger type or drilling bucket to suit conditions

in cohesive soils toe of pile can be underreamed up to 3 × pile shaft diameter to increase end bearing capacity

Figure 7.22 Underreaming piles

Pile foundations

Continuous flight auger bored piles (CFA)

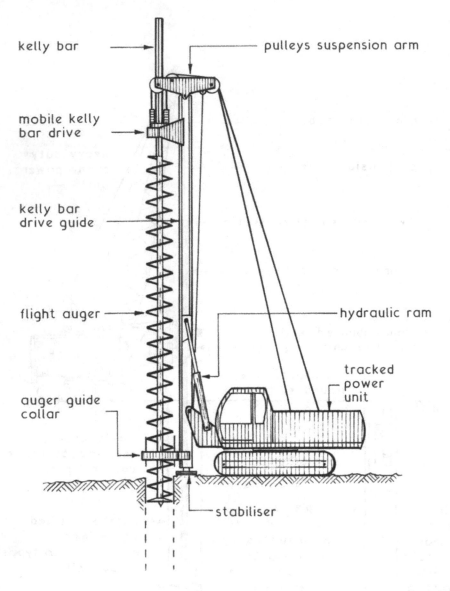

Figure 7.23 Continuous flight auger piling rig

Standard pile diameters are 300, 450 and 600mm. Depth of borehole to about 15m.

A variation of continuous flight auger bored piling that uses an open-ended hollow core to the flight. After boring to the required depth, high slump concrete is pumped through the hollow stem as the auger is retracted. Spoil is displaced at the surface and removed manually. In most applications there is no need to line the boreholes, as the subsoil has little time to be disturbed. A preformed reinforcement cage is pushed into the wet concrete.

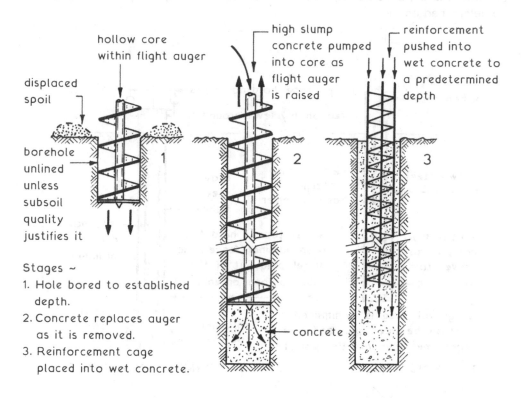

displaced spoil

hollow core within flight auger

high slump concrete pumped into core as flight auger is raised

reinforcement pushed into wet concrete to a predetermined depth

borehole unlined unless subsoil quality justifies it

concrete

Stages ~
1. Hole bored to established depth.
2. Concrete replaces auger as it is removed.
3. Reinforcement cage placed into wet concrete.

Figure 7.24 CFA installation sequence

Pile foundations

Driven in-situ piles

Also known as Franke Piles, there are used on medium to large contracts as an alternative to preformed piles, particularly where final length of pile is a variable to be determined on site.

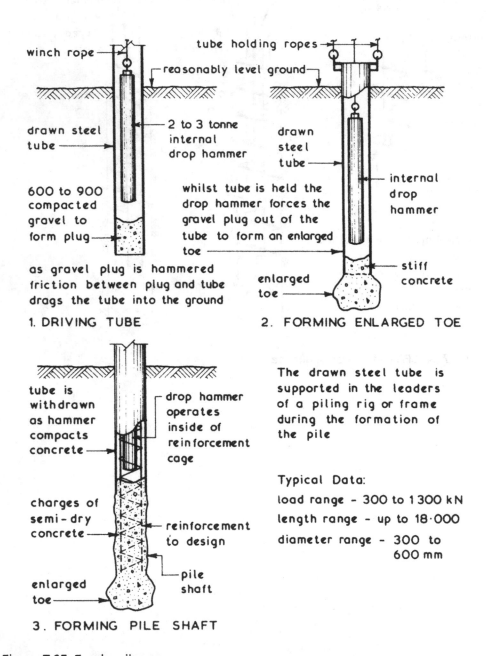

Figure 7.25 Franke piles

Displacement piles

These are precast concrete or steel pile sections that are also known as driven piles since they are usually driven into the ground, displacing the earth around the pile shaft. The pile is driven into the ground to a predetermined depth or to the required 'set', which is a measure of the subsoil's resistance to the penetration of the pile and hence its bearing capacity, by noting the amount of penetration obtained by a fixed number of hammer blows.

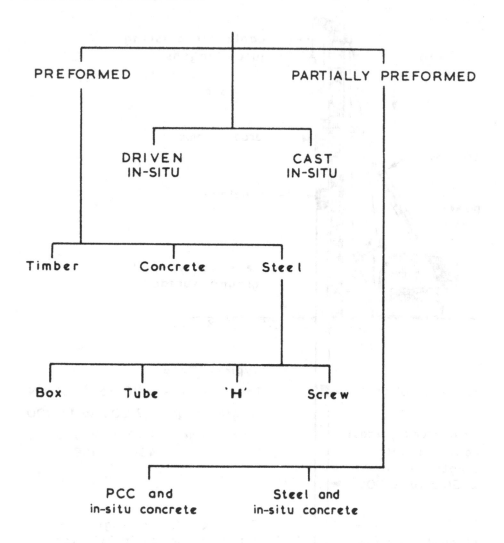

Figure 7.26 Displacement pile types

Pile foundations

Precast concrete pile

A variety of lengths/sizes/sections/types are available to suit a variety of load/ ground situations, used where soft soil deposits overlie firmer strata. These piles are percussion driven using a drop or single-acting hammer.

Typical Example [West's Hardrive Precast Modular Pile] ~

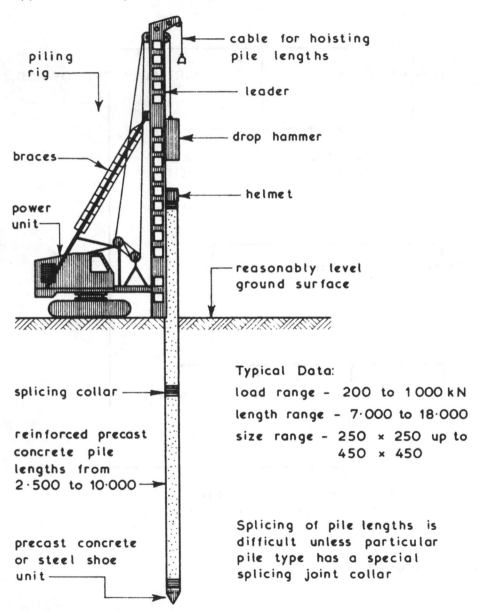

piling rig

cable for hoisting pile lengths

leader

drop hammer

braces

power unit

helmet

reasonably level ground surface

splicing collar

reinforced precast concrete pile lengths from 2·500 to 10·000

precast concrete or steel shoe unit

Typical Data:

load range – 200 to 1 000 kN

length range – 7·000 to 18·000

size range – 250 × 250 up to 450 × 450

Splicing of pile lengths is difficult unless particular pile type has a special splicing joint collar

Figure 7.27 Displacement pile driving

Precast sections can be mechanically connected together with steel pins and collars.

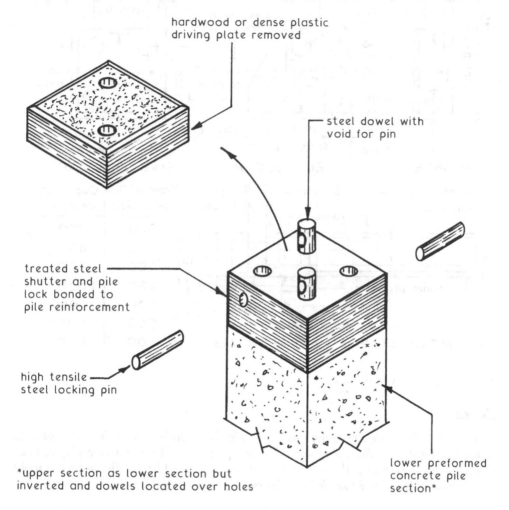

hardwood or dense plastic
driving plate removed

steel dowel with
void for pin

treated steel
shutter and pile
lock bonded to
pile reinforcement

high tensile
steel locking pin

*upper section as lower section but
inverted and dowels located over holes

lower preformed
concrete pile
section*

Figure 7.28 Precast concrete pile connection pieces

Pile caps

Piles can be used singly to support the load but often it is more economical to use piles in groups or clusters linked together with a reinforced concrete cap. The pile caps can also be linked together with reinforced concrete ground beams.

The usual minimum spacing for piles is shown in Table 7.3.

Table 7.3 Typical pile spacings

1. Friction piles	–	1.100 or not less than 3 whichever is the greater	×	pile diameter
2. Bearing piles	–	750mm or not less than 2 whichever is the greater	×	pile diameter

Pile foundations

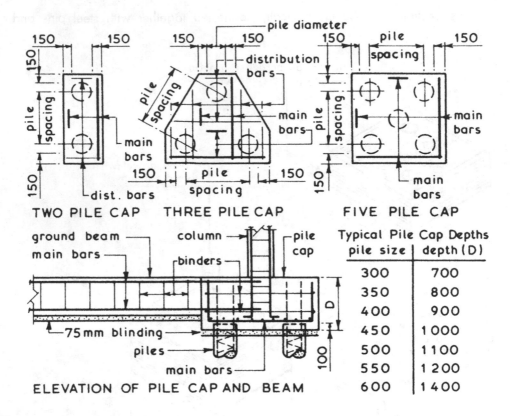

Figure 7.29 Types of pile cap

Typical Pile Cap Depths

pile size	depth (D)
300	700
350	800
400	900
450	1 000
500	1 100
550	1 200
600	1 400

Pile testing

To ensure that the piles satisfy the design criteria they can be tested. It is advisable to test load at least one pile per scheme. The test pile should be overloaded by at least 50% of its working load and this load should be held for 24 hours. The test pile should not form part of the actual foundations. Suitable testing methods are:

1. Jacking against kentledge placed over test pile.
2. Jacking against a beam fixed to anchor piles driven in on two sides of the test pile.

8 BASEMENT CONSTRUCTION

Opening remarks

Basement ~ a building floor which is below the ground floor of the building and is therefore constructed below ground level.

On sloping sites the basement may be only partially below ground level. Most basements can be classified into one of three groups.

1. Retaining Wall and Raft Basements - this is the general format for basement construction and consists of a slab raft foundation which forms the basement floor and helps to distribute the structural loads transmitted down the retaining walls.

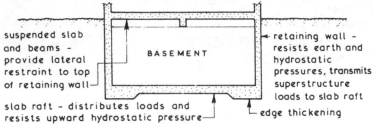

suspended slab and beams - provide lateral restraint to top of retaining wall⌐

BASEMENT

retaining wall - resists earth and hydrostatic pressures, transmits superstructure loads to slab raft

slab raft - distributes loads and resists upward hydrostatic pressure⌐

edge thickening

2. Box and Cellular Raft Basements - similar method to above except that internal walls are used to transmit and spread loads over raft as well as dividing basement into cells.

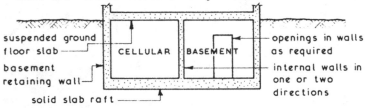

suspended ground floor slab

basement retaining wall⌐

CELLULAR | BASEMENT

openings in walls as required

internal walls in one or two directions

solid slab raft

3. Piled Basements - the main superstructure loads are carried to the basement floor level by columns where they are finally transmitted to the ground via pile caps and bearing piles. This method can be used where low bearing capacity soils are found at basement floor level.

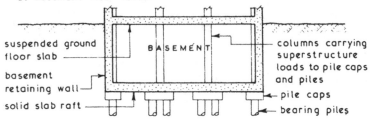

suspended ground floor slab

basement retaining wall⌐

solid slab raft

BASEMENT

columns carrying superstructure loads to pile caps and piles

pile caps

bearing piles

Figure 8.1 Basement construction methods

8.1 Secant pile walls

Secant piling forms a permanent structural wall of interlocking bored piles. Alternate piles are bored and cast by traditional methods and before the concrete has fully hardened the interlocking piles are bored using a toothed flight auger. This system is suitable for most types of subsoil and has the main advantages of being economical on small and confined sites; it is capable of being formed close to existing foundations and can be installed with the minimum of vibration and noise. Ensuring a complete interlock of all piles over the entire length may be difficult to achieve in practice; therefore the exposed face of the piles is usually covered with a mesh or similar fabric and faced with rendering or sprayed concrete. Alternatively a reinforced concrete wall could be cast in front of the contiguous piling. This method of ground water control is suitable for structures such as basements, road underpasses and underground car parks.

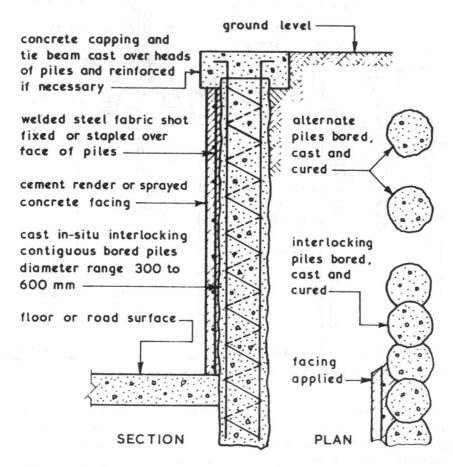

Figure 8.2 Secant piling

8.2 Diaphragm walls

These are structural concrete walls which can be cast in situ (usually by the bentonite slurry method) or constructed using precast concrete components. They are suitable for most subsoils and their installation generates only a small amount of vibration and noise, making them suitable for works close to existing buildings. The high cost of these walls makes them uneconomic unless they can be incorporated into the finished structure. Diaphragm walls are suitable for basements, underground car parks and similar structures.

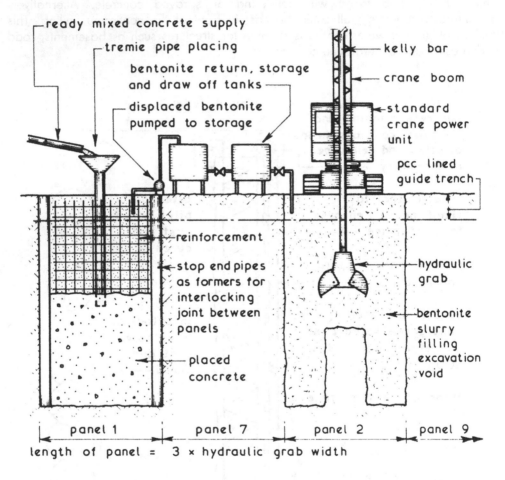

NB. Bentonite is a controlled mixture of fullers earth and water which produces a mud or slurry which has thixotropic properties and exerts a pressure in excess of earth + hydrostatic pressure present on sides of excavation.

Figure 8.3 Diaphragm wall construction

8.3 Top down basement construction

Basement Construction with Permanent Lateral Support

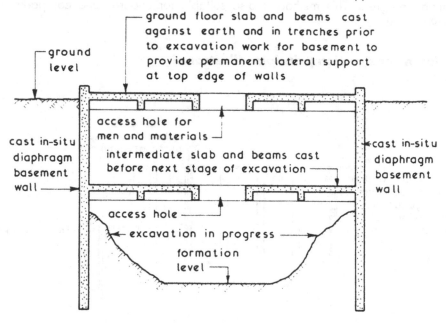

Basement Construction with Temporary Lateral Support

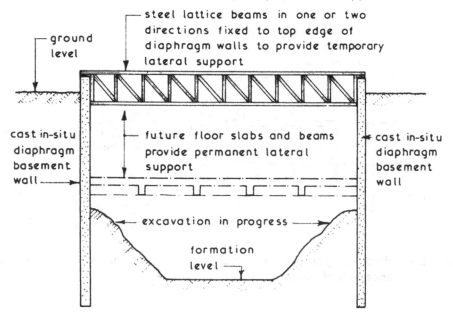

Figure 8.4 Top down basement construction

Top down basement construction

Top down construction is a method associated with deep, multi-level basement construction where intermediate level basement floors are constructed as excavation proceeds downwards. The main benefit of this method is that construction above ground can be conducted simultaneously with construction below ground, thereby optimizing progress. This method is also suitable for underground car parks and railway stations.

Top down construction principles illustrated

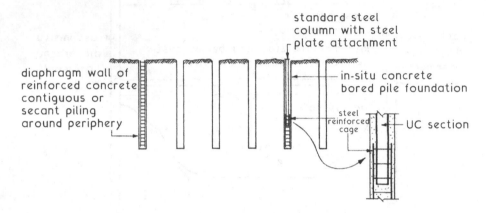

Figure 8.5 Diaphragm walls and plunge column piled foundation

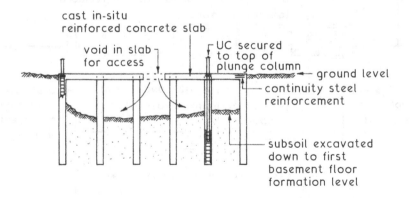

Figure 8.6 Reinforced ground floor slab cast with access for excavation and commencement of superstructure

Sub-floors concreted whilst superstructure proceeds

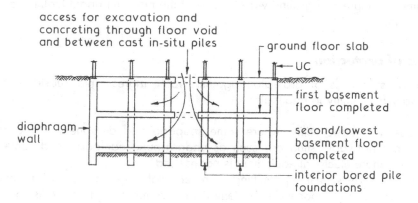

Figure 8.7 Access for excavation

Top down sequence

- Ground water controls/dewatering installed and activated (subject to sub-surface composition and water table height).
- Cast in-situ concrete diaphragm perimeter walls installed, possibly using contiguous or secant piling methods.
- Holes bored for piled foundations.
- Steel reinforcement cage positioned in boreholes.
- Load-bearing plunge columns (standard steel UC sections) lowered inside reinforcement cage.
- Boreholes concreted to ground level.
- In-situ reinforced concrete slab cast at ground level with shuttering to create suitable size opening(s) for excavation plant access. This and subsequent sub-floors act as lateral bracing to perimeter walls.
- Superstructure commences.
- Subsoil extracted through slab opening. Possible use of crane and bucket or long arm excavator. After initial soil removal, small backacters may be able to operate below ground floor slab.
- First basement formation level established and in-situ concrete floor slab cast with suitable size opening to access the next level down.
- Excavation procedure repeated to formation level of second basement floor and slab cast with access void if further subfloors required.
- Continue until design depth and number of sub-floors is reached.
- Basement construction completed with waterproofing (tanking) and finishes.
- Continue with superstructure construction.

8.4 Waterproofing basements

Preventing the ingress of ground water is one of the most important features of basement design.

Grades of protection

Grades or standards of protection against water ingress into substructures (BS 8102) vary from utility to habitable.

- Grade 1 – *basic utility*, where some seepage and dampness is acceptable. E.g. non-habitable situations such as underground car parks and some plant rooms where electrical equipment is not housed.
- Grade 2 – *better utility*, permitting a presence of moisture vapour but not water penetration. E.g. spaces in residential and commercial basements for equipment storage in general (non-electrical), possibly for retail storage of non-perishable goods.
- Grade 3 – *habitable*, i.e. for human occupation, therefore totally dry. Includes areas that are used in conjunction with habitable accommodation such as material and equipment storage.

Note: Both *utility* grades will require drainage by gravitational means or a sump collection point with pump discharge.

Basements can be waterproofed by one of three basic methods, namely:

1. Use of dense monolithic concrete walls and floor.
2. Tanking techniques.
3. Drained cavity system.

Dense monolithic concrete

The main objective is to form a watertight basement using dense, high quality, reinforced or pre-stressed concrete by a combination of high quality materials, good workmanship, attention to design detail and on-site construction methods. If strict control of all aspects is employed, a sound, watertight structure can be produced, but it should be noted that such structures are not always water vapour proof. If the latter is desirable some waterproof coating, lining or tanking should be used. The water tightness of dense concrete mixes depends primarily upon three factors:

1. Water/cement ratio.
2. Degree of compaction.
3. Water proofing additives.

Joints

In general these are formed in basement constructions to provide for movement accommodation (expansion joints) or to create a convenient stopping point in the construction process (construction joints). Joints are lines of weakness which will leak unless carefully designed and constructed; therefore they should be simple in concept and easy to construct.

Basement slabs are usually designed to span in two directions and as a consequence have relatively heavy top and bottom reinforcement. To enable them to fulfil their basic functions they usually have a depth in excess of 250mm. The joints, preferably of the construction type, should be kept to a minimum and if waterbars are specified they must be placed to ensure that complete compaction of the concrete is achieved.

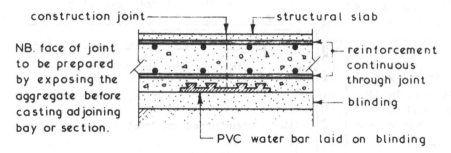

Basement Walls ~ joints can be horizontal and/or vertical according to design requirements. A suitable waterbar should be incorporated in the joint to prevent the ingress of water. The top surface of a kicker used in conjunction with single-lift pouring if adequately prepared by exposing the aggregate should not require a waterbar but if one is specified it should be either placed on the rear face or consist of a centrally placed mild steel strip inserted into the kicker whilst the concrete is still in a plastic state.

Typical Basement Wall Joint Details ~

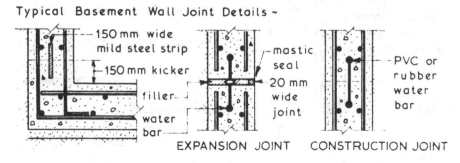

Figure 8.8 Basement wall joint details

Waterproofing basements

Tanking

The objective of tanking is to provide a continuous waterproof membrane, which is applied to the base slab and walls with complete continuity between the two applications.

The tanking can be applied externally or internally according to the circumstances prevailing on site. Mastic asphalt, bituminous compounds and epoxy resin compounds are commonly used for tanking membranes.

External mastic asphalt tanking

This is the preferred method since it not only prevents the ingress of water but it also protects the main structure of the basement from aggressive sulphates that may be present in the surrounding soil or ground water.

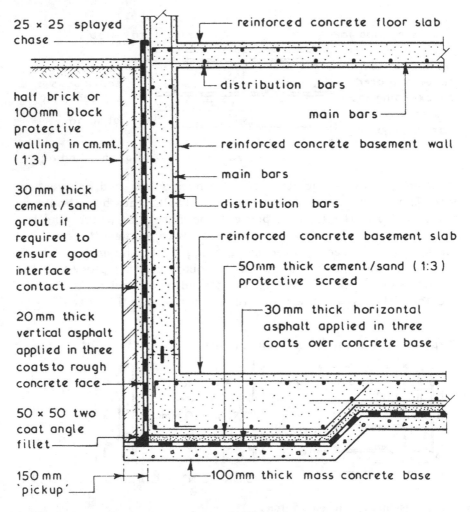

Figure 8.9 Basement external waterproof tanking

Internal mastic asphalt tanking

This method should only be adopted if external tanking is not possible, since it will not give protection to the main structure and, unless adequately loaded, may be forced away from the walls and/or floor by hydrostatic pressure. To be effective the horizontal and vertical coats of mastic asphalt must be continuous.

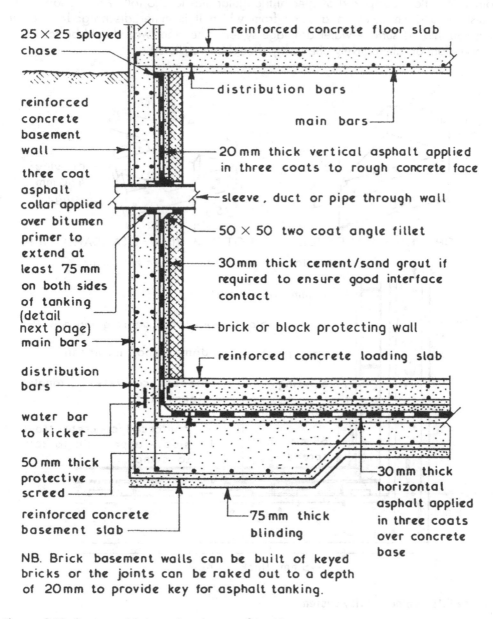

25 × 25 splayed chase

reinforced concrete floor slab

reinforced concrete basement wall

distribution bars

main bars

three coat asphalt collar applied over bitumen primer to extend at least 75 mm on both sides of tanking (detail next page) main bars

20 mm thick vertical asphalt applied in three coats to rough concrete face

sleeve, duct or pipe through wall

50 × 50 two coat angle fillet

30 mm thick cement/sand grout if required to ensure good interface contact

brick or block protecting wall

reinforced concrete loading slab

distribution bars

water bar to kicker

50 mm thick protective screed

reinforced concrete basement slab

75 mm thick blinding

30 mm thick horizontal asphalt applied in three coats over concrete base

NB. Brick basement walls can be built of keyed bricks or the joints can be raked out to a depth of 20mm to provide key for asphalt tanking.

Figure 8.10 Basement internal waterproof tanking

Waterproofing basements

Drained cavity system

This method of waterproofing basements can be used for both new and refurbishment work. The basic concept is very simple in that it accepts that a small amount of water seepage is possible through a monolithic concrete wall and the best method of dealing with such moisture is to collect it and drain it away. This is achieved by building an inner non-load-bearing wall to form a cavity which is joined to a floor composed of special triangular tiles laid to falls which enables the moisture to drain away to a sump from which it is either discharged direct or pumped into the surface water drainage system. The inner wall should be relatively vapour tight or, alternatively, the cavity should be ventilated.

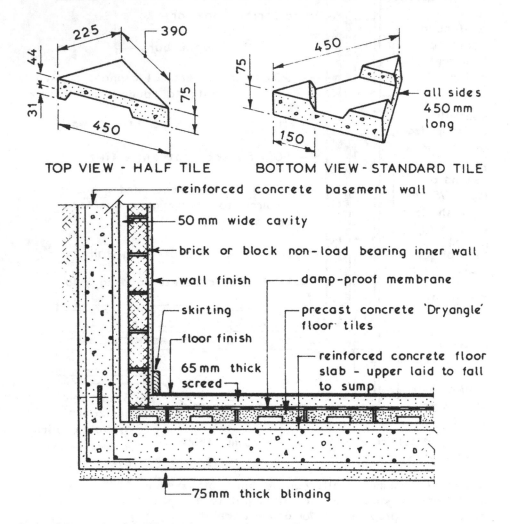

Figure 8.11 Drained cavity system

Service pipes and ducts

The unbroken continuity of tanking is essential to prevent ground water penetration. Therefore, if possible, pipework, ducting and cables should be located internally above ground level with vertical access into and within a basement. In some situations this will be impractical and penetration of the tanking is unavoidable. Here, particular attention to a watertight fit between pipe and tanking will be required. Two possible methods are:

1. Asphalt sleeve.
2. Sheet lead collar.

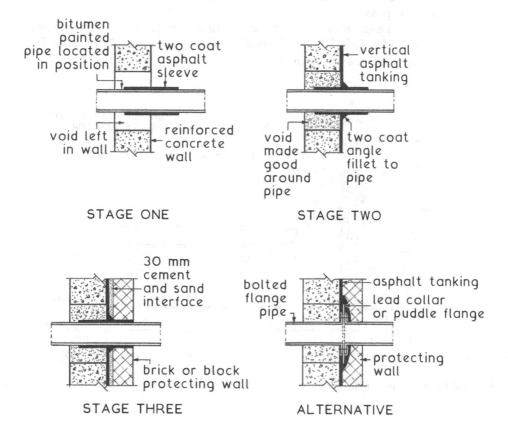

Figure 8.12 Service pipe passing through basement wall

8.5 Basement insulation

Basements benefit considerably from the insulating properties of the surrounding soil. However, that alone is insufficient to satisfy the typical requirements for wall and floor U-values of 0.35 and $0.25 W/m^2K$, respectively.

Refurbishment of existing basements may include insulation within dry lined walls and under-the-floor screed or particle board overlay. This should incorporate an integral vapour control layer to minimise risk of condensation.

External insulation of closed cell rigid polystyrene slabs is generally applied to new construction. These slabs combine low thermal conductivity with low water absorption and high compressive strength. The external face of insulation is grooved to encourage moisture run-off. It is also filter faced to prevent clogging of the grooves. Backfill is granular.

Tables and calculation methods to determine heat energy transfer for basements are provided in BS EN ISO 13370: *Thermal Performance of Buildings. Heat Transfer via the Ground. Calculation Methods.*

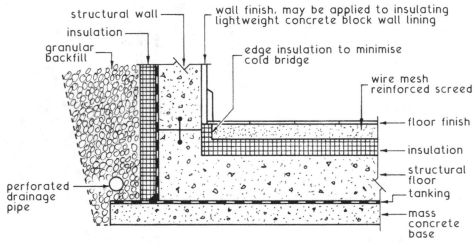

Note: reinforcement in concrete omitted, see details on previous pages.

Figure 8.13 Basement insulation

9 GROUND SUPPORTED FLOORS

Opening remarks

There are two primary types of ground floor:

- Ground supported concrete slabs.
- Suspended floors.

Ground floor functions:

- Support the imposed loads of people and furniture.
- Prevent ingress of water from the ground.
- Limit heat loss through the floor into the ground.
- Provide a flat, level surface to receive the chosen finish.

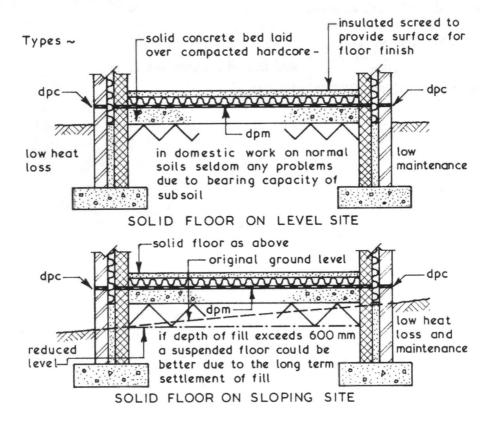

Figure 9.1 Types of ground floor

9.1 Ground supported concrete floors

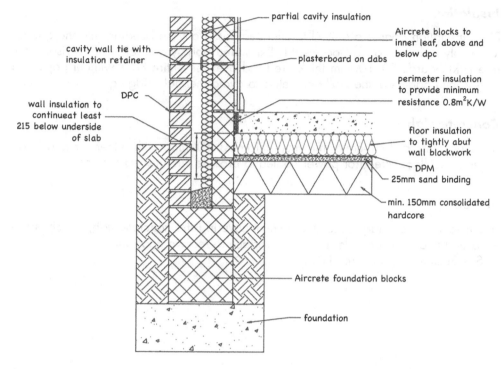

Figure 9.2 Ground supported concrete floor

Hardcore

Suitable materials for hardcore are graded: stone, quarry waste, gravel, crushed concrete or brick – all of which are non-organic, chemically inert and have low permeability. Hardcore is used to provide a stable base for the floor and is installed in layers not exceeding 150mm thick, which are compacted by roller or plate compactors so that no settlement will occur.

Blinding layer

Used when the DPM is installed above the hardcore to prevent the damp proof membrane (DPM) being punctured by sharp pieces of aggregate; it also helps level out the hardcore surface. A 25mm layer of sand or stone dust is used for the blinding layer.

Damp proof membrane (DPM)

This is a layer of heavy duty polythene sheeting that prevents moisture from the ground passing into the floor and into the interior of the building. All joints in the

Ground supported concrete floors

DPM must be lapped by 150mm and sealed with PVC tape. The DPM must lap into the DPC all around the building.

Insulation

Rigid insulation boards are used to limit heat loss from the building into the ground to comply with Approved Document L. Expanded polystyrene (EPS) or Polyisocyanurate (PIR) boards 50–100mm thick are fitted over the entire floor without gaps and with 25mm boards at the wall perimeters to avoid thermal bridging.

Concrete slab

Mass un-reinforced concrete, minimum thickness 100mm, with either tamped finish to receive a floor screed or power floated to provide the finished surface.

Screed

A semi-dry mix of sharp sand and cement covering the concrete slab, which is levelled and floated to provide a flat level surface to receive floor finishes.

See Section 9.3 for screed details.

9.2 Industrial concrete floors

These are floors designed to carry medium to heavy loadings, such as those used in factories, warehouses, shops, garages and similar buildings. Floors of this type are usually laid in alternate 4.5m wide strips running the length of the building. Transverse joints will be required to control the tensile stresses due to the thermal movement and contraction of the slab. The spacing of these joints will be determined by the design and the amount of reinforcement used. Such joints can either be formed by using a crack inducer or by sawing a 20 to 25mm deep groove into the upper surface of the slab within 20 to 30 hours of casting.

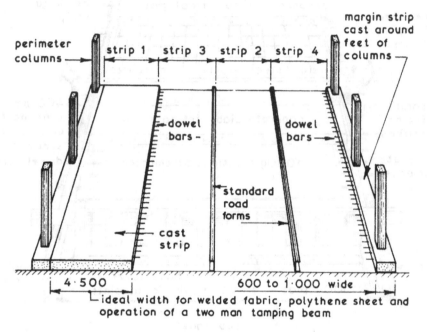

Figure 9.3 Concreting large floors in bays

Surface finishing

The surface of the concrete may be finished by power floating or trowelling, which is carried out when the concrete has set sufficiently to support the weight of machine and operator without deformation but has not yet hardened. Vacuum dewatering can be used to shorten the time delay between tamping the concrete and power floating the surface by extracting the surplus water. This process causes a reduction in slab depth of approximately 2%; therefore, packing strips are placed on the side forms before tamping to allow for sufficient surcharge of concrete.

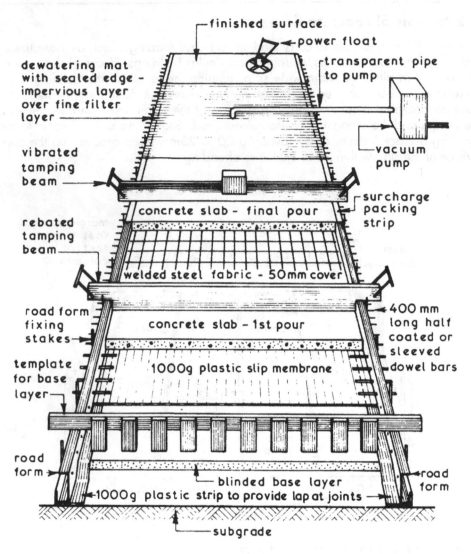

finished surface

power float

transparent pipe
to pump

dewatering mat
with sealed edge –
impervious layer
over fine filter
layer

vacuum
pump

vibrated
tamping
beam

concrete slab – final pour

surcharge
packing
strip

rebated
tamping
beam

welded steel fabric – 50mm cover

road form
fixing
stakes

concrete slab – 1st pour

400 mm
long half
coated or
sleeved
dowel bars

template
for base
layer

1000g plastic slip membrane

road
form

blinded base layer

road
form

1000g plastic strip to provide lap at joints

subgrade

Figure 9.4 Concrete slab elements

9.3 Concrete floor screeds

These are used to give an uneven concrete floor a suitable finish or to provide a flat finish to a precast concrete floor.

Typical screed mixes

Table 9.1 Typical screed thickness and mixes

Screed thickness	Cement	Dry fine aggregate <5mm	Coarse aggregate >5mm <10mm
up to 40mm	1	3 to 4 1/2	-
	1	3 to 4 1/2	-
40 to 75mm	1	1 1/2	3

Monolithic screeds

These are screeds laid directly on the concrete floor slab within three hours of placing concrete. Work should be carried out from scaffold board runways to avoid walking on the 'green' concrete slab.

Separate screeds

The screed is laid onto the concrete floor slab after it has cured. The floor surface must be clean and rough enough to ensure an adequate bond, unless the floor surface is prepared by applying a suitable bonding agent or by brushing with a cement/water grout of a thick cream-like consistency just before laying the screed.

Unbonded screeds

The screed is laid directly over a damp-proof membrane or over a damp-proof membrane and insulation. A rigid form of floor insulation is required where the concrete floor slab is in contact with the ground. Care must be taken during this operation to ensure that the damp-proof membrane is not damaged.

Floating screeds

A resilient quilt of 25mm thickness is laid with butt joints and turned up at the edges against the abutment walls, the screed being laid directly over the resilient quilt. The main objective of this form of floor screed is to improve the sound insulation properties of the floor.

Latex floor screed

A modified screed with liquid resin or synthetic polymer additives to modify the cement and sand commonly used for industrial applications. The screed is spread in very thin layers and is ideal for levelling uneven and irregular surfaces of sound composition.

Concrete floor screeds

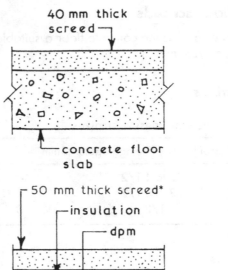

40 mm thick screed

concrete floor slab

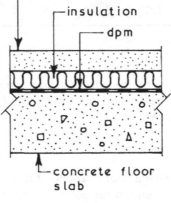

50 mm thick screed*

insulation

dpm

concrete floor slab

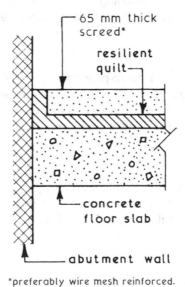

65 mm thick screed*

resilient quilt

concrete floor slab

abutment wall

*preferably wire mesh reinforced.

Figure 9.5 Screeds

9.4 Resistance to contaminants

Approved Document C, *Site Preparation and Resistance to Contaminants and Moisture*, deals with potential hazards to health and safety from the ground to the building and its occupants. Potential hazards include vegetation, methane and radon gasses, ingress of water and water-borne contaminants.

Methane

Methane is produced by deposited organic material decaying in the ground. It often occurs with carbon dioxide and traces of other gases to form a cocktail known as landfill gas. It has become an acute problem in recent years, as planning restrictions on 'green-field' sites have forced development of derelict and reclaimed 'brown-field' land.

The gas would normally escape into the atmosphere, but under a building it pressurises until percolating through cracks, cavities and junctions with services. Being odourless, it is not easily detected until coming into contact with a naked flame; then the result is devastating!

Radon

Radon is a naturally occurring colour/odourless gas produced by radioactive decay of radium. It originates in uranium deposits of granite subsoils as far apart as the south-west and north of England and the Grampian region of Scotland. Concentrations of radon are considerably increased if the building is constructed of granite masonry. The combination of radon gas and the tiny radioactive particles known as radon daughters are inhaled. In some people with several years' exposure, research indicates a high correlation with cancer-related illness and death.

Protection of buildings and the occupants from subterranean gases can be achieved by passive or active measures incorporated within the structure:

1. Passive protection consists of a complete airtight seal integrated within the ground floor and walls. A standard LDPE damp-proof membrane of 0.3mm thickness should be adequate if carefully sealed at joints, but thicknesses of up to 1mm are preferred, combined with foil and/or wire reinforcement.
2. Active protection requires installation of a permanently running extractor fan connected to a gas sump below the ground floor. It is an integral part of the building services system and will incur operating and maintenance costs throughout the building's life.

Resistance to contaminants

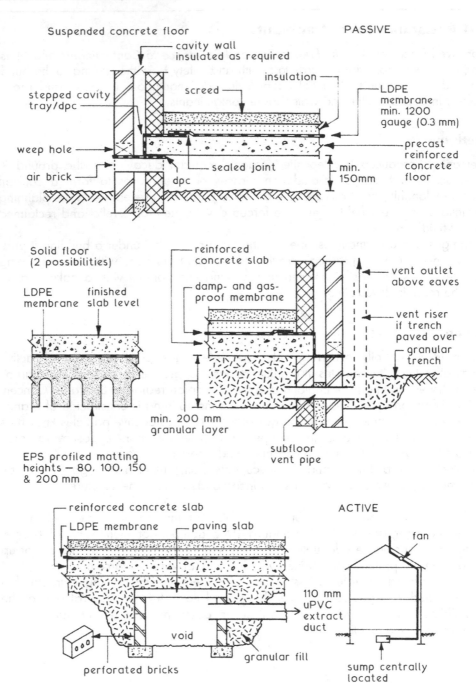

Figure 9.6 Gas resistant construction

9.5 Ground floor insulation

Unlike walls and roofs, the heat loss through a ground floor varies with its size and shape. The approved U value calculation methodology from BS EN ISO 13,370: 1998 uses the ratio of the exposed floor perimeter to the floor area to take account of the variation in heat loss due to floor size and shape. The measurement of the perimeter and area should be to the finished inside surfaces of the perimeter walls that enclose the heated space.

Ground-bearing floors can include insulation either below or above the concrete slab. If the insulation is below the slab the thermal capacity of the building is increased, helping to maintain steady internal temperatures. If the insulation is above the slab, the building will respond much more quickly to space heating.

Rigid insulation materials are used under and above concrete slabs that can support the applied loads with the minimum of compression.

9.6 Ground-bearing concrete

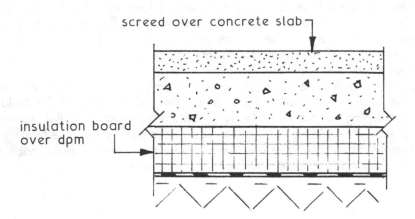

Figure 9.7 Insulation under floor slab

Table 9.2 Typical solid floor insulation thicknesses and U values for P/A ratios

P/A ratio	Typical insulation thickness (mm)		
1.0	105	140	175
0.8	100	135	170
0.6	90	130	160
0.4	75	110	130
0.2	30	60	80
U value	*0.25*	*0.20*	*0.18*

Note: P/A ratio is the exposed perimeter of a ground floor divided by the floor area.

9.7 Suspended beam and block

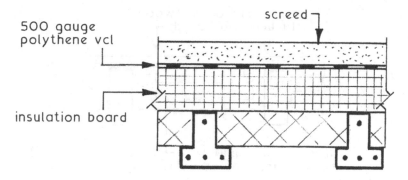

Figure 9.8 Insulation over beam and block floor

Table 9.3 Typical beam and block floor insulation thicknesses and U values for P/A ratios

P/A ratio	Typical insulation thickness (mm)		
1.0	110	145	175
0.8	105	140	165
0.6	100	130	155
0.4	90	120	145
0.2	60	95	125
U value	0.25	0.20	0.18

Note: Figures apply to mineral wool insulation board, $\lambda = 0.038$W/mK. Closed cell extruded polystyrene can also be used, $\lambda = 0.029$W/mK.

9.8 Suspended timber

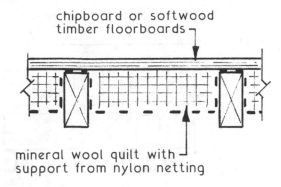

Figure 9.9 Insulating between timber floor joists

Table 9.4 Typical insulation thicknesses between timber joists and U values for P/A ratios

P/A ratio	Typical insulation thickness (mm)		
1.0	140	150	180
0.8	140	140	180
0.6	120	140	180
0.4	120	140	150
0.2	75	90	120
U value	0.25	0.22	0.20

9.9 Thermal bridging ground floors

The cavity wall inner leaf interrupts the insulating envelope at the ground floor and wall interface. To prevent thermal bridging, condensation and possible mould growth, the cavity insulation should overlap the floor insulation by at least 150mm.

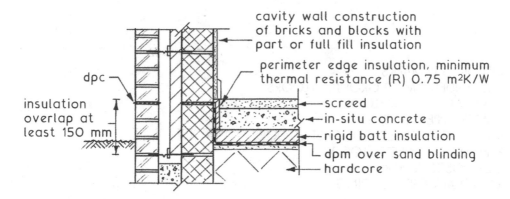

Figure 9.10 Perimeter edge insulation at abutment with wall

10 SUSPENDED FLOORS

OPENING REMARKS
SUSPENDED TIMBER GROUND FLOORS
BEAM AND BLOCK FLOORS
TIMBER UPPER FLOORS
IN-SITU REINFORCED CONCRETE FLOORS
PRECAST CONCRETE FLOORS
COMPOSITE FLOORS
RAISED ACCESS FLOORING

Opening remarks

Suspended floors transfer their dead and live loads to the building structure and not directly to the ground. Two types of suspended ground floors are common in domestic construction; timber joist and beam and block. All floors above ground level are suspended floors.

10.1 Suspended timber ground floors

A suspended timber floor will meet the requirements of Approved Document C if:

- The subfloor is covered with oversite concrete to prevent plant growth and resist moisture.
- There is a minimum 150mm ventilated air space between the sub floor and the underside of the timber joists with through ventilation to prevent dampness and fungal attack.
- The timber joists are protected from moisture from the supporting structure by a DPC.

The depth of the floor joist depends upon the span. To reduce the span intermediate sleeper walls are built on the oversite concrete to support the joists at mid span or one-third spans. The sleeper walls are of honeycomb construction to permit through ventilation.

Note: In old, solid wall buildings timber joists were built into the walls without a DPC and consequently the joist ends rot away over time.

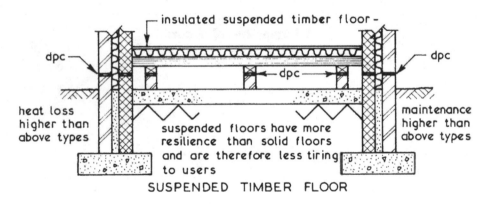

Figure 10.1 Types of ground floor

Timber floor insulation

Traditionally timber floors were not insulated. However, to comply with Approved Document L, all new floors (other than industrial floors) must be insulated. There are two methods available: insulation between the joists or insulation above the joists.

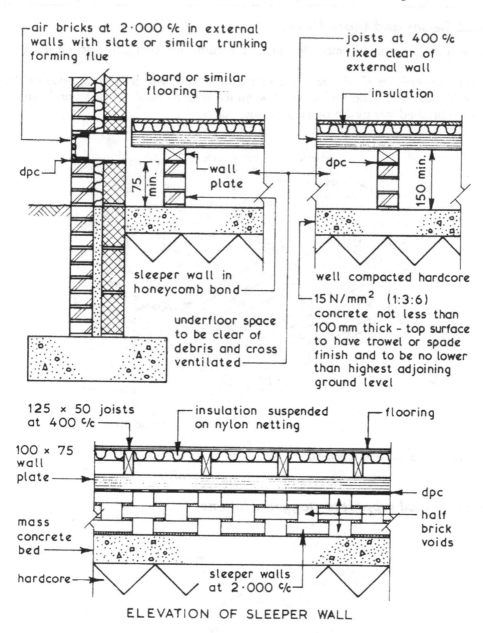

air bricks at 2·000 c/c in external walls with slate or similar trunking forming flue

board or similar flooring

dpc

75 min.

wall plate

sleeper wall in honeycomb bond

underfloor space to be clear of debris and cross ventilated

joists at 400 c/c fixed clear of external wall

insulation

dpc

150 min.

well compacted hardcore

15 N/mm² (1:3:6) concrete not less than 100 mm thick – top surface to have trowel or spade finish and to be no lower than highest adjoining ground level

125 × 50 joists at 400 c/c

insulation suspended on nylon netting

flooring

100 × 75 wall plate

dpc

mass concrete bed

half brick voids

hardcore

sleeper walls at 2·000 c/c

ELEVATION OF SLEEPER WALL

Figure 10.2 Sleeper walls

10.2 Beam and block floors

These have superseded timber in domestic construction and can also be used on upper floors.

The floor consists of pre-stressed concrete 'T' beams, which are supported on the external walls and concrete blocks between the beams. Semi dry sand and cement grout is brushed into all the beam and block joints.

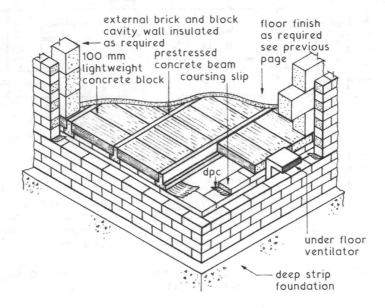

Figure 10.3 Beam and block floor

Internal wall support to staggered beams

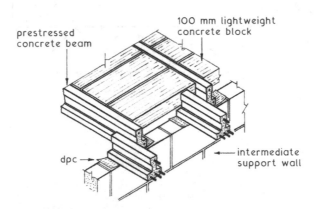

Figure 10.4 Intermediate support walls

BEAMS PARALLEL WITH EXTERNAL WALL

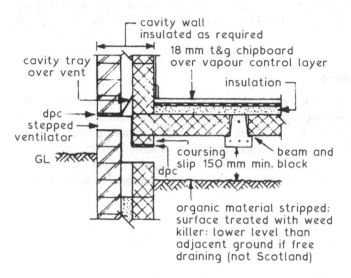

POLYPROPYLENE VENTILATOR

stepped telescopic sleeve

cavity wall insulated as required

18 mm t&g chipboard over vapour control layer

cavity tray over vent

insulation

dpc

stepped ventilator

GL

coursing slip

beam and block

150 mm min.

dpc

organic material stripped; surface treated with weed killer: lower level than adjacent ground if free draining (not Scotland)

grill clips to sleeve: 1500 mm²/m run of wall OR 500 mm²/m² of floor area (take greater value)

Figure 10.5 Sub floor ventilation

Figure 10.6 Telescopic vent

BEAMS BEARING ON EXTERNAL WALL

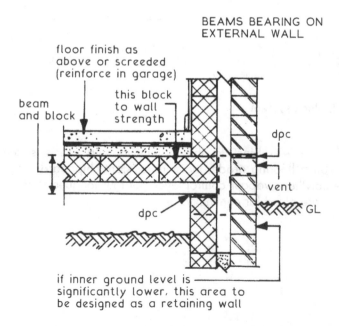

floor finish as above or screeded (reinforce in garage)

this block to wall strength

beam and block

dpc

dpc

vent

GL

if inner ground level is significantly lower, this area to be designed as a retaining wall

Figure 10.7 Floor beams built into external walls

Beam and block floors

Beam and block insulation

Rigid board insulation (EPS or PIR) is laid on a vapour control layer (DPM) over the beams and blocks. Two floor finishing options are available: screed or chipboard.

Beam and EPS block floors

These have evolved from the principles of beam and block floor systems, the lightweight and easy to cut properties of the blocks provide for speed of construction and a high level of thermal insulation is possible. The EPS blocks fit under the PCC beams, preventing thermal bridging to the screed. These systems require a reinforced screed/structural topping in accordance with the manufacturers' requirements. A low-density polyethylene (LDPE) methane/radon gas membrane can be incorporated under the screed if local conditions require it.

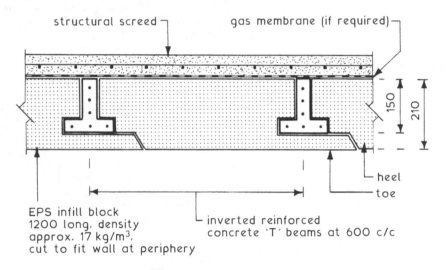

Figure 10.8 EPS block system

Floating floor finish ~ subject to the system manufacturer's specification and accreditation, 18mm flooring grade, moisture-resistant chipboard can be used over a 1000 gauge polythene vapour control layer. All four tongued and grooved edges of the chipboard are glued for continuity.

10.3 Timber upper floors

Also known as intermediate floors, these are supported on the building structure and load-bearing walls.

Primary functions

- Provide a level surface with sufficient strength to support the imposed loads of people and furniture, plus the dead loads of flooring and ceiling.
- Reduce heat loss from lower floor as required.
- Provide required degree of sound insulation.
- Provide required degree of fire resistance

Basic construction

A timber suspended upper floor consists of a series of beams or joists supported by load-bearing walls sized and spaced to carry all the dead and imposed loads without undue deflection. Traditionally solid timber joists were used and the section size required was determined from TRADA span tables. However, engineered timber joists have superseded solid timber in use.

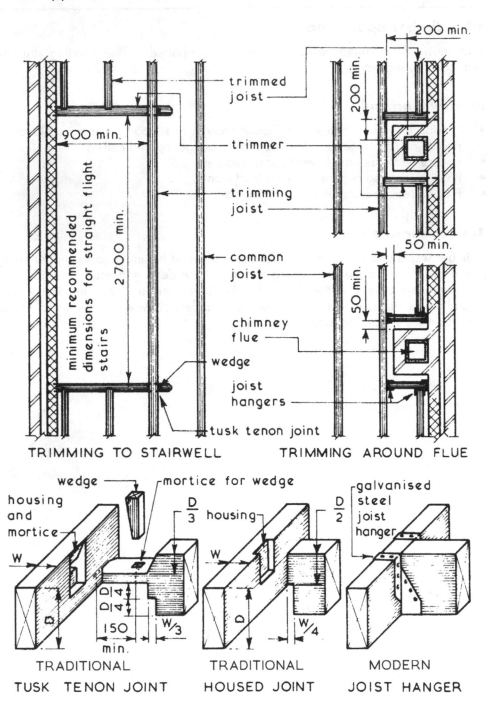

Figure 10.9 Trimming joists

Structural design of floors will be satisfied for most situations by using the minimum figures given for uniformly distributed loading (UDL). These figures provide for static loading and for the dynamics of occupancy. The minimum figures given for concentrated or point loading can be used where these produce greater stresses.

Table 10.1 Typical UDL loadings for timber floors

Application	UDL (kN/m^2)	Concentrated (kN)
Dwellings		
Communal areas	1.5	1.4
Bedrooms	1.5	1.8
Bathroom/WC	2.0	1.8
Balconies (use by one family)	1.5	1.4
Commercial/industrial		
Hotel/motel bedrooms	2.0	1.8
Communal kitchen	3.0	4.5
Offices and general work Areas	2.5	2.7
Kitchens/laundries/ laboratories	3.0	4.5
Factories and workshops	5.0	4.5
Balconies – guest houses	3.0	1.5/m run at outer edge
Balconies – communal areas in flats	3.0	1.5/m run at outer edge
Balconies – hotels/motels	4.0	1.5/m run at outer edge
Warehousing/storage		
General use for static items	2.0	1.8
Reading areas/libraries	4.0	4.5
General use, stacked items	2.4/m height	7.0
Filing areas	5.0	4.5
Paper storage	4.0/m height	9.0
Plant rooms	7.5	4.5
Book storage	2.4/m height (min. 6.5)	7.0

See also: BS EN 1991-1-1: *Densities, Self-Weight, Imposed Loads for Buildings.*

Timber upper floors

Typical spans and loading for floor joists of general structural (GS) grade are shown in Table 10.2.

Table 10.2 Timber floor joist span tables

Sawn size (mm)	Dead weight of flooring and ceiling, excluding the self-weight of the joists (kg/m^2)								
	< 25			25–50			50–125		
	Spacing of joists (mm ^{c}/c)								
	400	450	600	400	450	600	400	450	600
	Maximum clear span (m)								
50 × 75	1.45	1.37	1.08	1.39	1.30	1.01	1.22	1.11	0.88
50 × 100	2.18	2.06	1.76	2.06	1.95	1.62	1.82	1.67	1.35
50 × 125	2.79	2.68	2.44	2.67	2.56	2.28	2.40	2.24	1.84
50 × 150	3.33	3.21	2.92	3.19	3.07	2.75	2.86	2.70	2.33
50 × 175	3.88	3.73	3.38	3.71	3.57	3.17	3.30	3.12	2.71
50 × 200	4.42	4.25	3.82	4.23	4.07	3.58	3.74	3.53	3.07
63 × 100	2.41	2.29	2.01	2.28	2.17	1.90	2.01	1.91	1.60
63 × 125	3.00	2.89	2.63	2.88	2.77	2.52	2.59	2.49	2.16
63 × 150	3.59	3.46	3.15	3.44	3.31	3.01	3.10	2.98	2.63
63 × 175	4.17	4.02	3.66	4.00	3.85	3.51	3.61	3.47	3.03
63 × 200	4.73	4.58	4.18	4.56	4.39	4.00	4.11	3.95	3.43

Solid timber joists

Floor joists can either be built into the external walls or suspended on galvanised steel joist hangers that are built into the walls. The joist hanger method is quicker and easier to install and prevents gaps around joist and the wall which allow air leakage.

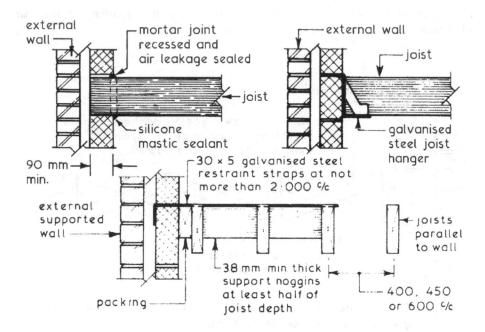

external
wall

mortar joint
recessed and
air leakage sealed

joist

silicone
mastic sealant

90 mm
min.

external wall

joist

galvanised
steel joist
hanger

external
supported
wall

30 × 5 galvanised steel
restraint straps at not
more than 2·000 c/c

38 mm min. thick
support noggins
at least half of
joist depth

packing

joists
parallel
to wall

400, 450
or 600 c/c

Figure 10.10 Support and restraint for timber joists

Timber upper floors

To restrict the floor joists from twisting, which could damage ceiling finishes, strutting is needed.

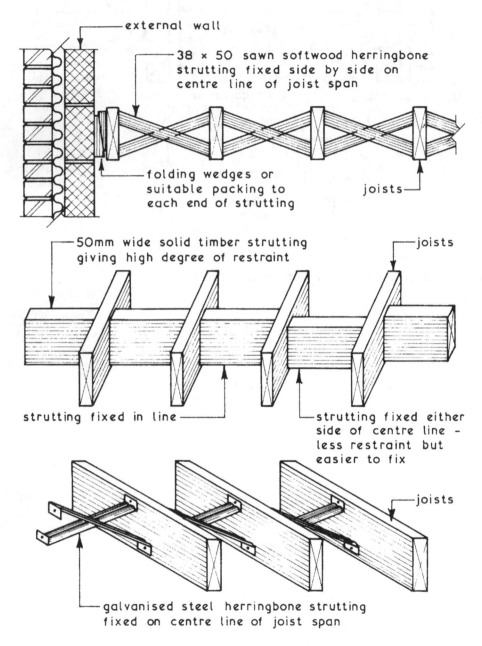

external wall

38 × 50 sawn softwood herringbone strutting fixed side by side on centre line of joist span

folding wedges or suitable packing to each end of strutting

joists

50mm wide solid timber strutting giving high degree of restraint

joists

strutting fixed in line

strutting fixed either side of centre line – less restraint but easier to fix

joists

galvanised steel herringbone strutting fixed on centre line of joist span

Figure 10.11 Strutting

Engineered timber joists

More economical in terms of material and predictable in terms of engineering are I joists or open web joists and these have superseded solid timber joists with volume house developers. I joists, aka TJI, joists consist of softwood or laminated timber strips forming the top and bottom of the joists, with a sheet of OSB rebated into the top and bottom sections. Open web joists consist of solid or laminated timber top and bottom sections connected by light gauge galvanised steel V braces. The spaces between can accommodate a wide range of building services within the floor structure.

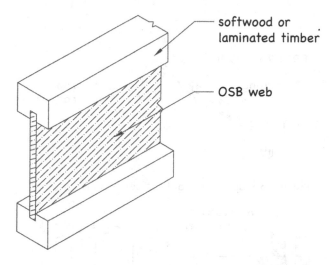

softwood or
laminated timber

OSB web

Figure 10.12 Engineered timber I joist

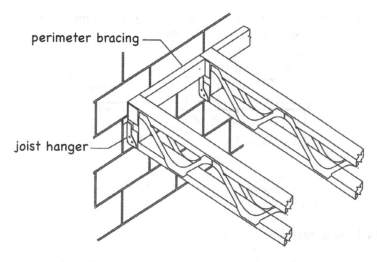

perimeter bracing

joist hanger

Figure 10.13 Open steel web engineered joists

Timber upper floors

Lateral restraint

External, compartment (fire), separating (party) and internal load-bearing walls must have horizontal support from adjacent floors, to restrict movement. Exceptions occur when the wall is less than 3m long.

Methods

- 90 mm end bearing of floor joists, spaced not more than 1.2m apart.
- Galvanised steel straps spaced at intervals not exceeding 2m and fixed square to joists.

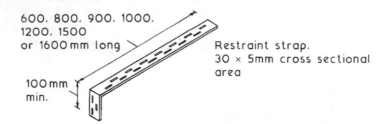

600, 800, 900, 1000, 1200, 1500 or 1600 mm long

100 mm min.

Restraint strap, 30 × 5mm cross sectional area

3. Joists carried by BS approved galvanised steel hangers.

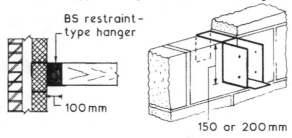

BS restraint-type hanger

100 mm

150 or 200mm

4. Adjacent floors at or about the same level, contacting with the wall at no more than 2 m intervals.

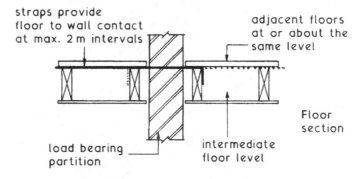

straps provide floor to wall contact at max. 2 m intervals

adjacent floors at or about the same level

load bearing partition

intermediate floor level

Floor section

Figure 10.14 Lateral restraint straps

Trimming members

These are the edge members of an opening in a floor and are the same depth as common joists but are usually 25mm wider.

Notches and holes for services

Timber joists can be weakened by notching and drilling for pipes and cables and should be limited.

Hole limitations: Diameter maximum: 0.25 × joist depth.
Spacing minimum: 3 × diameter apart, measured centre to centre. Position in the neutral axis: between 0.25 and 0.40 × clear span, measured from the support.

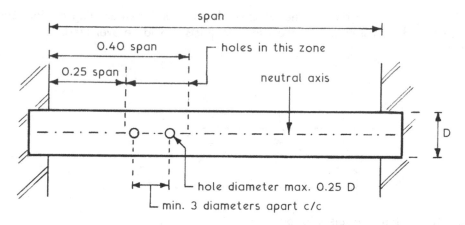

Figure 10.15 Permitted joist notching zones

Notching is the only practical way of accommodating rigid pipes and conduits in joisted floors.

Depth maximum: 0.125 × joist depth.
Position between: 0.07 and 0.25 × clear span measured from the support.

Timber upper floors

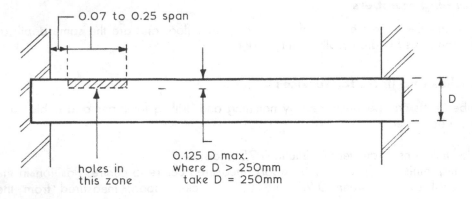

Figure 10.16 Permitted floor joist notching zones

Notching of a joist reduces the effective depth, thereby weakening the joist and reducing its design strength. To allow for this, joists should be oversized.

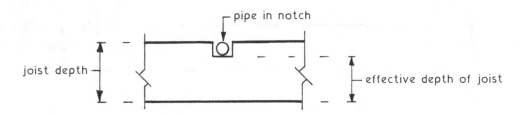

Figure 10.17 Notching joists for pipes

Floor boarding

Traditionally, softwood tongue and groove boarding was the standard method, laid at right angles to the joists and fixed with two 65mm-long cut floor boards per joist. The ends of board lengths are butt jointed on the centre line of the supporting joist.

Chipboard or particle board, which is made from particles of wood bonded with a synthetic resin and/or other organic binders, is a faster, more economic and sustainable material to use. Standard floorboards are 2400mm x 600mm at 18 or 22mm thickness with tongued and grooved joints to all edges. Boards are laid at right angles to joists with all cross joints directly supported. May be specified as unfinished or waterproof quality, indicated with a dull green dye.

Fire protection

For fire protection, floors are categorised depending on their height relative to adjacent ground (see Table 10.3).

Tests for fire resistance relate to load-bearing capacity, integrity and insulation, and are determined by BS 476–21: *Fire Tests on Building Materials and Structures*.

Table 10.3 Fire resistance requirements for height above ground level

Height of top floor above ground	Fire resistance (load-bearing)
Less than 5m	30 minutes
More than 5m	60 minutes (30 min. for a three-storey dwelling)

Methods for determining the fire resistance of load-bearing elements of construction:

- Integrity, the ability of an element to resist fire penetration.
- Insulation, the ability to resist heat penetration so that fire is not spread by radiation and conduction.

MODIFIED 30-MINUTE FIRE RESISTANCE

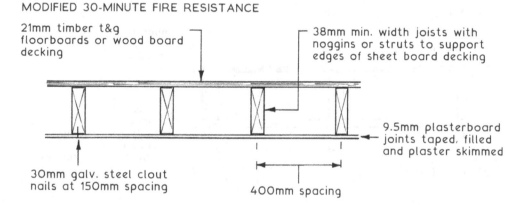

Figure 10.18 Fire resistance

- Collapse – 30 minutes minimum.
- Integrity – 15 minutes minimum.
- Insulation – 15 minutes minimum.

Note 1: Full 30-minute fire resistance must be used for floors over basements.

Note 2: Where a floor provides support or stability to a wall, or vice versa, the fire resistance of the supporting element must not be less than the fire resistance of the other element.

Building Regulations, *AD B Fire safety, Volume 1 – Dwelling houses*.

Timber upper floors

Modified 30-minute fire resistance:

A lower specification than the full 30-minute fire resistance often found during refurbishment of housing constructed during the latter part of the 20th century. Here, 9.5mm plasterboard lath (rounded edge board 1200mm long × 406mm wide – now obsolete) was used with floor joists at 400mm spacing. It is now general practice to use a 12.5mm plasterboard ceiling lining to dwelling house floors as this not only provides better fire resistance, but is more stable.

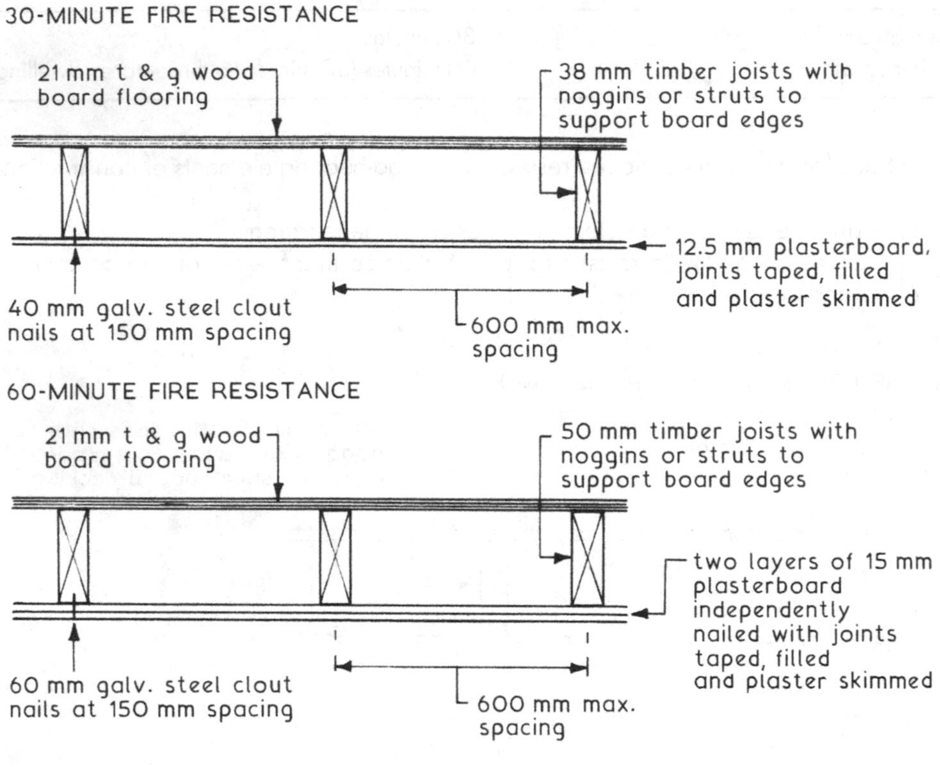

30-MINUTE FIRE RESISTANCE

21 mm t & g wood board flooring

38 mm timber joists with noggins or struts to support board edges

12.5 mm plasterboard, joints taped, filled and plaster skimmed

40 mm galv. steel clout nails at 150 mm spacing

600 mm max. spacing

60-MINUTE FIRE RESISTANCE

21 mm t & g wood board flooring

50 mm timber joists with noggins or struts to support board edges

two layers of 15 mm plasterboard independently nailed with joints taped, filled and plaster skimmed

60 mm galv. steel clout nails at 150 mm spacing

600 mm max. spacing

Figure 10.19 Fire resistance

10.4 In-situ reinforced concrete floors

A simple reinforced concrete flat slab cast to act as a suspended floor is not usually economical for spans over 5m. To overcome this problem beams can be incorporated into the design to span in one or two directions. Such beams usually span between columns that transfer their loads to the foundations. The disadvantages of introducing beams are the greater overall depth of the floor construction and the increased complexity of the formwork and reinforcement. To reduce the overall depth of the floor construction flat slabs can be used where the beam is incorporated with the depth of the slab. This method usually results in a deeper slab with complex reinforcement, especially at the column positions.

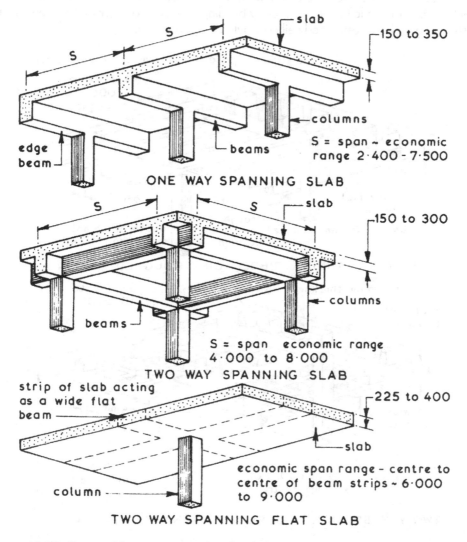

Figure 10.20 One and two way spanning floor slabs

In-situ reinforced concrete floors

Ribbed floors

To reduce the overall depth of a traditional, cast in-situ, reinforced concrete beam and slab suspended floor a ribbed floor could be used. The basic concept is to replace the wide spaced deep beams with narrow spaced shallow beams or ribs, which will carry only a small amount of slab loading.

Ribbed floors can be designed as one or two way spanning floors. One way spanning ribbed floors are sometimes called troughed floors whereas the two way spanning ribbed floors are called coffered or waffle floors. Ribbed floors are usually cast against metal, glass-fibre or polypropylene preformed moulds which are temporarily supported on plywood decking, joists and props.

The advantage of this method is that they have greater span and load potential per unit weight than flat slab construction due to the considerable reduction of dead load. The regular pattern of voids created with waffle moulds produces a honeycombed effect, which may be left exposed in utility buildings such as car parks. Elsewhere, such as shopping malls, a suspended ceiling would be appropriate.

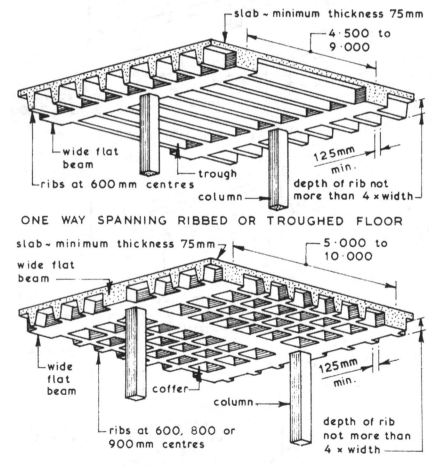

ONE WAY SPANNING RIBBED OR TROUGHED FLOOR

TWO WAY SPANNING COFFERED OR WAFFLE FLOOR

Figure 10.21 Examples of ribbed floor construction

Hollow pot floors

These are, in essence, a ribbed floor with permanent formwork in the form of hollow clay or concrete pots. The main advantage of this type of cast in-situ floor is that it has a flat soffit, which is suitable for the direct application of a plaster finish or an attached dry lining. The voids in the pots can be utilised to house small diameter services within the overall depth of the slab. These floors can be designed as one or two way spanning slabs, the common format being the one way spanning floor.

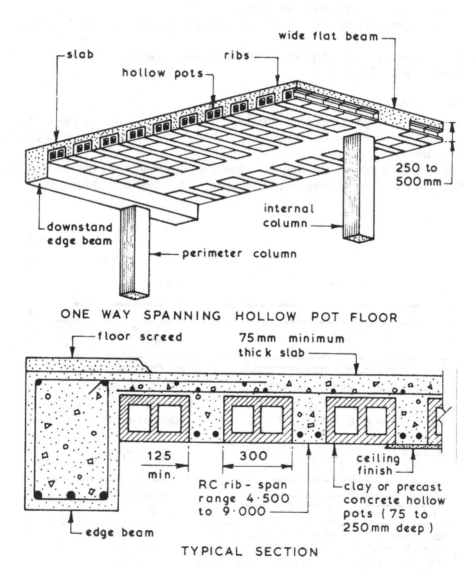

Figure 10.22 Hollow pot floors

10.5 Precast concrete floors

These are available in several basic formats and provide an alternative form of floor construction to suspended timber floors and in-situ reinforced concrete suspended floors. The main advantages of precast concrete floors are:

- Elimination of the need for formwork, except for nominal propping which is required with some systems.
- Curing time of concrete is eliminated; therefore the floor is available for use as a working platform at an earlier stage.
- Superior quality control of product is possible with factory-produced components.

The main disadvantages of precast concrete floors when compared with in-situ reinforced concrete floors are:

- Less flexible in design terms.
- Formation of large openings in the floor for ducts, shafts and stairwells usually have to be formed by casting an in-situ reinforced concrete floor strip around the opening position.
- Higher degree of site accuracy is required to ensure that the precast concrete floor units can be accommodated without any alterations or making good.

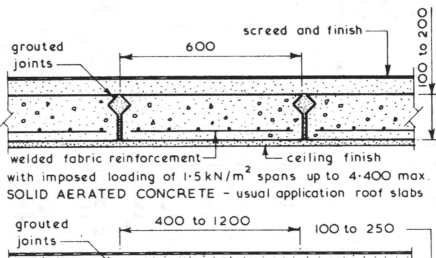

Figure 10.23 Solid and hollow pre–cast concrete floor formats

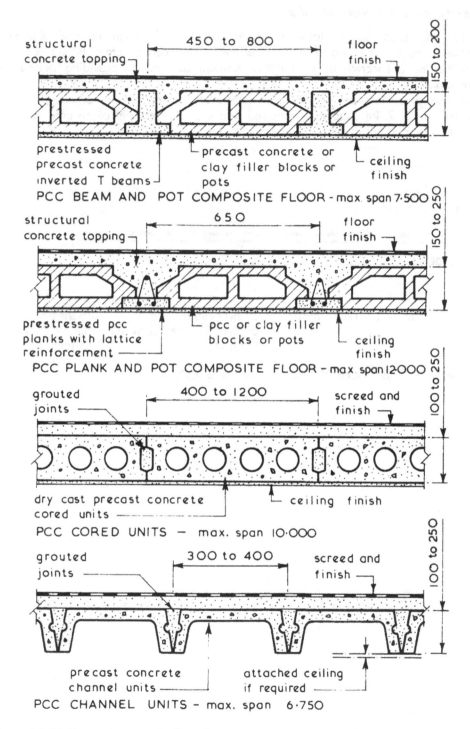

Figure 10.24 Precast concrete floor types

Precast concrete floors

Pre-stressed concrete planks

Precast units produced in widths of 600 to 1200mm. These are relatively thin and require a structural topping of in-situ, steel mesh reinforced concrete. Planks combine with the in-situ concrete as a structural element as well as providing permanent formwork.

During placement of wet concrete temporary support from vertical props at a maximum of 2.4m spacing is required and should remain in place until the concrete has set and cured.

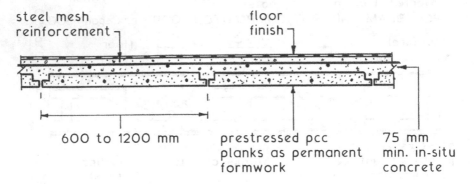

Figure 10.25 Prestressed PCC planks – spans to 10.000

10.6 Composite floors

Comprising of galvanised, light gauge, steel profiled sheets as a permanent form-work for in-situ concrete. Used in steel frame construction the sheets are fixed to the beams with welded steel sheer studs. The steel decking is quick and easy to install and provides a working platform as work progresses. Propping is generally not required during concrete placement. Steel mesh 'anti-crack' reinforcement is cast into the upper section of the slab.

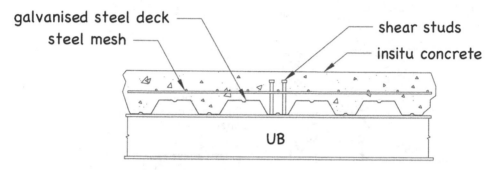

Figure 10.26 Composite floor section

10.7 Raised access flooring

Used in commercial buildings with open plan office spaces, this floor system allows for the horizontal distribution of building services. The adjustable floor pedestals allow for levelling on uneven concrete decks and can support a variety of decking materials. Cavity fire stops are required between decking and structural floor at appropriate intervals (see Building Regulations, *A.D. B, Volume 2, Section 9*).

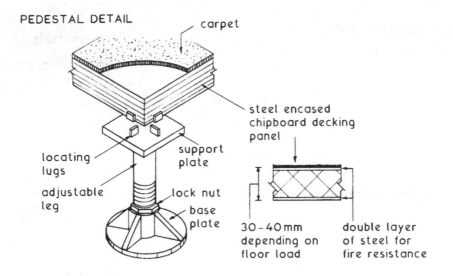

PEDESTAL DETAIL

carpet

steel encased chipboard decking panel

locating lugs

support plate

adjustable leg

lock nut

base plate

30–40 mm depending on floor load

double layer of steel for fire resistance

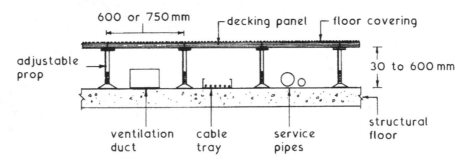

FLOOR SECTION

600 or 750 mm

decking panel

floor covering

adjustable prop

30 to 600 mm

ventilation duct

cable tray

service pipes

structural floor

Figure 10.27 Raised access flooring

11 EXTERNAL WALLS

Opening remarks

Solid masonry wall construction was the standard form of building construction in the UK until the introduction of cavity wall construction circa the 1920s. Apart from stone rubble walls, all stone faced walls incorporated brickwork/blockwork to the interior side.

11.1 Stone walls

These are solid walls constructed of stones used either in a random rough or uneven form as extracted from the quarry, or roughly dressed to a square or rectangular shape. Walls of this type are common in rural areas, particularly those close to quarry locations. Many walls are dry bonded as estate and farm boundaries. Otherwise, stones are laid with a relatively deep bed joint of mortar to accommodate irregularities, often as garden features and external walls to dwellings. Lime based mortar is preferred with long runs of stone walling as it is less prone to thermal movement cracking than cement mortar.

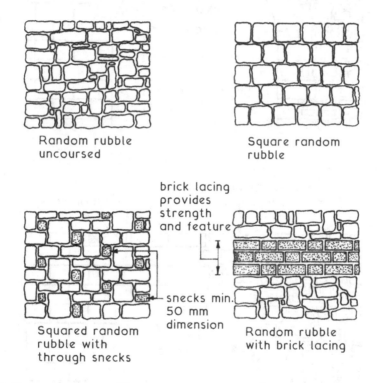

Random rubble
uncoursed

Square random
rubble

brick lacing
provides
strength
and feature

snecks min.
50 mm
dimension

Squared random
rubble with
through snecks

Random rubble
with brick lacing

Figure 11.1 Types of stone walling

Stone walls

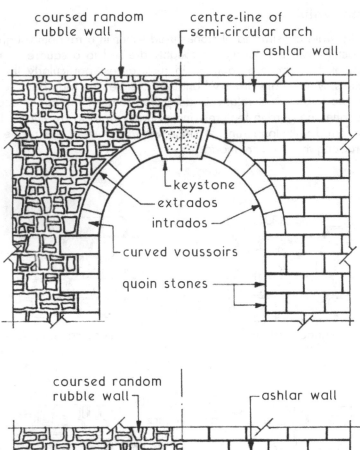

coursed random rubble wall

centre-line of semi-circular arch

ashlar wall

keystone

extrados

intrados

curved voussoirs

quoin stones

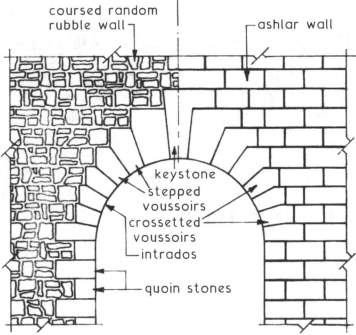

coursed random rubble wall

ashlar wall

keystone

stepped voussoirs

crossetted voussoirs

intrados

quoin stones

Figure 11.2 Stone arch construction

Arch construction

By the arrangement of the bricks or stones in an arch over an opening it will be self-supporting once the jointing material has set and gained adequate strength. The arch must therefore be constructed over a temporary support until the arch becomes self-supporting. The traditional method is to use a framed timber support called a centre. Permanent arch centres are also available for small spans and simple formats.

The UK and Ireland have many sources of stone to include the ones listed in Table 11.1.

Table 11.1   Types of natural stone in the UK

Type	Location	Colour
Granite	Cornwall	Light grey
	Highlands and Western Isles	Bright grey to black
	Leicestershire	Various, light to dark
	Westmorland, Cumbria	Reddish brown
	Down, Ireland	Greenish grey
Limestone	Portland, Dorset	White to light brown
	Bath, West Country	White to cream
	Lincolnshire	Cream to buff
	Wiltshire	Light brown
	Somerset	Yellowish grey
	Kent	Blue grey
	Leicestershire	Cream to yellow
	Nottinghamshire	Yellowish brown
	Oxfordshire	Buff
Sandstone	Yorkshire	Grey and light brown
	Forest of Dean, Glos.	Grey/blue, grey/pink
	Gloucestershire	Dull blue
	Glamorganshire	Dull blue
	Derbyshire	Buff and light grey
	Sussex	Buff, brown specs
Marble	Dorset	Green/blue or grey
	Derbyshire	Light grey
	Argyllshire	Pale green/white
	Galway, Ireland	Green

Stone walls

Stonework is often a preferred alternative to brickwork, particularly where local planning requirements stipulate that vernacular characteristics are maintained.

Naturally quarried stones

These vary marginally in appearance, even when extracted in the same location from one quarry. Texture and slight colour variation adds to the character and visual attractiveness, along with the presence of veins, crystals, fossils and other visible features that manifest with wear and exposure.

Artificial stone

Otherwise known as reconstituted, reconstructed or cast stone, these are factory manufactured by casting natural stone aggregates of 15mm maximum size with white or coloured cement into moulds of specific dimensions. The usual ratio is three to four parts of aggregate to one part cement. These defect-free blocks produced under quality controlled conditions are an economical substitute for natural stone. Block sizes can vary to suit specific applications, but colour and texture are uniform. A variation is a 20 to 25mm facing of natural stone aggregate, fine sand aggregate and cement (possibly colour pigmented) set over a base of wet concrete.

Further working of cast stone can produce a high quality surface finish to very accurate overall dimensions. In this precise form, blocks can be used as an artificial type of ashlar facing over brickwork or standard blockwork backgrounds (see below for a definition of ashlar and for a summary of ashlar walling).

The long-term weathering qualities of cast stone differ from that of natural stone. Natural stone weathers slowly with a gradual and fairly consistent colour change. Artificial stone has the characteristics of concrete. With time it may crack, become dull and attract dirt staining at joints, corners and projecting features such as copings and sills.

Ashlar ~ hewn and dressed stone facing block.
Ashlaring ~ stonework comprising blocks of stone finely squared and dressed to a precise finish and dimensions, laid to courses of not less than 300mm in height. It is a classic façade treatment to many existing prestigious buildings, but now rarely used due to the cost of manufacturing the stone to fine tolerances and the time-consuming craftsmanship required for construction. Examples are included here not just for historical reference; many buildings constructed with ashlar walling are the subject of ongoing refurbishment, conversion, modification, adaptation and repair. Limestone is popular for ashlar, produced in thicknesses between 100 and 300mm and bedded in a mortar known as mason's putty. The traditional mix for this mortar comprises stone dust, lime putty and portland cement in the ratio of 7:5:2.

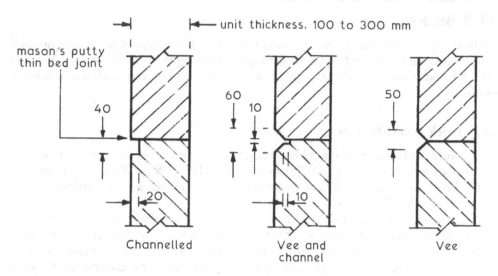

Figure 11.3 Examples of joint profiles to ashlar walling units (dimensions in mm)

Lime mortars

Traditional mortars are a combination of lime, sand and water. These mixes are very workable and have sufficient flexibility to accommodate a limited amount of wall movement due to settlement, expansion and contraction. The long-term durability of lime mortars is poor as they can break down in the presence of atmospheric contaminants and surface growths. Nevertheless, lime is frequently specified as a supplementary binder with cement, to increase mix workability and to reduce the possibility of joint shrinkage and cracking–a characteristic of stronger cement mortars. Lime mortars are still used today in conservation work.

BS 1217: *Cast Stone. Specification.*
BS EN 771–5: *Specification for Masonry Units. Manufactured Stone Masonry Units.*

11.2 Bricks

Bricks in the UK are predominantly made from clay. The composition of clay deposits vary around the UK, providing the wide range of brick types and colours seen. Calcium silicate bricks manufactured from sand, lime and additives can also be found to a lesser extent.

Clay brick manufacture

Bricks were originally made by hand kneading clay into an open wooden box, with dimensions typically 250mm × 125mm × 75mm (10" × 5" × 3") with a protrusion in the bottom if a *frog* or indent was required as a mortar key. Oversized dimensions allowed for shrinkage.

Machine manufacture began during the industrial revolution (mid 18th to 19th centuries) as greater demand for bricks led to brick production on an industrial scale. Machinery was introduced to mass produce pressed bricks in moulds and for wire cutting extruded clay. Both techniques are the basis of production today. Wire cutting to size is used where clay is mechanically forced through a brick sized opening. Circular bars in the central area of the opening leave two or three perforations to provide for a mortar key when laid.

Kiln drying and firing

In 1858 the German Friedrich Hoffmann devised the circular or oval kiln system containing several chambers. The chambers have loading doors, fire holes and flues connecting to a common chimney. Each chamber represents a different stage of drying, pre-heating, firing using natural gas, fuel oil or solid fuel (1100°C) and cooling, all of which are undertaken on a large scale without transportation. A later development is the straight tunnel kiln with a central firing zone. With this, bricks stacked on trolleys run on a narrow railway line through slowly increasing and decreasing heat zones. Both types of kiln are used today, although the tunnel is generally preferred as this is easier to control.

Calcium silicate brick manufacture

This originated during the 19th century as a form of cast artificial stone. Caustic lime and sand were mixed with water into a workable consistency and left until the lime was completely hydrated. It was then pressed into moulds for steam curing and initial hardening. After dry stacking for several weeks the bricks matured to a usable strength.

By the end of the 19th century, processing was considerably improved. Steam at 200°C was pressurised in an *autoclave* (a sealed vessel used to create a chemical reaction between the lime and sand under high pressure; approximately 15 bar or $1.5MN/m^2$). The reaction formed hydrosilicates (hydrated calcium silicates), resulting in an extremely durable and strong building brick.

This principle is the basis for calcium silicate brick production today.

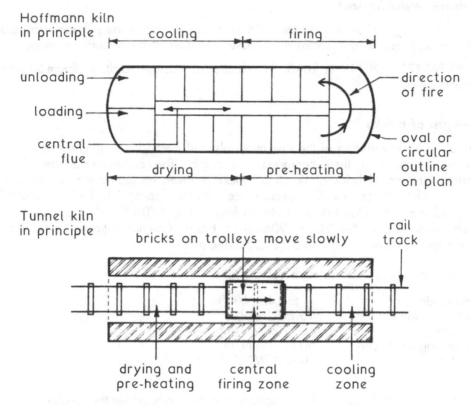

Figure 11.4 Hoffmann brick manufacture process

Typical composition:

- Silica (sand) 84%
- Lime (quicklime or hydrated lime) 7%
- Alumina and oxide of lime 2%
- Water, manganese and alkalis 7%

Properties:

- Similar production costs to clay brick manufacture.
- High crushing strength.
- Dimensionally accurate, i.e. edges straight, even and without defects, warping or twisting.
- Uniformity of dull white colour and of texture (perceived as uninteresting by some).
- A variation is flint-lime, produced by adding powdered flint colouring agents; options for buffs, blues, mauves, etc.

Bricks

Environmental impact

Due to less energy fuel use in production and without generating the air pollutants that occur with clay brick manufacture, considered to have a relatively low environmental impact.

Ref. BS EN 771–2:2011+A1:2015: *Specification for Masonry Units. Calcium Silicate Masonry Units.*

Strength of bricks

Due to the wide variation of the raw materials and methods of manufacture, bricks can vary greatly in their compressive strength. The compressive strength of a particular type of brick or batch of bricks is taken as the arithmetic mean of a sample of ten bricks tested in accordance with the appropriate British Standard. A typical range for clay bricks would be from 20 to 170MN/m^2, the majority of which would be in the 20 to 90MN/m^2 band. Calcium silicate bricks have a compressive strength between 20 and 50MN/m^2.

Typical brick compressive strength comparisons:

Hand made clay	6 to 14MN/m^2
Machine made clay	15 to 20MN/m^2
Stock (clay)	10 to 20MN/m^2
Engineering brick (clay)	50 to 70MN/m^2
Sand-lime	20 to 50MN/m^2

Brick classification ~ clay bricks are specified in three categories for durability:

F0 – suitable for passive exposure.
F1 – suitable for moderate exposure.
F2 – suitable for severe exposure.

Bricks made from clay (BS EN 771–1) or from sand and lime (BS EN 771–2) and are available in a wide variety of strengths, types, textures, colours and special shaped bricks to BS 4729. The standard brick weighs between 2 and 4kg, and is easily held in one hand.

Imperial brick dimensions

Imperial sizes varied significantly over time and by manufacturer but typically were 9" × 4 3/8" xx 2 7/8" (228mm × 110mm × 73mm). For new work bonding to old imperial work, 73mm high bricks are available but they retain the metric length of 215mm.

Metric brick dimensions

The UK government policy in 1965 adopted the metric system from the imperial system of weights and measures. Accordingly, the standardised dimensions for bricks was revised to: length 215mm, width 102.5mm and height 65mm. The length (215mm) is equal to twice its width (102.5mm) plus one standard 10mm joint, and three times its height (65mm) plus two standard joints and, like blocks, they must be laid in a definite pattern or bond if they are to form a structural wall.

National Annex (informative) to BS EN 771–1: 2011, and BS 4729: 2005.

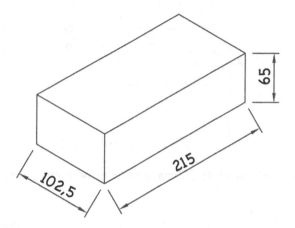

Figure 11.5 Standard metric brick dimensions (mm)

Gauged brickwork

The modular coordination system for construction is based on a module (M) of 100mm and multi-modules of 3M, 6M, 12M, 15M, 30M and 60M. For standard metric gauged brickwork, the base unit is 3M or 300mm. Vertically, four courses of 65mm brickwork with 10mm joints coordinates to a height of 300mm. Horizontally, four 215mm bricks with 10mm joints coordinate to 900mm or 9M.

BS 6750: 1986.

Brick cut specials

Bonding is not solely for aesthetic enhancement. In many applications, e.g. English bonded manhole walls, the disposition of bricks is to maximise wall strength and integrity. In a masonry wall the amount of overlap should not be less than one quarter of a brick length. Specials may be machine or hand cut from standard bricks, or they may be purchased as purpose-made. These purpose-made bricks are relatively expensive as they are individually manufactured in hardwood moulds.

BS 4729: *Clay and Calcium Silicate Bricks of Special Shapes and Sizes.*

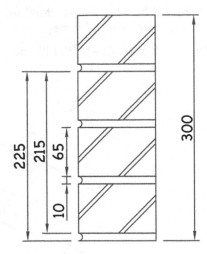

Figure 11.6 Gauged brickwork to 300mm module using 10mm joints

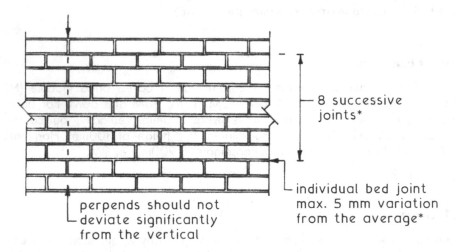

Figure 11.7 Vertical alignment of brick perpend joints

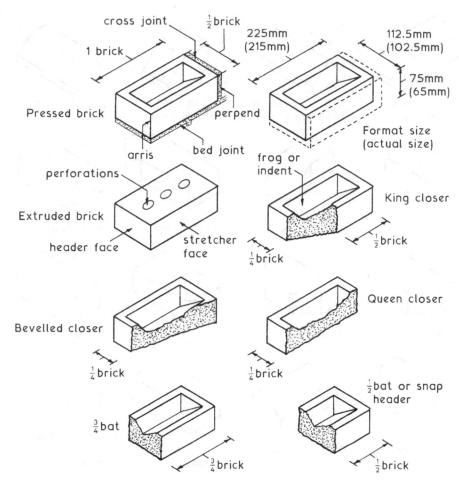

Figure 11.8 Standard bricks and cut specials

Bricks

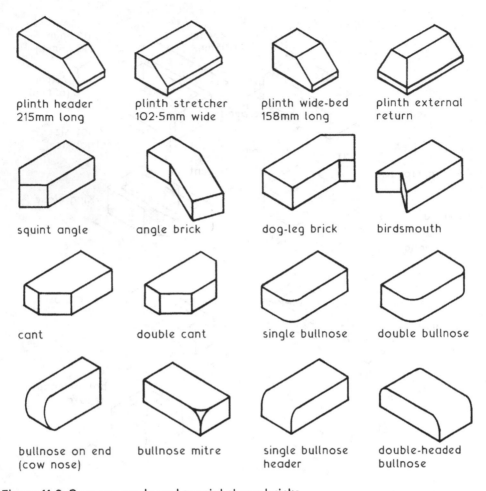

plinth header
215mm long

plinth stretcher
102·5mm wide

plinth wide-bed
158mm long

plinth external
return

squint angle

angle brick

dog-leg brick

birdsmouth

cant

double cant

single bullnose

double bullnose

bullnose on end
(cow nose)

bullnose mitre

single bullnose
header

double-headed
bullnose

Figure 11.9 Purpose-made and special shape bricks

11.3 Blocks

Concrete blocks are manufactured in a wide variety of dimensions, strengths and types. The standard walling block, which coordinates to three courses of brickwork allowing for 10mm mortar joints: length 440mm, height 215mm and 100mm wide, although other widths are available.

Generally used for the inner leaf of cavity walls and internal walls and plastered, fair faced block walling can be used as the external or internal wall finish with a wide variety of decorative surface finishes available. Blocks suitable for external solid walls are classified as load bearing and are required to have a minimum declared compressive strength of not less than $2.9N/mm^2$.

Concrete blocks can be solid, cellular or hollow and can incorporate insulation within the hollows. Solid blocks can be further classified by density: aerated, lightweight, medium or dense. Available in a range of compressive strengths of 2.9, 3.6, 7.3, 8.7, 10.4, 17.5, 22.5, 30.0 and 40.0.

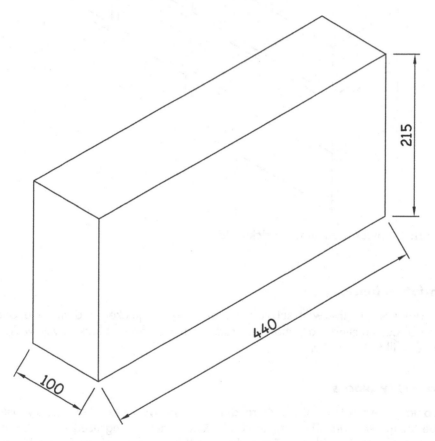

Figure 11.10 Standard metric block dimensions (mm)

Blocks

Aerated and lightweight blocks have good thermal insulation qualities but low compressive strength and sound insulation performance; conversely, dense concrete blocks have low thermal insulation properties but high compressive strength and sound insulation.

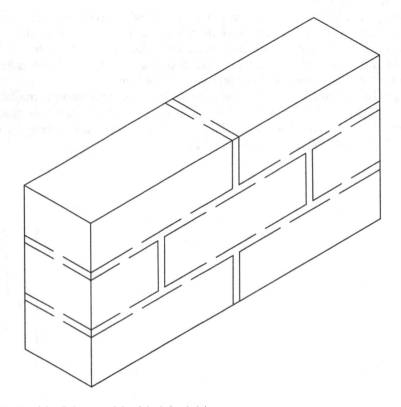

Figure 11.11 1 block is equal to 6 bricks laid

Foundation blocks

These are used in the wall substructure and are a quicker and more economic method of constructing walls than the traditional method of block cavity walls with concrete infill to the cavity.

Hollow clay blocks

Horizontally perforated units with rendering keyways have been widely used in Europe for many years. These types of blocks are now being used in the UK as an alternative to concrete blocks. Typically, vertically perforated units with horizontal tongue and groove connections are laid on a special thin bed mortar. Construction using this system is quicker than traditional blockwork construction.

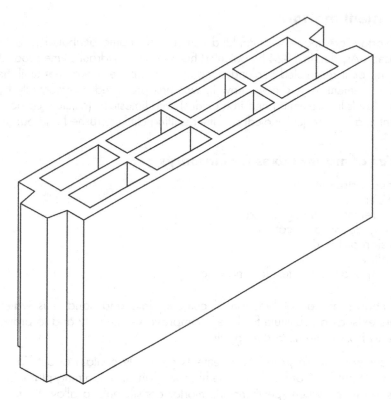

Figure 11.12 Hollow clay interlocking block

BS 6073–2: *Precast Concrete Masonry Units.*
BS EN 772–16: *Methods of Test for Masonry Units*
BS EN 771–3:2011+A1:2015: *Specification for masonry units. Aggregate concrete masonry units (Dense and lightweight aggregates).*
BS EN 771–3: *Specification for masonry units. Aggregate concrete masonry units (Dense and lightweight aggregates).*

11.4 Cement mortars

Modern mortars are made with Portland cement, the name attributed to a bricklayer named Joseph Aspdin. In 1824 he patented his improved hydraulic lime product as Portland cement, as it resembled Portland stone in appearance. It was not until the 1920s that Portland cement, as we now know it, was first produced commercially by mixing a slurry of clay (silica, alumina and iron oxides) with limestone (calcium carbonate). The mix is burnt in a furnace (calcinated) and the resulting clinker crushed and bagged.

Properties of mortar mixes for masonrys

- Adequate strength.
- Workability.
- Water retention during laying.
- Plasticity during application.
- Adhesion or bond.
- Durability.
- Good appearance ~ texture and colour.

Modern mortars are a combination of cement, lime and sand plus water. Liquid plasticisers exist as a substitute for lime, to improve workability and to provide some resistance to frost when used during winter.

Masonry cement ~ these proprietary cements generally contain about 75% Portland cement and about 25% of fine limestone filler with an air entraining plasticiser. Allowance must be made when specifying the mortar constituents to allow for the reduced cement content. These cements are not suitable for concrete.

BS 6463–101, 102 and 103: *Quicklime, Hydrated Lime and Natural Calcium Carbonate.*

BS EN 197–1: *Cement. Composition, Specifications and Conformity Criteria for Common Cements.*

Ready mixed mortar ~ this is delivered dry for storage in purpose-made silos with integral mixers as an alternative to site blending and mixing.

This ensures:

- Guaranteed factory quality controlled product.
- Convenience.
- Mix consistency between batches.
- Convenient facility for satisfying variable demand.
- Limited wastage.
- Optimum use of site space.
- Mortar and cement strength.

Strength of mortars

Mortars consist of an aggregate (sand) and a binder which is usually cement; cement plus additives to improve workability; or cement and lime. The factors controlling the strength of any particular mix are the ratio of binder to aggregate plus the water: cement ratio. The strength of any particular mix can be ascertained by taking the arithmetic mean of a series of test cubes.

Test samples are made in prisms of 40 × 40mm cross section, 160mm long. After 28 days samples are broken in half to test for flexural strength. The broken pieces are subject to a compression test across the 40mm width. An approximate comparison between mortar strength (MN/m^2 or N/mm^2), mortar designations (i to v) and proportional mix ratios is shown in Table 11.2. Included is guidance on application.

Proportional mixing of mortar constituents by volume is otherwise known as a prescribed mix, or simply a recipe.

Mortar classification

Table 11.2 Mortar classifications

Traditional designation	BS EN 998–2 strength	Proportions by volume		Application
		cement/lime/sand	cement/sand	
i	12	1:0.25:3	1:3	Exposed external
ii	6	1:0.5:4–4.5	1:3–4	General external
iii	4	1:1:5–6	1:5–6	Sheltered internal
iv	2	1:2:8–9	1:7–8	General internal
v	-	1:3:10–12	1:9–10	Internal, grouting

Relevant standards

BS EN 1996: *Design of Masonry Structures.*
BS EN 196: *Methods of Testing Cement.*
BS EN 998–2: *Specification for Mortar for Masonry. Masonry Mortar.*
PD 6678: *Guide to the Specification of Masonry Mortar.*
BS EN 1015: *Methods of Test for Mortar for Masonry.*

11.5 Damp proof courses (DPC)

Building Regulation C2 states that the walls (and floors) of a building shall be adequately protected from the harmful effects of ground moisture; this is remedied by installing a damp proof course (DPC) in the wall at a minimum of 150mm above ground level. To prevent ground moisture from the floor transferring to the abutting walls, it is necessary to 'lap' the floor Damp Proof Membrane (DPM) with the wall DPC.

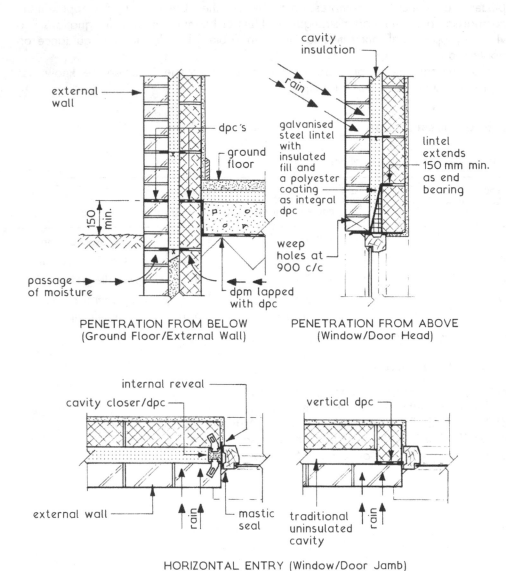

Figure 11.13   Damp proof course locations

The primary function of a DPC:

- Resist moisture penetration from below (rising damp).
- Resist moisture penetration from above.
- Resist moisture penetration from horizontal entry.

DPC and DPM

A damp proof membrane prevents damp entering floors from below.

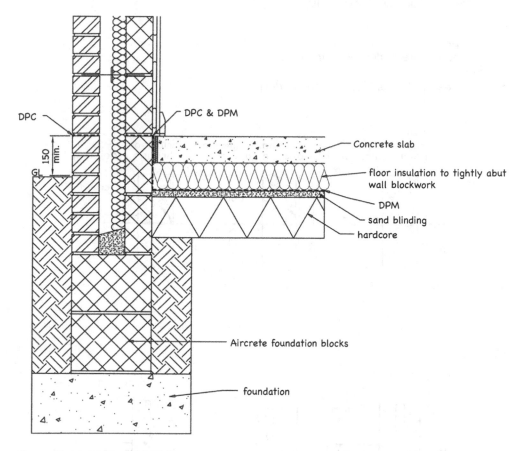

DPC

DPC & DPM

150 min.

GL

Concrete slab

floor insulation to tightly abut wall blockwork

DPM

sand blinding

hardcore

Aircrete foundation blocks

foundation

Figure 11.14 DPC and DPM

Damp proof courses (DPC)

DPC failure

In addition to damp proof courses failing due to deterioration or damage, they may be bridged as a result of:

- Faults occurring during construction.
- Work undertaken after construction, with disregard for the damp proof course.

BS 743, 6398, 6515, 8102 and 8215.
Building Regulations, Approved Document C2, Section 5:

Examples of legacy DPC materials:

- Lead
- Copper
- Bitumen impregnated hessian fibres
- Slate
- Engineering bricks (1:3 cement mortar)

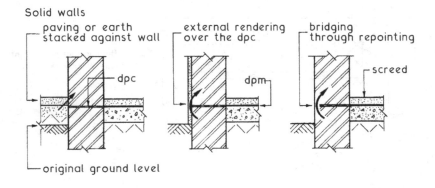

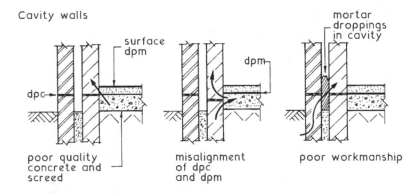

Figure 11.15 DPC and DPM defects

DPC materials in common use today:

- Low Density Polyethylene (LDPE)
- Bitumen polymer
- Polypropylene

All DPCs should be lapped at least by 100mm and adhesive sealed. All DPC must be continuous with any DPM in the floor.

BS 743: *Specification for Materials for Damp-Proof Courses.*
BS 8102: *Code of practice for Protection of Structures Against Water from the Ground.*
BS 8215: *Code of Practice for Design and Installation of Damp-Proof Courses in Masonry Construction.*
BRE Digest 380: *Damp-Proof Courses.*

11.6 Solid brick walls

Walls are classified by their thickness in terms of brick lengths; solid brick construction for external walls is a minimum of 1 brick thick for two-storey housing and 1½ to 2 bricks thick for taller buildings.

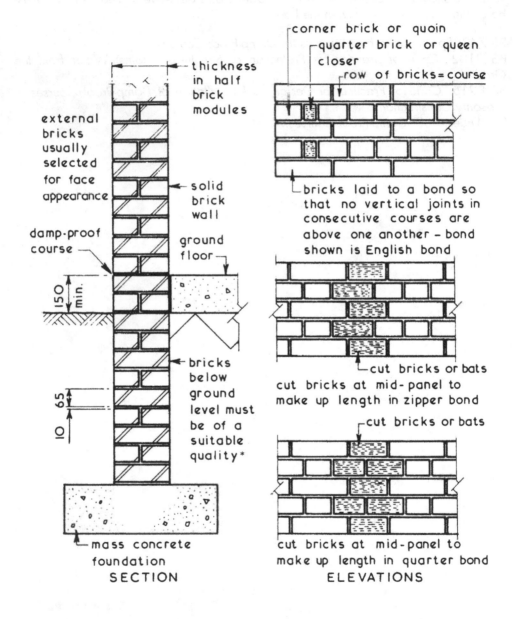

Figure 11.16 Solid brick wall construction methods

Bonding

This is the arrangement of bricks in a wall, column or pier laid to a set pattern to maintain an adequate lap.

Purposes of brick bonding:

1. Obtain maximum strength whilst distributing the loads to be carried throughout the wall, column or pier.
2. Ensure lateral stability and resistance to side thrusts.
3. Create an acceptable appearance.

Simple bonding rules:

1. Bond is set out along length of wall working from each end to ensure that no vertical joints are above one another in consecutive courses.
2. Walls that are not in exact bond length can be set out as in Figure 11.18.
3. Transverse or cross joints continue unbroken across the width of wall unless stopped by a face stretcher.

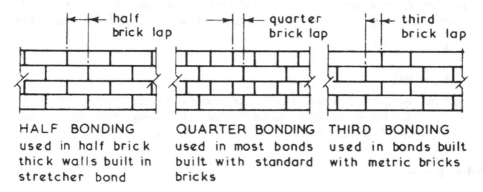

HALF BONDING
used in half brick
thick walls built in
stretcher bond

QUARTER BONDING
used in most bonds
built with standard
bricks

THIRD BONDING
used in bonds built
with metric bricks

Figure 11.17 Lap forms

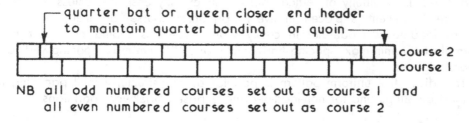

NB all odd numbered courses set out as course 1 and
all even numbered courses set out as course 2

Figure 11.18 Maintaining half lap bond with Queen Closers

319

Solid brick walls

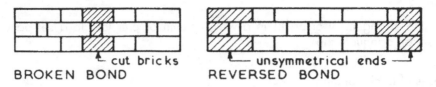

BROKEN BOND   REVERSED BOND

Figure 11.19 Methods of maintaining bonds

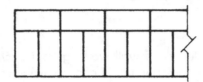

Figure 11.20 One and a half brick thick wall bond in plan view

English bond

Formed by laying alternate courses of stretchers and headers, it is one of the strongest bonds but it will require more facing bricks than other bonds (89 facing bricks per m^2).

Flemish bond

Formed by laying headers and stretchers alternately in each course. Not as strong as English bond but is considered to be aesthetically superior and uses less facing bricks (79 facing bricks per m^2).

Other bonds

STACK BOND

Stack bonding is the quickest, easiest and most economical bond to lay, as there is no need to cut bricks or to provide special sizes. Visually the wall appears unbonded as continuity of vertical joints is structurally unsound, unless wire bed-joint reinforcement is placed in every horizontal course, or alternate courses where loading is moderate. In cavity walls, wall ties should be closer than normal at 600mm max. spacing horizontally and 225mm max. spacing vertically and staggered.

This distinctive uniform pattern is popular as non-structural infill panelling to framed buildings and for non-load-bearing exposed brickwork partitions.

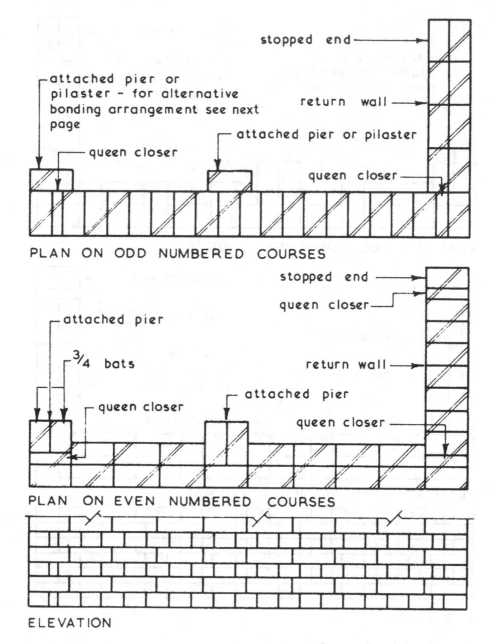

PLAN ON ODD NUMBERED COURSES

PLAN ON EVEN NUMBERED COURSES

ELEVATION

Figure 11.21 Building English bond walls

Solid brick walls

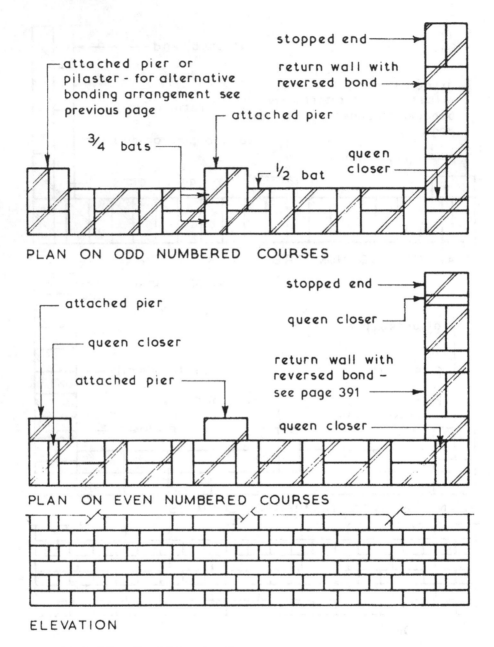

attached pier or pilaster - for alternative bonding arrangement see previous page

¾ bats

attached pier

½ bat

stopped end

return wall with reversed bond

queen closer

PLAN ON ODD NUMBERED COURSES

attached pier

queen closer

attached pier

stopped end

queen closer

return wall with reversed bond - see page 391

queen closer

PLAN ON EVEN NUMBERED COURSES

ELEVATION

Figure 11.22 Building Flemish Bond walls

I course of headers to 3 courses of stretchers

I header to 3 stretchers in each course

ENGLISH GARDEN WALL BOND - gives quick lateral spread of load - uses less facings than English bond.

FLEMISH GARDEN WALL BOND - enables a fair face to be kept on both sides of a one brick thick wall.

ENGLISH CROSS BOND - header placed next to end stretcher in every other stretcher course which thus staggers stretchers enabling patterns or diapers to be picked out in different texture or coloured bricks.

PLAN ON ODD COURSES

2/3 bats   voids

PLAN ON EVEN COURSES

RAT TRAP BOND - uses brick on edge courses - hollow pockets or voids reduce total weight of wall and by the bricks on edge there is an overall saving of materials.

Figure 11.23 Brick bonding variations

Solid brick walls

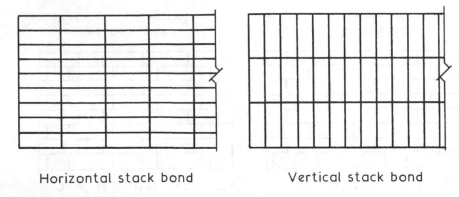

Horizontal stack bond   Vertical stack bond

Figure 11.24 Stack bonding

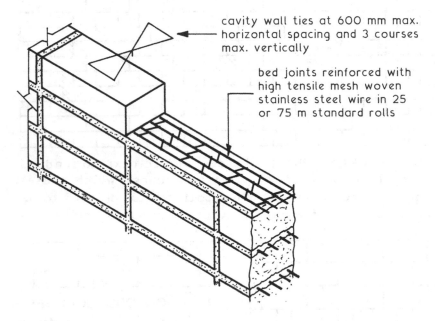

cavity wall ties at 600 mm max. horizontal spacing and 3 courses max. vertically

bed joints reinforced with high tensile mesh woven stainless steel wire in 25 or 75 m standard rolls

Figure 11.25 Reinforced stack bond

11.7 Design of masonry walls

For small and residential buildings up to three storeys high the sizing of load-bearing brick walls can be taken from data given in Section 2C of Approved Document A. The alternative methods for these and other load-bearing brick walls are given in:

BS EN 1996: *Design of Masonry Structures.*
BS 8103–2: *Structural Design of Low Rise Buildings. Code of Practice for Masonry Walls for Housing.*

The main factors governing the load bearing capacity of brick walls and columns are:

1. Thickness of wall.
2. Strength of bricks used.
3. Type of mortar used.
4. Slenderness ratio of wall or column.
5. Eccentricity of applied load.

Thickness of wall

This must always be sufficient throughout its entire body to carry the design loads and induced stresses. Other design requirements such as thermal and sound insulation properties must also be taken into account when determining the actual wall thickness to be used.

Effective thickness ~ this is the assumed thickness of the wall or column used for the purpose of calculating its slenderness ratio.

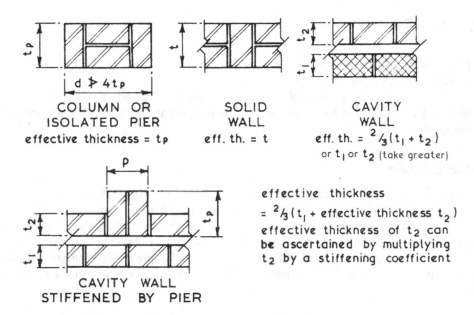

COLUMN OR
ISOLATED PIER
effective thickness = t_p

SOLID
WALL
eff. th. = t

CAVITY
WALL
eff. th. = $\frac{2}{3}(t_1 + t_2)$
or t_1 or t_2 (take greater)

CAVITY WALL
STIFFENED BY PIER

effective thickness
= $\frac{2}{3}(t_1 +$ effective thickness $t_2)$
effective thickness of t_2 can be ascertained by multiplying t_2 by a stiffening coefficient

Figure 11.26 Wall thicknesses
Note: All dimensions are maximum.

Design of masonry walls

Slenderness ratio

This is the relationship of the effective height to the effective thickness thus:

$$\text{Slenderness ratio} = \frac{\text{effective height}}{\text{effective thickness}} = \frac{h}{t} > 27$$

Effective height

This is the dimension taken to calculate the *slenderness ratio as opposed to the actual height.*

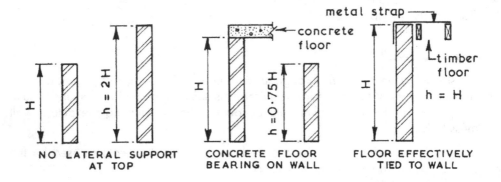

Figure 11.27 Effective thickness = tK

Effective thickness ~ this is the dimension taken to calculate the slenderness ratio as opposed to the actual thickness.

Stress reduction, the permissible stress for a wall, is based on the basic stress multiplied by a reduction factor related to the slenderness ratio and the eccentricity of the load.

Attached piers

The main function of an attached pier is to give lateral support to the wall of which it forms a part from the base to the top of the wall. It also has the subsidiary function of dividing a wall into distinct lengths whereby each length can be considered as a wall. Generally walls must be tied at the end to an attached pier, buttressing or return wall.

Requirements for the external wall of a small, single-storey, non-residential building or annex exceeding 2.5m in length or height and of floor area not exceeding 36m²:

- Minimum thickness, 90mm, i.e. 102.5mm brick or 100mm block.
- Built solid of bonded brick or block masonry and bedded in cement mortar.

- Surface mass of masonry, minimum 130kg/m² where floor area exceeds 10m².
- No lateral loading permitted excepting wind loads.
- Maximum length or width not greater than 9m.
- Maximum height as shown in Figure 2.7.
- Lateral restraint provided by direct bearing of roof.
- Maximum of two major openings in one wall of the building. Height maximum 2.1m, width maximum 5m (if two openings, total width maximum 5m).
- Bonded or connected to piers of a minimum size of 390 × 190mm at maximum 3m centres for the full wall height as shown in Figure 11.31. Pier connections are with pairs of wall ties of 20 × 3mm flat stainless steel type at 300mm vertical spacing.

- Major openings A and B are permitted in one wall only. Aggregate width is 5m maximum. Height not greater than 2.1m. No other openings within 2m.
- Other walls not containing a major opening can have smaller openings of maximum aggregate area 2.4m².
- Maximum of only one opening between piers.
- Distance from external corner of a wall to an opening at least 390mm unless the corner contains a pier.
- The minimum pier dimension of 390 × 190mm can be varied to 327 × 215mm to suit brick sizes.

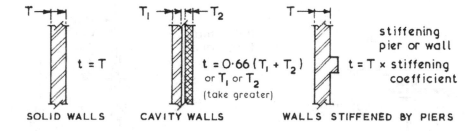

Figure 11.28 Stress reduction

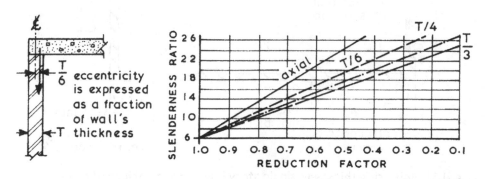

Figure 11.29 Reduction factor

Design of masonry walls

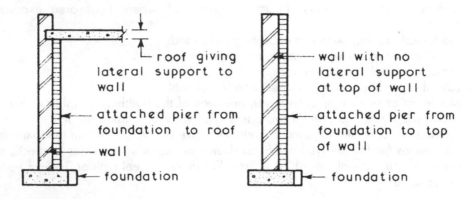

Figure 11.30 Half brick thick wall single story building with attached piers

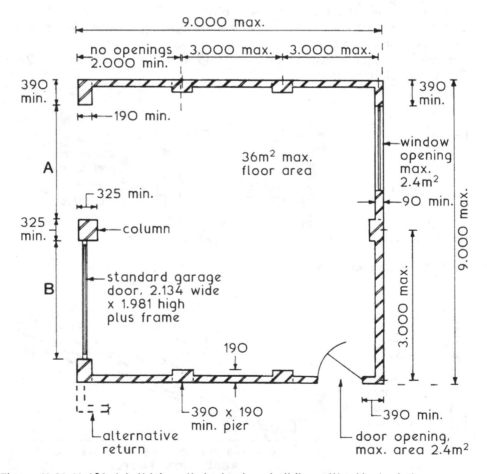

Figure 11.31 Half brick thick wall single story building with attached piers

Construction of half brick and 100mm thick solid concrete block walls (90mm min.) with attached piers has height limitations to maintain stability. The height of these buildings will vary depending on the roof profile; it should not exceed the lesser value in the following examples.

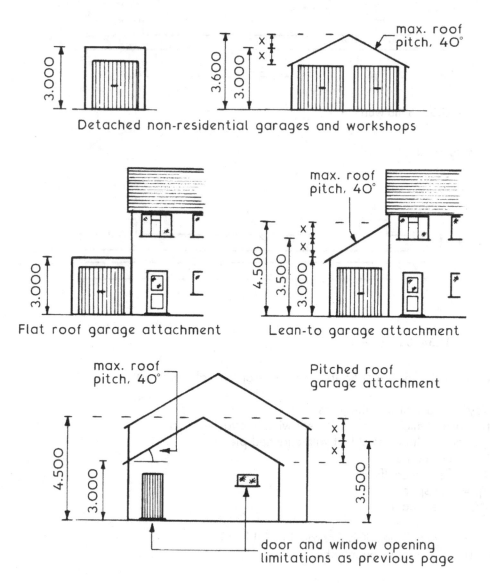

Figure 11.32 Small non-residential buildings or annexes
Note: All dimensions are maximum.

Height is measured from the top of the foundation to the top of the wall, except where shown at an intermediate position. Where the underside of the floor slab provides an effective lateral restraint, measurements may be taken from here.

Design of masonry walls

Piers, effective thickness

Solid wall

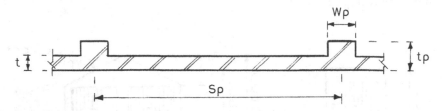

Figure 11.33 Solid wall

Cavity wall

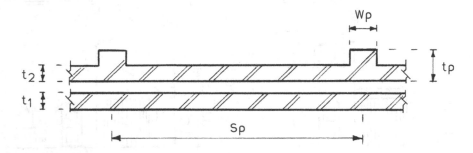

Figure 11.34 Cavity wall

Effective thickness = 0.66 $(t_1 + t_2K)$ or t_1 or t_2K (take greater)

Key : t = actual wall thickness
t_1 = actual thickness of cavity leaf without attached pier
t_2 = actual thickness of leaf with attached pier
t_p = pier thickness
K = stiffening coefficient
S_p = pier spacing
W_p = pier width

Stiffening coefficient K is illustrated in Table 11.3.
    Linear interpolation may be applied.

Table 11.3 Stiffening coefficient K

$Sp \div Wp$	$tp \div t = 1$	2	3
≤6	K=1.0	1.4	2.0
10	1.0	1.2	1.4
≥20	1.0	1.0	1.0

Linear interpolation may be applied

E.g.

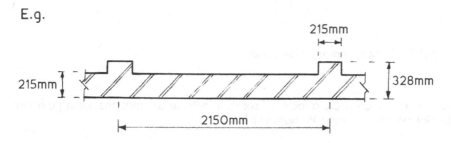

$Sp \div Wp = 2150 \div 215 = 10$
$tp \div t = 328 \div 215 = 1.523$
By interpolation, K = 1.104
Effective thickness, tK = 215 x 1.104 = 237.4 mm

Figure 11.35 Example with pier centres at 2.15m

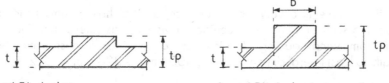

tp ≤ 1.5t design
as a pier

tp > 1.5t design as a column
effective thickness = tp or b
depending on direction of bending

Figure 11.36 Pier effective thickness

Piers and columns, effective height calculations

E.g. A one-and-a-half brick column of 4.000m effective height/length.

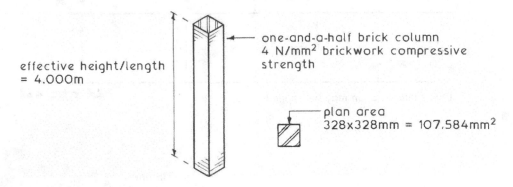

effective height/length = 4.000m

one-and-a-half brick column 4 N/mm² brickwork compressive strength

plan area
328x328mm = 107,584mm²

Figure 11.37 Free standing brick column

It is usual to incorporate a factor of safety into these calculations. Using a figure of 2, the brickwork safe bearing strength will be:

$$4N\,/\,mm^2 \div 2 = 2N\,/\,mm^2$$

Permissible load will be:

$$2N\,/\,mm^2 \times 107,584 = 215,168 \text{ Newtons}$$

$$\text{Newtons} \div \text{gravity}\left(9.81\,m/s^2\right) = 215,168\,kg \div 9.81 = 21,933\,kg\ (22 \text{ tonnes approx})$$

Note: This example is a very simple application that has assumed that there is no axial loading with bending or eccentric forces such as lateral loading from the structure or from wind loading. It also has a low slenderness ratio (effective height/length to least lateral dimension), i.e. 4000 mm ÷ 328 mm = 12.2.

Free standing piers/columns

Piers or columns ~ free standing piers comprise bricks and mortar bonded in accordance with the principles applied to walls. This is essential to maximise strength and distribution of loads.

Alternate courses are arranged for bricks to bridge upper and lower vertical joints as shown in the examples given in Figure 11.38.
Note: Some bricks may require slight trimming to maintain a 10mm mortar width and to align alternating courses.

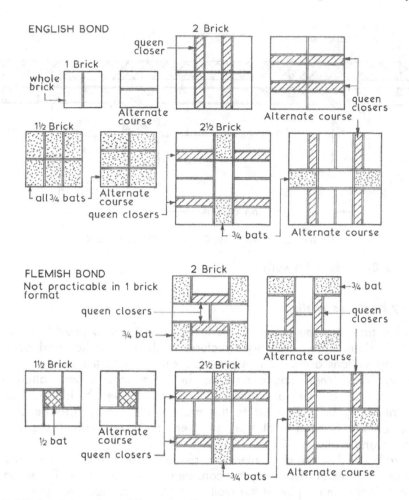

Figure 11.38 Brick bonding of freestanding columns

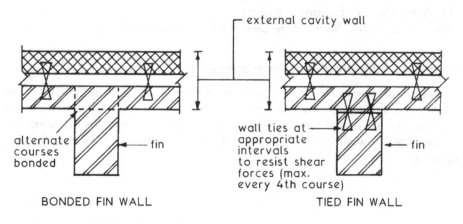

Figure 11.39 Fin wall bonding methods

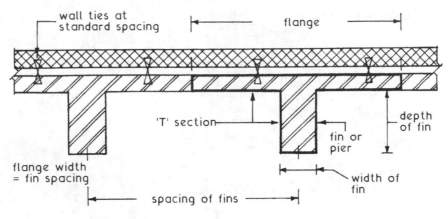

FIN WALL AS A STRUCTURAL 'T' SECTION

Figure 11.40 Buttress or fin walls

Masonry buttress walls

Historically, finned or buttressed walls have been used to provide lateral support to tall, single-storey masonry structures such as churches and cathedrals. Modern applications are similar in principle and include theatres, gymnasiums, warehouses, etc. Where space permits, they are an economic alternative to masonry cladding of steel or reinforced concrete framed buildings. The fin or pier is preferably brick bonded to the main wall. It may also be connected with horizontally bedded wall ties, sufficient to resist vertical shear stresses between fin and wall.

Structurally, the fins are deep piers that reinforce solid or cavity masonry walls. For design purposes the wall may be considered as a series of 'T' sections composed of a flange and a pier. If the wall is of cavity construction, the inner leaf is not considered for bending moment calculations, although it does provide stiffening to the outer leaf or flange.

BS EN 1996: *Design of Masonry Structures.*
PD 6697: *Recommendations for the Design of Masonry Structures.*

Vertical and horizontal wall tolerances

Individual masonry units may not be uniform in dimensions and shape. This characteristic is particularly applicable to natural stone masonry and to kiln manufactured clay bricks. Dimensional variations and other irregularities must be accommodated within an area of walling. The following (shown in Figures 11.41 and 11.42) is acceptable.

Vertical alignment and straightness

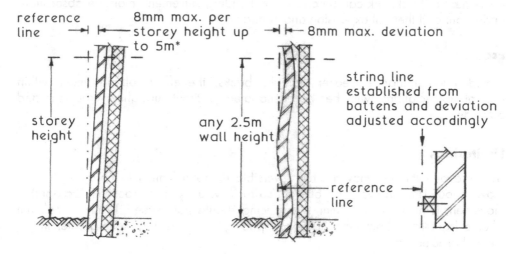

*Walls > 5m height, max. 12mm out of plumb. Limited to 8mm/storey height. E.g. typical storey height of 2.5m = 3.2mm/m.

Figure 11.41 Vertical tolerance for straightness of walls

Horizontal alignment

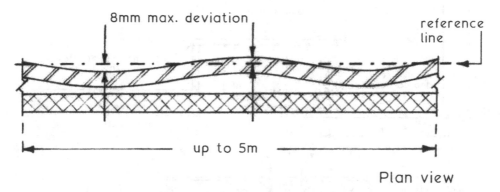

Figure 11.42 Horizontal tolerance for straightness of walls

Design of masonry walls

Movement control

Movement of brickwork can be caused by building settlement, moisture absorption, drying out and thermal expansion and contraction.

Effects

Provided the mortar is weaker than the bricks, the effects of movement within normal expectations will be accommodated without unsightly cracking and damage.

Limitations

Moisture and thermal movement is reversible up to a limit. Clay brick walls can move about 1mm in every 1m of wall, so for a visually and practically acceptable movement joint width of 12mm, the spacing should not exceed 12m. For calcium silicate (sand-lime) bricks the maximum spacing is 9m as they have greater movement characteristics.

Applications

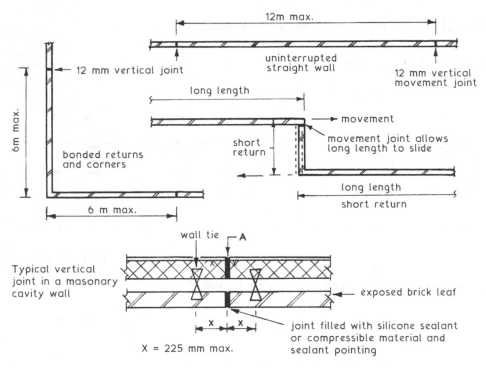

Figure 11.43 Allowing for horizontal movement in a wall

11.8 Jointing and pointing

The appearance of a building can be significantly influenced by the mortar finishing treatment to masonry. Finishing may be achieved by jointing or pointing.

Jointing ~ the finish applied to mortar joints as the work proceeds.
Pointing ~ the process of removing semi-set mortar to a depth of about 20mm and replacing it with fresh mortar. Pointing may contain a colouring pigment to further enhance the masonry.

Finish profiles, typical examples shown pointed

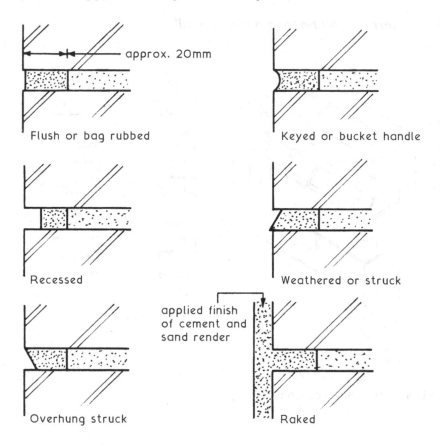

Figure 11.44 Pointing/jointing methods

Note: Recessed and overhung finishes should not be used in exposed situations, as rainwater can be detained. This could encourage damage by frost action and growth of lichens.

11.9 Feature brickwork

To enhance the appearance of plain brick walls, standard bricks in patterns or special bricks are used to create architectural features

Brick plinths

These are used as a projecting feature to enhance external wall appearance at its base. The exposed projection determines that only frost-proof quality bricks are suitable and that recessed or raked-out joints, which could retain water, must be avoided.

Typical external wall base in a cavity wall

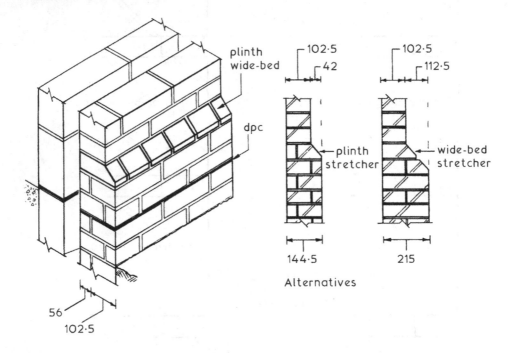

Figure 11.45 Brick plinth construction

Brick corbels

These are a projecting feature at higher levels of a building. This may be created by using plinth bricks laid upside-down with header and stretcher formats maintaining bond. For structural integrity, the amount of projection (P) must not exceed one-third of the overall wall thickness (T).

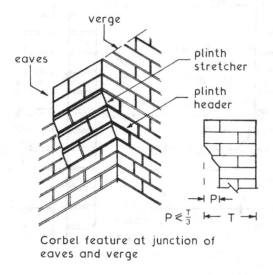

verge

eaves

plinth stretcher

plinth header

$P \leqslant \frac{T}{3}$

Corbel feature at junction of eaves and verge

Figure 11.46 Brick corbelling at the eaves

Parapet walls

A parapet is a low wall projecting above the level of a roof, bridge or balcony forming a guard or barrier at the edge. Parapets are exposed to the elements, justifying careful design and construction for durability.

Feature brickwork

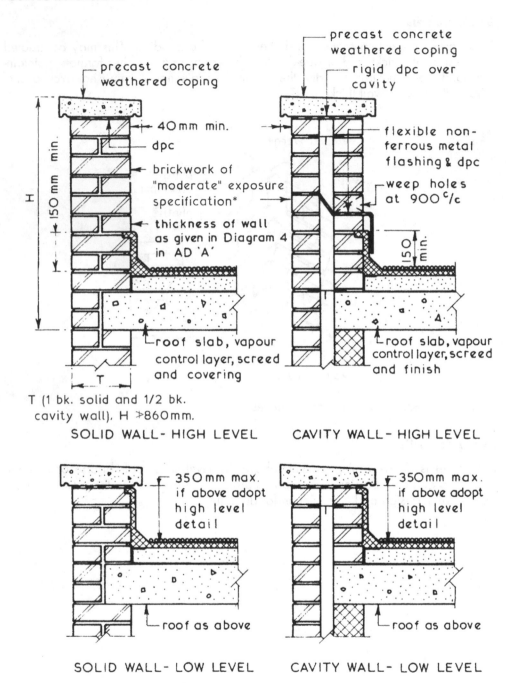

precast concrete
weathered coping

precast concrete
weathered coping

rigid dpc over
cavity

40mm min.

dpc

brickwork of
"moderate" exposure
specification*

thickness of wall
as given in Diagram 4
in AD 'A'

flexible non-
ferrous metal
flashing & dpc

weep holes
at 900 c/c

roof slab, vapour
control layer, screed
and covering

roof slab, vapour
control layer, screed
and finish

150 mm min.

H

150 min.

T

T (1 bk. solid and 1/2 bk.
cavity wall), H ⩾860mm.

SOLID WALL- HIGH LEVEL

CAVITY WALL- HIGH LEVEL

350mm max.
if above adopt
high level
detail

350mm max.
if above adopt
high level
detail

roof as above

roof as above

SOLID WALL- LOW LEVEL

CAVITY WALL- LOW LEVEL

Figure 11.47 Typical parapet wall details

11.10 Cavity wall construction

This has been the standard form of wall construction in the UK from the 1920s onwards. Cavity walls consist of an outer brick or block leaf or skin separated from an inner brick or block leaf or skin by an air space called a cavity. Over time, to improve the thermal performance of cavity walls, insulation was installed in the cavity. For structural stability the two leaves of a cavity wall are tied together with wall ties. Cavity walls have better weather-resistance properties than a comparable solid brick or block walls. Cavities are not normally ventilated and are closed by roof insulation at eaves level.

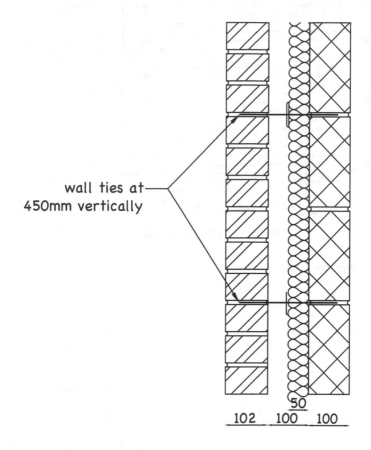

Figure 11.48 Partial fill cavity wall

Minimum requirements:

- Thickness of each leaf, 90mm.
- Width of cavity, 50mm.
- Wall ties at $2.5/m^2$.
- Compressive strength of bricks, $6N/mm^2$ up to two storeys.
- Compressive strength of blocks, $2.9N/mm^2$ up to two storeys.

See Section 2C of Approved Document A – Building Regulations for design limitations

Cavity wall construction

Wall ties

Located at a rate of 2.5/m², or at equivalent spacings shown below and as given in Section 2C of Approved Document A – Building Regulations.

Stainless steel or non-ferrous vertical twist-type ties are used for cavity widths between 75 and 300mm.

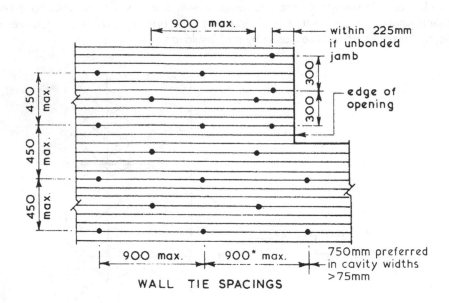

WALL TIE SPACINGS

Figure 11.49 Wall tie spacing

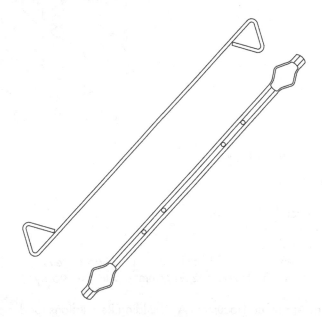

Figure 11.50 Ancon stainless steel cavity wall ties. Ref: www.ancon.co.uk

11.11 Cavity wall insulation

The minimum performance standards for exposed walls, set out in Approved Document L to meet the requirements of Part L of the Building Regulations, can be achieved in several ways, but all require careful specification, detail and construction of the wall fabric, insulating material(s) and/or applied finishes. Cavity walls can have the insulation on the outer face, in the cavity or the inner face of the wall; the most common is within the cavity.

Full fill cavity wall batts

Rockwool and Fibreglass insulation in rigid slabs or 'batts' have been used to provide full fill cavity wall insulation for many years. Available in thicknesses between 50 and 150mm, batts are typically 455mm wide (to fit between two courses of blocks and cavity wall ties) and 1200mm long. A high level of workmanship is required to ensure that the cavity is free from mortar and that all slabs are butted tightly together with no gaps.

Partial fill cavity boards

These are high performance insulation boards which means they can provide the same level of insulation of a full filled cavity of Rockwool batts but only partially filling the cavity, allowing for a clear air gap in the cavity between the insulation and the inner face of the brickwork. The insulation is held against the block wall with clips on the cavity wall ties.

Typical thermal conductivity values of insulation materials are shown in Table 11.4.

Table 11.4 Typical thermal conductivities of insulation materials

Material	Thermal conductivity (W/m K)
Phenolic foam	0.018–0.031
Polyurethane foam (rigid)	0.019–0.023
Foil-faced Polyisocyanurate foam	0.021
Polyisocyanurate foam	0.023–0.025
Extruded polystyrene	0.028–0.036
Expanded PVC	0.030
Mineral wool	0.031–0.040
Glass wool	0.031–0.040
Expanded polystyrene	0.033–0.040
Cellulose (recycled paper)	0.035–0.040
Sheep's wool	0.037–0.039
Rigid foamed glass	0.037–0.055

Cavity wall insulation

Typical examples of existing construction that would require upgrading to satisfy contemporary UK standards:

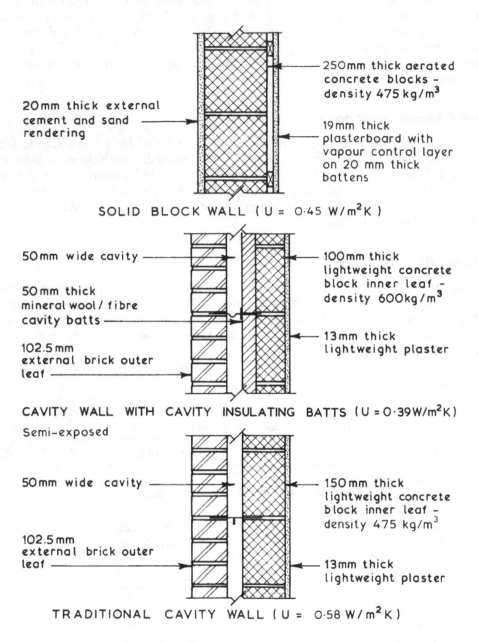

20 mm thick external cement and sand rendering

250 mm thick aerated concrete blocks – density 475 kg/m³

19 mm thick plasterboard with vapour control layer on 20 mm thick battens

SOLID BLOCK WALL (U = 0·45 W/m²K)

50 mm wide cavity

50 mm thick mineral wool / fibre cavity batts

102.5 mm external brick outer leaf

100 mm thick lightweight concrete block inner leaf – density 600 kg/m³

13 mm thick lightweight plaster

CAVITY WALL WITH CAVITY INSULATING BATTS (U = 0·39 W/m²K)

Semi-exposed

50 mm wide cavity

102.5 mm external brick outer leaf

150 mm thick lightweight concrete block inner leaf – density 475 kg/m³

13 mm thick lightweight plaster

TRADITIONAL CAVITY WALL (U = 0·58 W/m²K)

Figure 11.51 Older wall constructions

Typical examples of contemporary construction practice that achieve a thermal transmittance or U-value below 0.30W/m²K:

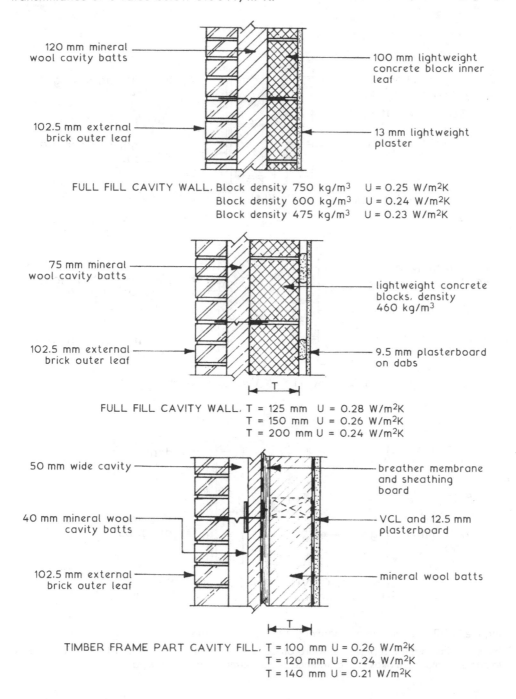

120 mm mineral wool cavity batts

100 mm lightweight concrete block inner leaf

102.5 mm external brick outer leaf

13 mm lightweight plaster

FULL FILL CAVITY WALL. Block density 750 kg/m³   U = 0.25 W/m²K
Block density 600 kg/m³   U = 0.24 W/m²K
Block density 475 kg/m³   U = 0.23 W/m²K

75 mm mineral wool cavity batts

lightweight concrete blocks, density 460 kg/m³

102.5 mm external brick outer leaf

9.5 mm plasterboard on dabs

FULL FILL CAVITY WALL. T = 125 mm   U = 0.28 W/m²K
T = 150 mm   U = 0.26 W/m²K
T = 200 mm   U = 0.24 W/m²K

50 mm wide cavity

breather membrane and sheathing board

40 mm mineral wool cavity batts

VCL and 12.5 mm plasterboard

102.5 mm external brick outer leaf

mineral wool batts

TIMBER FRAME PART CAVITY FILL. T = 100 mm   U = 0.26 W/m²K
T = 120 mm   U = 0.24 W/m²K
T = 140 mm   U = 0.21 W/m²K

Figure 11.52 Contemporary wall constructions
Note: Mineral wool insulating batts have a thermal conductivity (A) value of between 0.025 and 0.040W/mK depending on density.

Cavity wall insulation

Further examples of contemporary construction that achieve a thermal transmittance or U-value below 0.30W/m²K:

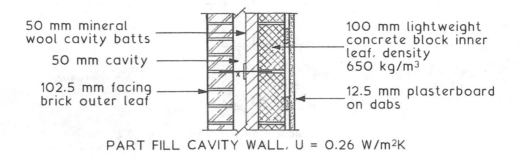

50 mm mineral wool cavity batts

50 mm cavity

102.5 mm facing brick outer leaf

100 mm lightweight concrete block inner leaf, density 650 kg/m³

12.5 mm plasterboard on dabs

PART FILL CAVITY WALL, U = 0.26 W/m²K

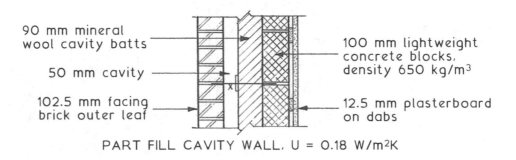

90 mm mineral wool cavity batts

50 mm cavity

102.5 mm facing brick outer leaf

100 mm lightweight concrete blocks, density 650 kg/m³

12.5 mm plasterboard on dabs

PART FILL CAVITY WALL, U = 0.18 W/m²K

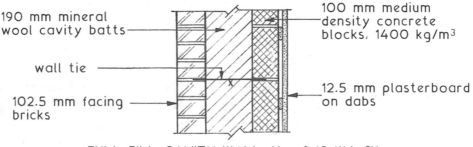

190 mm mineral wool cavity batts

wall tie

102.5 mm facing bricks

100 mm medium density concrete blocks, 1400 kg/m³

12.5 mm plasterboard on dabs

FULL FILL CAVITY WALL, U = 0.18 W/m²K
with 140 mm insulation batts, U=0.22 W/m²K

Figure 11.53 Further contemporary wall constructions

Note: An internal finish of plasterboard on dabs, or 13mm lightweight plaster applied directly to concrete blocks, makes little difference to the thermal transmittance value.

Masonry with external insulation

Improvements to various forms of existing construction with typical thermal transmittance U-values (W/m^{2}K):

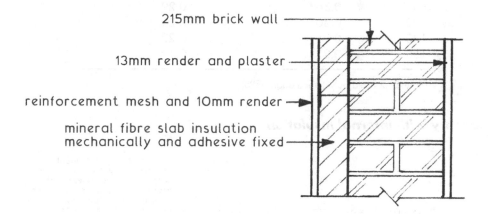

215mm brick wall

13mm render and plaster

reinforcement mesh and 10mm render

mineral fibre slab insulation mechanically and adhesive fixed

Figure 11.54 Retrofit external wall insulation to old solid wall building

Table 11.5 Typical U values for retro fit external insulation to solid brick walls

Insulation mm	U-value
0	2.17
80	0.36
100	0.30
120	0.26
150	0.21

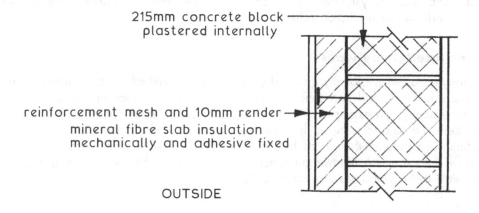

215mm concrete block plastered internally

reinforcement mesh and 10mm render
mineral fibre slab insulation mechanically and adhesive fixed

OUTSIDE

Figure 11.55 External insulation to thick block walls

Cavity wall insulation

Table 11.6 Typical U values for block walls with external insulation

Insulation	U-values Med. dens blk. $\lambda = 0.5$	Ltwt. dens blk. $\lambda = 0.2$
80	0.36	0.29
100	0.30	0.25
120	0.26	0.22
150	0.21	0.19

Note: Expanded polystyrene board (EPS)

Masonry with internal insulation

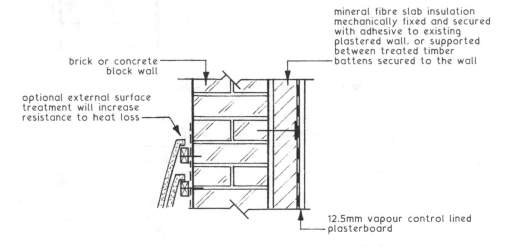

mineral fibre slab insulation mechanically fixed and secured with adhesive to existing plastered wall, or supported between treated timber battens secured to the wall

brick or concrete block wall

optional external surface treatment will increase resistance to heat loss

12.5mm vapour control lined plasterboard

Figure 11.56 Retrofit internal insulation to solid brick wall building

Note: Expanded polystyrene board (EPS) is an alternative insulation material that will provide similar insulation values.

Stone faced cavity walls

In many areas where locally quarried stone is the established building material, traditions are maintained and specified as part of planning compliance requirements. Dressed stone is an example of an established external feature and can be used as an alternative to brickwork as the outer leaf of cavity walls. Stone widths of 150 to 225mm are usual, although as little as 100mm is possible with mechanically cut and finished stone. A lightweight concrete block inner leaf provides lateral support and stability with wall ties positioned at the rate of $2.5/m^2$.

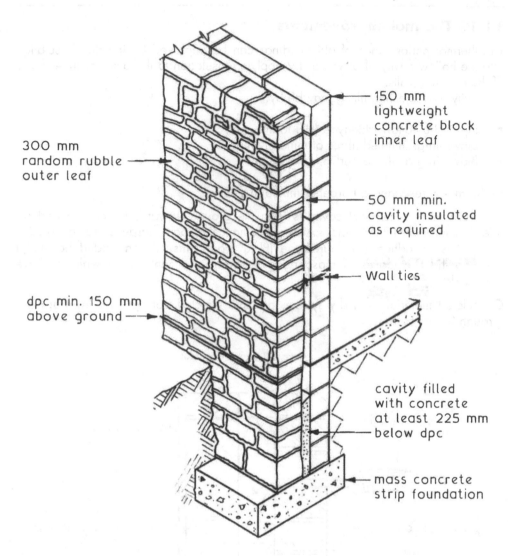

300 mm
random rubble
outer leaf

150 mm
lightweight
concrete block
inner leaf

50 mm min.
cavity insulated
as required

Wall ties

dpc min. 150 mm
above ground

cavity filled
with concrete
at least 225 mm
below dpc

mass concrete
strip foundation

Figure 11.57 Typical stone faced cavity wall detail
Note: Solid walls of stone need to be at least 300mm wide to ensure stability, and
in excess of 400mm (depending on quality) to resist penetration of rainfall.

11.12 Thermal improvements

The thermal performance of old buildings can be improved by injecting insulation into the hollow cavity of a wall or by applying insulation to the outer or inner faces of the external walls.

Cavity wall injection filling may be by:

- Injected urea-formaldehyde (UF) foam.
- Blown in granulated fibres of mineral wool.
- Blown in granulated perlite.

UF foam ~ a resin and a hardener mixed in solution.

Placement into the wall cavity is from the ground upward, by compressed air injection through 12mm holes. After placement the foam hardens. An alternative material is an adhesive polyurethane (PUR) foam. This has an added benefit of bonding the two leaves of masonry, a remedial treatment where wall ties have corroded.

Granulated materials ~ a dry system, pressure injected from ground level upward through 20mm holes.

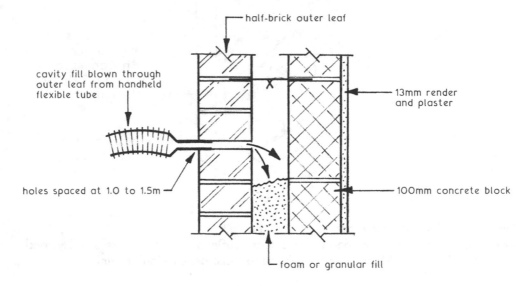

Figure 11.58 Blown cavity wall insulation

BS 5617: *Specification for Urea-Formaldehyde Foam Systems Suitable for Thermal Insulation of Cavity Walls with Masonry or Concrete Inner and Outer Leaves.*

BS 5618: *Code of Practice for Thermal Insulation of Cavity Walls (with Masonry or Concrete Inner and Outer Leaves) by Filling with Urea-Formaldehyde Foam Systems.*

Approved Document D – Toxic Substances, D1: Cavity Insulation.

External insulation

This is feasible only if space and access permits.

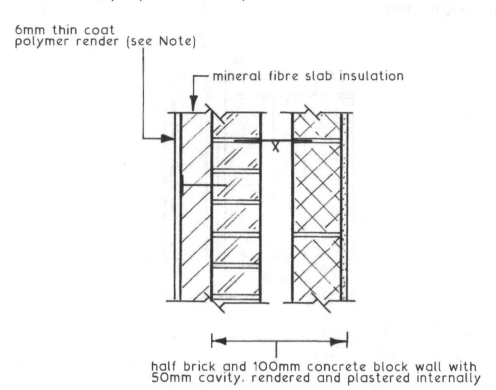

6mm thin coat
polymer render (see Note)

mineral fibre slab insulation

half brick and 100mm concrete block wall with
50mm cavity, rendered and plastered internally

Figure 11.59 External insulation to cavity wall
Note: Polymer renders as standard, vapour permeable with good weatherproofing qualities.

Table 11.7   Typical U values for hollow cavity walls with external insulation

	Typical U values	
Ins. (mm)	Med. dens. blk. $\lambda = 0.5$	Low dens. blk. $\lambda = 0.3$
50	0.51	0.48
75	0.38	0.36
100	0.30	0.29

Thermal improvements

Internal insulation

This is problematic because of the loss of room volume with possible rewiring and plumbing necessary.

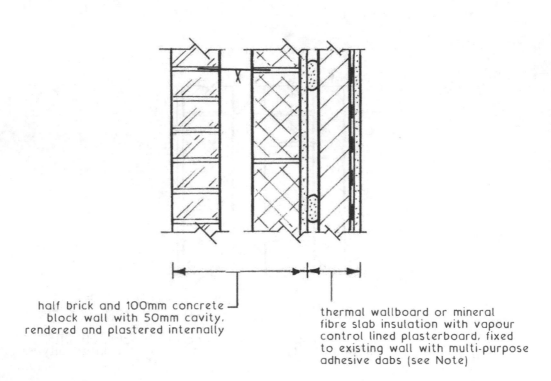

half brick and 100mm concrete block wall with 50mm cavity, rendered and plastered internally

thermal wallboard or mineral fibre slab insulation with vapour control lined plasterboard, fixed to existing wall with multi-purpose adhesive dabs (see Note)

Figure 11.60 Internal insulation to cavity walls

Note: Vapour control layer is an impervious lining of polythene or metal foil placed to the warm side of insulation. It prevents warm moisture from the interior of a house condensing on the cold parts of construction.

Table 11.8 Typical U values for hollow cavity walls with internal insulation

Ins. (mm)	Typical U values	
	Med. dens. blk. $\lambda = 0.5$	Low dens. blk. $\lambda = 0.3$
50	0.44	0.42
75	0.34	0.33
100	0.28	0.27

11.13 Thermal or cold bridging

This is heat loss and possible condensation, occurring mainly around window and door openings and at the junction between ground floor and wall. Other opportunities for thermal bridging occur where uniform construction is interrupted by unspecified components, e.g. occasional use of bricks and/or tile slips to make good gaps in thermal block inner leaf construction.

Note: This practice was quite common, but is no longer acceptable by current legislative standards in the UK.

Prime areas for concern

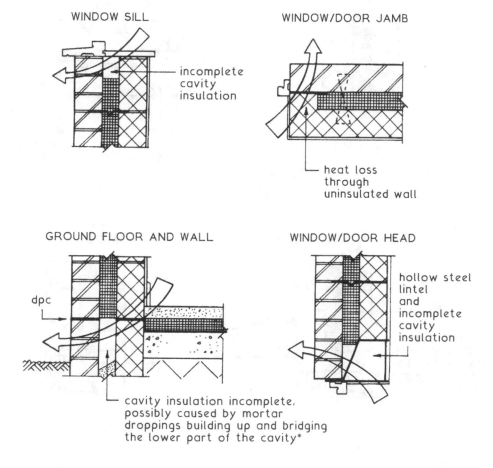

WINDOW SILL

incomplete cavity insulation

WINDOW/DOOR JAMB

heat loss through uninsulated wall

GROUND FLOOR AND WALL

dpc

cavity insulation incomplete, possibly caused by mortar droppings building up and bridging the lower part of the cavity*

WINDOW/DOOR HEAD

hollow steel lintel and incomplete cavity insulation

*Cavity should extend down at least 225mm below the level of the lowest dpc (A.D. C: Section 5).

Figure 11.61 Examples where thermal bridging can occur

Thermal or cold bridging

Continuity of insulated construction in the external envelope is necessary to prevent thermal bridging. Nevertheless, some discontinuity is unavoidable where the pattern of construction has to change. For example, windows and doors have significantly higher U-values than elsewhere. Heat loss and condensation risk in these situations is regulated by limiting areas, effectively providing a trade-off against very low U-values elsewhere.

The following details should be observed around openings and at ground floor:

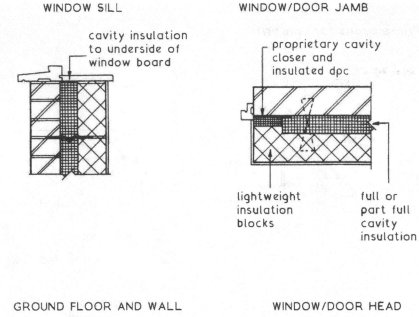

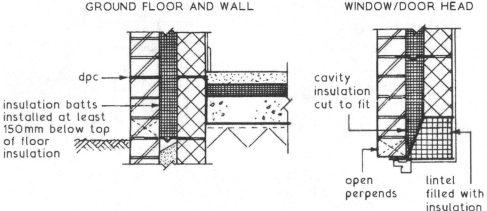

Figure 11.62 Methods of preventing thermal bridging

11.14 Lintels and arches

The primary function of any support over an opening is to carry the loads above the opening and transmit them safely to the abutments, jambs or piers on both sides. A support over an opening is usually required since the opening infilling, such as a door or window frame, will not have sufficient strength to carry the load through its own members.

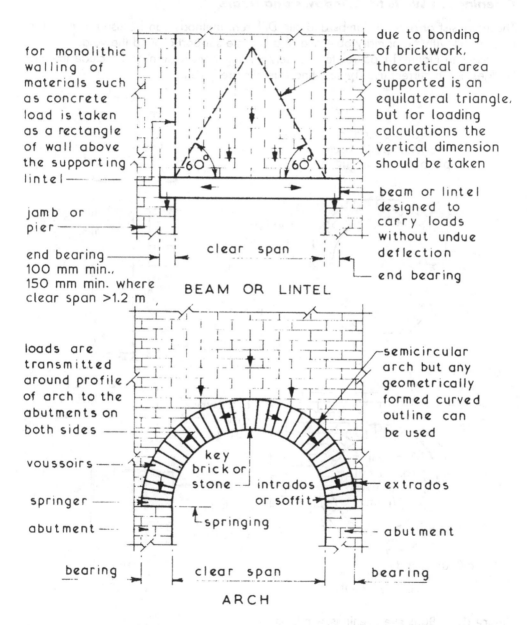

for monolithic walling of materials such as concrete load is taken as a rectangle of wall above the supporting lintel ——

jamb or pier ——

end bearing 100 mm min., 150 mm min. where clear span >1.2 m

due to bonding of brickwork, theoretical area supported is an equilateral triangle, but for loading calculations the vertical dimension should be taken

60°   60°

clear span

BEAM OR LINTEL

beam or lintel designed to carry loads without undue deflection

end bearing

loads are transmitted around profile of arch to the abutments on both sides ——

voussoirs ——

springer ——

abutment ——

bearing

key brick or stone —| intrados or soffit

springing

clear span

ARCH

semicircular arch but any geometrically formed curved outline can be used

extrados

abutment

bearing

Figure 11.63 Type of support

Lintels and arches

Arch cavity tray

The profile of an arch does not lend itself to simple positioning of a damp proof course. At best, it can be located horizontally at upper extrados level. This leaves the depth of the arch and masonry below the DPC vulnerable to dampness. Proprietary cavity trays resolve this problem.

Openings in Walls for Windows and Doors

These consist of a head, jambs and sill. Different methods can be used in their formation, all with the primary objective of adequate support around the void.

Details in Figure 11.65 relate to older/existing construction and include thermal bridges not in current use. Typical head details are given.

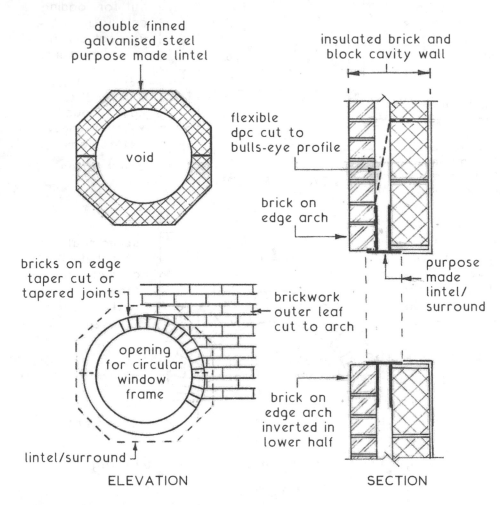

Figure 11.64 Bulls eye openings/windows

Note: Extruded polystyrene closed cell foam insulating panel 25 to 50mm in thickness, adhesive secured to the background.

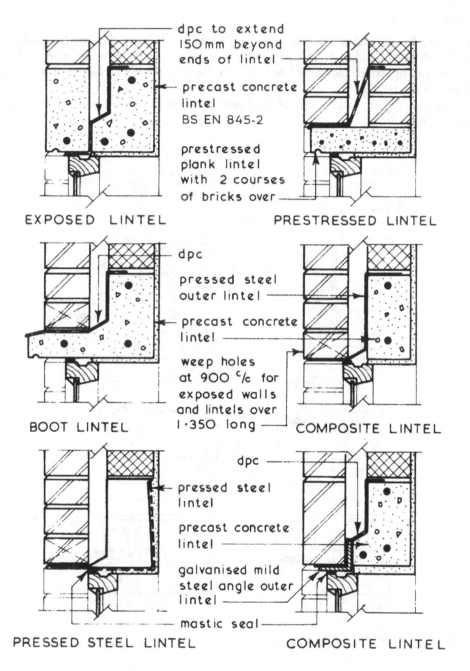

EXPOSED LINTEL

PRESTRESSED LINTEL

BOOT LINTEL

COMPOSITE LINTEL

PRESSED STEEL LINTEL

COMPOSITE LINTEL

dpc to extend 150 mm beyond ends of lintel

precast concrete lintel BS EN 845-2

prestressed plank lintel with 2 courses of bricks over

dpc

pressed steel outer lintel

precast concrete lintel

weep holes at 900 c/c for exposed walls and lintels over 1·350 long

dpc

pressed steel lintel

precast concrete lintel

galvanised mild steel angle outer lintel

mastic seal

Figure 11.65 Examples of older lintels over wall openings

BS EN 845–2: *Specification for Ancillary Components for Masonry. Lintels.*

Lintels and arches

Jambs

Jambs are the vertical sides of an opening. Traditionally these may be bonded as in solid walls or unbonded as in cavity walls. The latter must have some means of preventing the ingress of moisture from the outer leaf to the inner leaf and hence the interior of the building.

Details in Figure 11.66 are of past methods not in current use. Typical jamb details are given.

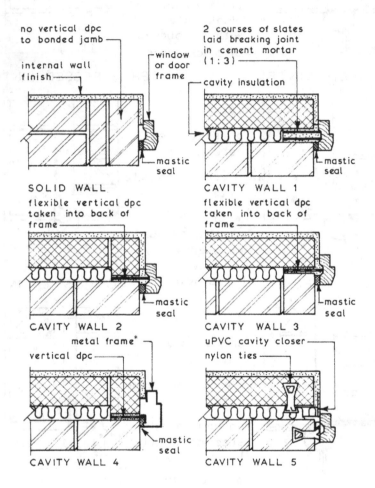

Figure 11.66 Plan view of jamb openings in walls

Sills

The sill is the bottom horizontal part of an opening, the primary function of any sill is to collect the rainwater which has run down the face of the window or door and shed it clear of the wall below.

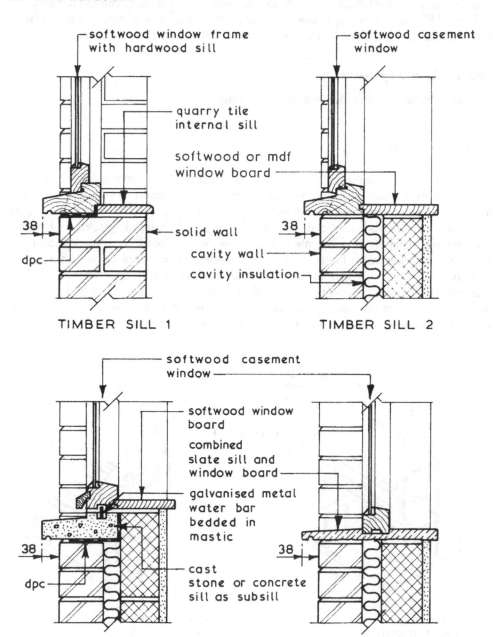

Figure 11.67 Typical sill details

11.15 Timber frame construction

This technique has a long history of conventional practice in Scandinavia and North America for residential construction where timber supplies are plentiful, but has only had limited use in the UK since the 1960s and relies upon imported timber supplies.

In the UK timber framing is used mainly for the inner leaf of cavity wall construction, maintaining the outer leaf of facing brickwork for aesthetic reasons.

Prefabricated systems

Assembly techniques are derived from two systems:

1. Balloon frame.
2. Platform frame.

Both systems are designed for rapid site construction, with considerably fewer site operatives than traditional construction. Factory-made panels are based on a stud framework of timber, normally metric equivalent (ex.) 100 × 50mm, an outer sheathing of plywood, particleboard or similar sheet material, insulation between the framing members and an internal lining of plasterboard. An outer cladding of brickwork weatherproofs the building and provides a traditional appearance.

A balloon frame consists of two-storey height panels with an intermediate floor suspended from the framework. In the UK, the platform frame is preferred with intermediate floor support directly on the lower panel. It is also easier to transport, easier to handle on site and has fewer shrinkage and movement problems.

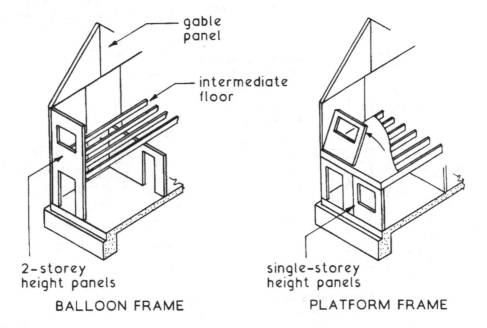

Figure 11.68 Timber framing systems

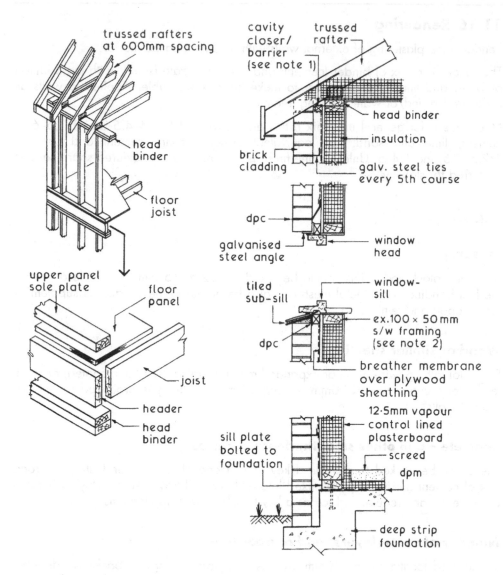

Figure 11.69 Typical details

Notes:

1. Cavity barriers prevent fire spread. The principal locations are between elements and compartments of construction (see Building Regulations A.D. B3).
2. Thermal bridging through solid framing may be reduced by using rigid EPS insulation and lighter 'I' section members of plywood or oriented strand board (OSB).

11.16 Rendering

Render ~ the plastering of external walls with a cement based mortar.

The render is a mix of binder (cement) and fine aggregate (sand), with the addition of water and lime or a plasticiser to make the mix workable. Applied to walls as a decorative and/or waterproofing treatment.

Mix ratios ~ for general use, mix ratios are between 1:0.5:4–4.5 and 1:1:5–6 of cement, lime and sand. The equivalent using masonry cement and sand is 1:2.5–3.5 and 1:4–5. Unless a fine finish is required, coarse textured sharp sand is preferred for stability.

Substrates

Masonry

Brick and block-work joints should be raked out 12 to 15 mm to provide a key for the first bonding coat. Metal mesh can also be nailed to the surface as supplementary support and reinforcing.

Wood or similar sheeting

Fix metal lathing, wire mesh or expanded metal of galvanised (zinc coated) or stainless steel secured every 300mm. A purpose-made lathing is produced for timber-framed walls.

Concrete – and other smooth, dense surfaces

These can be hacked to provide a key or spatter-dashed. Spatter-dash is a strong mix of cement and sand (1:2) mixed into a slurry and trowelled roughly, or thrown on, to leave an irregular surface as a key to subsequent applications.

Number of coats (layers) and composition

In sheltered locations, one 10mm layer is adequate for regular backgrounds. Elsewhere, two or possibly three separate applications are required to adequately weatherproof the wall and to prevent the brick or block-work joints from 'grinning' through. Render mixes should become slightly weaker towards the outer layer to allow for greater flexure at the surface, i.e. less opportunity for movement and shrinkage cracking.

Render finishes

Smooth, textured, rough-cast and pebble-dashed.

Smooth ~ fine sand and cement finished with a steel trowel (6 to 8mm).

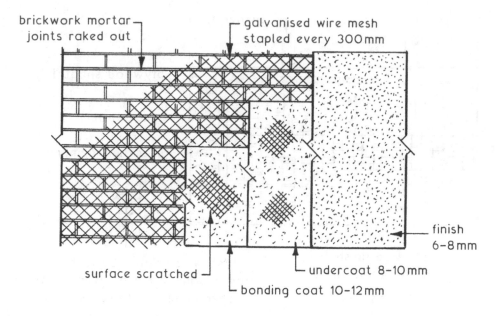

Figure 11.70 Three coat application to a masonry background

Textured ~ final layer finished with a coarse brush, toothed implement or a fabric roller (10 to 12mm with 3mm surface treated).

Rough-cast ~ irregular finish resulting from throwing the final coat onto the wall (6 to 10mm).

Pebble or dry dash ~ small stones thrown onto a strong mortar finishing coat (10 to 12mm).

Ref. BS EN 13914–1: *Design, Preparation and Application of External Rendering and Internal Plastering.*

Polymer renders

Polymer modified renders are either in factory-premixed 20 or 25kg bags for the addition of water prior to application or ready mixed in 20litre tubs. The composition is polymeric additives with the base materials of sand and cement, these may include latexes or emulsions, water soluble polymer powders and resins and anti-crack fibres as reinforcement. These renders are thinner than traditional renders and need a smooth substrate to achieve a flat appearance.

Rendering

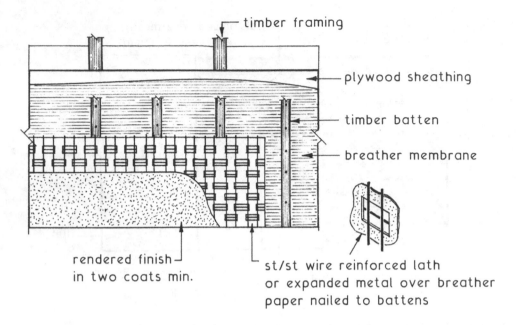

Figure 11.71 Render application to a timber framed background

Application

Two-coat work for irregular surfaces and refurbishment, 8–10mm base coat reinforced with galvanised or stainless steel mesh secured to the wall. Woven glass-fibre mesh can also be used, generally embedded in the base coat over existing cracks or damaged areas.

Finish

6–8mm over scratched base coat. Surface treatments as described for traditional renders.

Properties:

- Enhanced weatherproofing.
- Vapour permeable.
- Good resistance to impact.
- Inherent adhesion.
- Consistent mix produced in quality controlled factory conditions.
- Suitable for a variety of backgrounds.

11.17 Claddings

The external walls of block or timber frame construction can be clad with tiles, timber boards or plastic board sections.

Timber cladding

Timber boards such as match boarding and shiplap can be fixed vertically to horizontal battens or horizontally to vertical battens. Plastic moulded board claddings can be applied in a similar manner. The battens to which the claddings are fixed should be treated with a preservative against fungi and beetle attack and should be fixed with corrosion resistant nails.

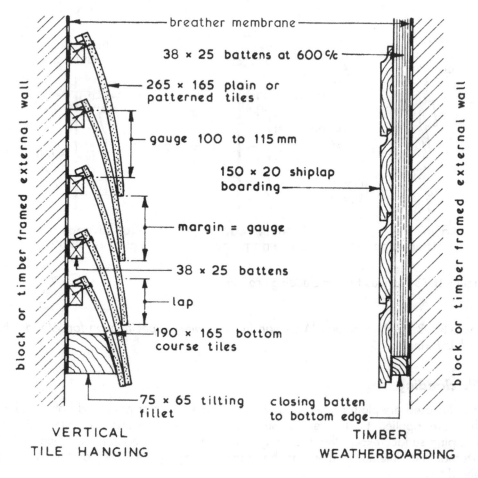

VERTICAL
TILE HANGING

TIMBER
WEATHERBOARDING

Figure 11.72 Typical details

Claddings

Shrinkage

Timber is subject to natural movement. With slender cladding sections, allowance must be made to accommodate this by drying to an appropriate moisture content before fixing, otherwise gaps will open between adjacent boards.

Fixing

Round head galvanised or stainless steel nails will avoid corrosion and metal staining. Lost head nails should be avoided as these can pull through. Annular ring shank nails provide extra grip. Dense timbers such as Siberian larch, Douglas fir and hardwoods should be pre-drilled 2mm over nail diameter. Double nailing may be required in very exposed situations.

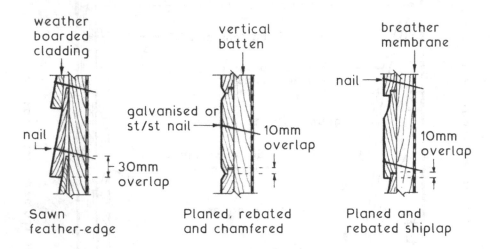

Figure 11.73 Various timber cladding profiles

BS 1186–3: *Timber for and Workmanship in Joinery. Specification for Wood Trim and its Fixing.*

Tile cladding

The tiles used are plain roofing tiles with either a straight or patterned bottom edge. They are applied to the vertical surface in the same manner as tiles laid on a sloping surface except that the gauge can be wider and each tile is twice nailed. External and internal angles can be formed using special tiles or they can be mitred.

Brick slips

These are thin bricks resembling tiles of 15 to 20mm thickness with face dimensions the same as standard bricks and a finish or texture to match. Originally slips were cut from whole bricks, now most brick manufacturers make slips to complement their stock items.

Other terminology includes brick tiles, brick cladding units, slip bricks and briquettes.

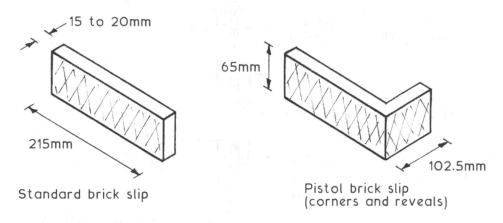

Standard brick slip

Pistol brick slip
(corners and reveals)

Figure 11.74 Typical brick slip details

Brick slips are used to maintain continuity of appearance where brickwork infill panels are supported by a reinforced concrete frame. Other uses are as a cost effective dressing to inexpensive or repaired backgrounds and for use where it is too difficult to build with standard bricks.

Claddings

Support system

Profiled galvanised or stainless steel mesh secured with screws and nylon discs 300mm horizontally and 150mm vertically apart into a sound background. Slips are stuck to the mesh with a structural adhesive of epoxy mortar, a two-part mix of resin and hardener combined with sand filler.

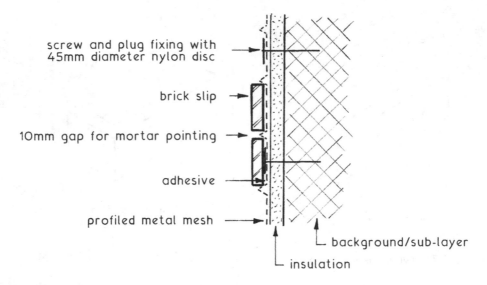

Figure 11.75 Brick slip support system

12 PITCHED ROOFS CLASSIFICATION

OPENING REMARKS
STRUCTURE
STABILITY
UNDERLAYS
DOUBLE LAP TILES
SINGLE LAP TILES
RIDGES AND HIPS
SLATES
FIBRE CEMENT SLATES
ROOF VENTILATION
ROOF INSULATION
VAPOUR CONTROL LAYER (VCL)
STRUCTURAL INSULATED PANELS (SIPS)
THATCHED ROOFS
STEEL ROOF TRUSSES
DOUBLE SKIN, ENERGY ROOF SYSTEMS
LONG SPAN ROOFS
MEMBRANE ROOF
FABRIC
ROOFLIGHTS

Opening remarks

Roofs with slopes greater than 10° are classed as pitched roofs. Pitches greater than 70° are classified as walls.

Roofs can be designed in many different forms and in combinations of these forms. Primary functions of any roof are to:

1. Provide an adequate barrier to the penetration of the elements.
2. Maintain the internal environment by providing an adequate resistance to heat loss.

A roof is in a very exposed situation and must therefore be designed and constructed in such a manner as to:

1. Safely resist all imposed loadings such as snow and wind.
2. Be capable of accommodating thermal and moisture movements.
3. Be durable so as to give a satisfactory performance and reduce maintenance to a minimum.

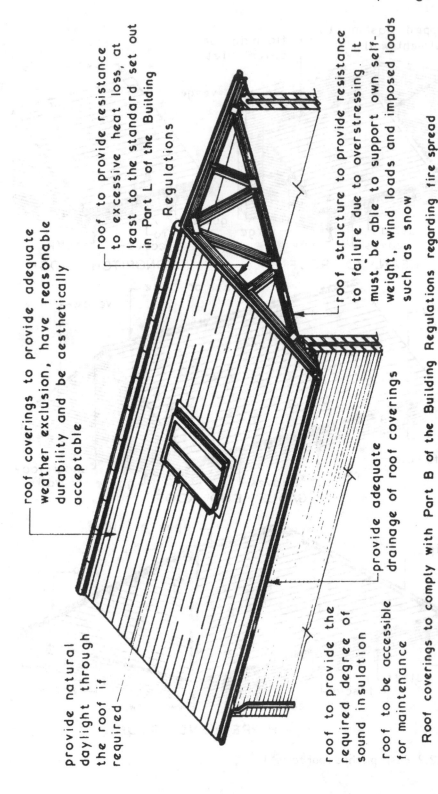

roof to provide resistance to excessive heat loss, at least to the standard set out in Part L of the Building Regulations

roof structure to provide resistance to failure due to overstressing. It must be able to support own self-weight, wind loads and imposed loads such as snow

roof coverings to provide adequate weather exclusion, have reasonable durability and be aesthetically acceptable

provide adequate drainage of roof coverings

Roof coverings to comply with Part B of the Building Regulations regarding fire spread

provide natural daylight through the roof if required

roof to provide the required degree of sound insulation

roof to be accessible for maintenance

Figure 12.1 Roof performance requirements

Opening remarks

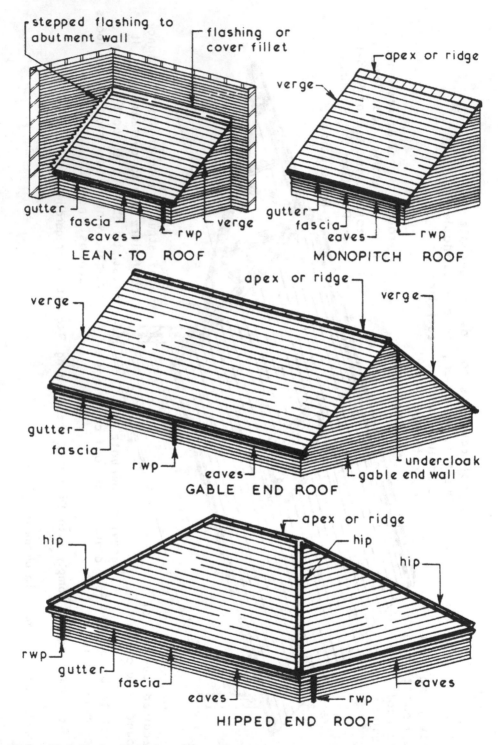

Figure 12.2 Basic pitched roof forms 1

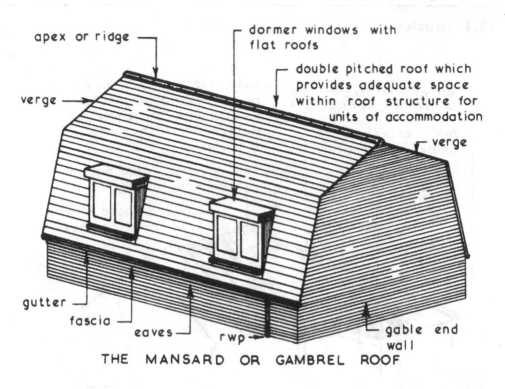

apex or ridge

verge

dormer windows with flat roofs

double pitched roof which provides adequate space within roof structure for units of accommodation

verge

gutter

fascia

eaves

rwp

gable end wall

THE MANSARD OR GAMBREL ROOF

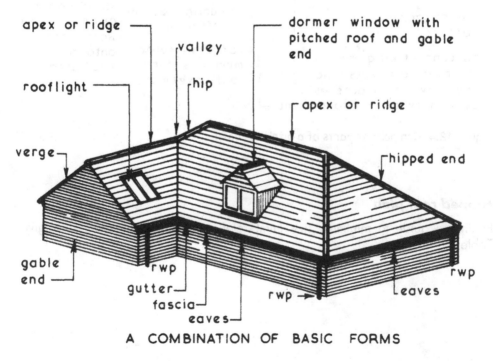

apex or ridge

valley

hip

rooflight

dormer window with pitched roof and gable end

apex or ridge

verge

hipped end

gable end

rwp

gutter

fascia

eaves

rwp

rwp

eaves

A COMBINATION OF BASIC FORMS

Figure 12.3 Basic roof forms 2

12.1 Structure

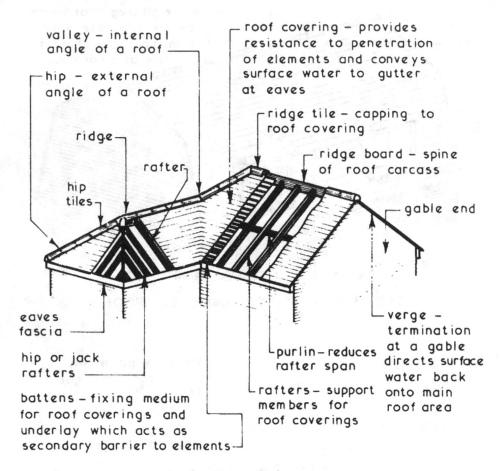

valley – internal angle of a roof

hip – external angle of a roof

ridge

rafter

hip tiles

roof covering – provides resistance to penetration of elements and conveys surface water to gutter at eaves

ridge tile – capping to roof covering

ridge board – spine of roof carcass

gable end

eaves fascia

hip or jack rafters

battens – fixing medium for roof coverings and underlay which acts as secondary barrier to elements

purlin – reduces rafter span

rafters – support members for roof coverings

verge – termination at a gable directs surface water back onto main roof area

Figure 12.4 Component parts of a pitched roof

Hipped roof plan

Purlins: a guide to minimum size (mm) relative to span and spacing is give in Table 12.1.

Table 12.1 Timber purlin span

Span (m)	Spacing (m)		
	1.75	2.25	2.75
2.0	125 × 75	150 × 75	150 × 100
2.5	150 × 75	175 × 75	175 × 100
3.0	175 × 100	200 × 100	200 × 125
3.5	225 × 100	225 × 100	225 × 125
4.0	225 × 125	250 × 125	250 × 125

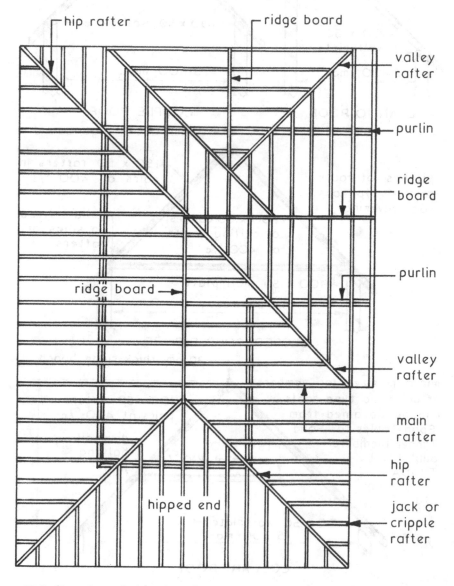

Figure 12.5 Plan view of hipped roof

Types of pitched roof

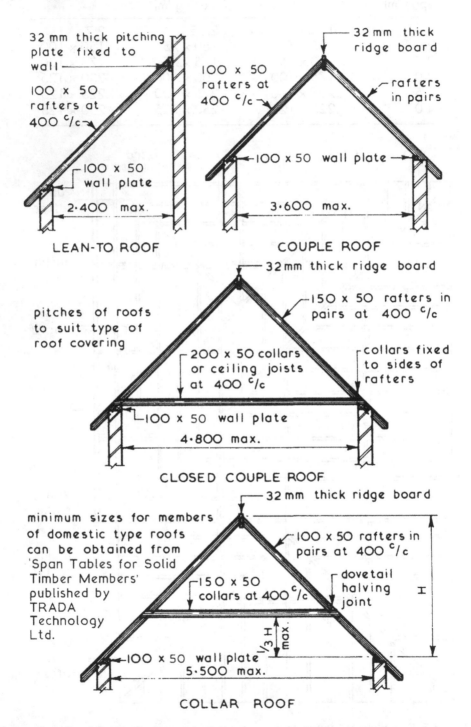

32 mm thick pitching plate fixed to wall

100 x 50 rafters at 400 ᶜ/c

100 x 50 wall plate

2·400 max.

LEAN-TO ROOF

32 mm thick ridge board

100 x 50 rafters at 400 ᶜ/c

rafters in pairs

100 x 50 wall plate

3·600 max.

COUPLE ROOF

32mm thick ridge board

150 x 50 rafters in pairs at 400 ᶜ/c

pitches of roofs to suit type of roof covering

200 x 50 collars or ceiling joists at 400 ᶜ/c

collars fixed to sides of rafters

100 x 50 wall plate

4·800 max.

CLOSED COUPLE ROOF

32 mm thick ridge board

minimum sizes for members of domestic type roofs can be obtained from 'Span Tables for Solid Timber Members' published by TRADA Technology Ltd.

100 x 50 rafters in pairs at 400 ᶜ/c

150 x 50 collars at 400 ᶜ/c

dovetail halving joint

H

1/3 H max.

100 x 50 wall plate

5·500 max.

COLLAR ROOF

Figure 12.6 Roof forms for spans to 5.5m

Traditional 'cut roof' elements

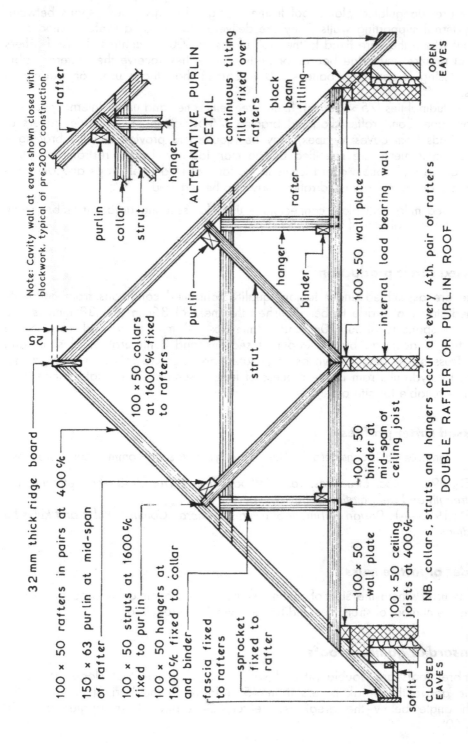

Note: Cavity wall at eaves shown closed with blockwork, typical of pre-2000 construction.

rafter

purlin

collar

strut

hanger

ALTERNATIVE PURLIN DETAIL

continuous tilting fillet fixed over rafters

block beam filling

OPEN EAVES

rafter

hanger

binder

100 × 50 wall plate

internal load bearing wall

32 mm thick ridge board

100 × 50 rafters in pairs at 400 c/c

150 × 63 purlin at mid-span of rafter

100 × 50 struts at 1600 c/c fixed to purlin

100 × 50 hangers at 1600 c/c fixed to collar and binder

fascia fixed to rafters

sprocket fixed to rafter

25

100 × 50 collars at 1600 c/c fixed to rafters

purlin

strut

100 × 50 binder at mid-span of ceiling joist

100 × 50 wall plate

100 × 50 ceiling joists at 400 c/c

soffit

CLOSED EAVES

NB. collars, struts and hangers occur at every 4th pair of rafters

DOUBLE RAFTER OR PURLIN ROOF

Figure 12.7 Section details of purlin and strutting

377

Structure

Roof trusses

These are triangulated plane roof frames designed to give clear spans between the external supporting walls. They are delivered to site as a prefabricated component where they are fixed to the wall plates at 600mm centres. Trussed rafters do not require any ridge boards or purlins since they receive their lateral stability by using larger tiling battens (50 × 25mm) than those used on traditional roofs.

Longitudinal ties (75 × 38mm), fixed over ceiling ties and under internal ties near to roof apex, and rafter diagonal bracing (75 × 38mm), fixed under rafters at gable ends from eaves to apex, may be required to provide stability bracing – actual requirements are specified by the manufacturer. Lateral restraint to gable walls at top and bottom chord levels in the form of mild steel straps at 2.0m maximum centres over two trussed rafters may also be required.

Galvanised mild steel plate connectors ~ used for securing adjacent timber members in trussed rafter roof frames.

Trussed rafter production

These are assembled under factory quality controlled conditions from uniformly planed structural grade timber. Finished thickness of 35mm (ex. 38 mm) is suitable for spans to about 10m and 47mm (ex. 50 mm) to around 15m. Some variation is achieved by the depth of sections and configuration of struts and ties. Truss plates (also known as nail plates) are pressure rolled or pressed into place by hydraulic ram in the process of truss assembly. Truss plate connections are not suitable for site assembly.

Trussed rafter profiles

The design potential is considerable. The following shows some common configurations.

BS EN 595: *Timber Structures. Test Methods. Test of Trusses for the Determination of Strength and Deformation Behaviour.*
BS EN 1995–1-1: *Design of Timber Structures. General. Common Rules and Rules for Buildings.*

Girder or lattice truss

This is used where the size of purlins would be uneconomically large and/or an excessive number of struts, ties and hangers would be required.

Mansard or gambrel roofs

Gambrel roofs are double pitched with a break in the roof slope. The pitch angle above the break is less than 45° relative to the horizontal, whilst the pitch angle below the break is greater. Generally, these angles are 30° and 60°.

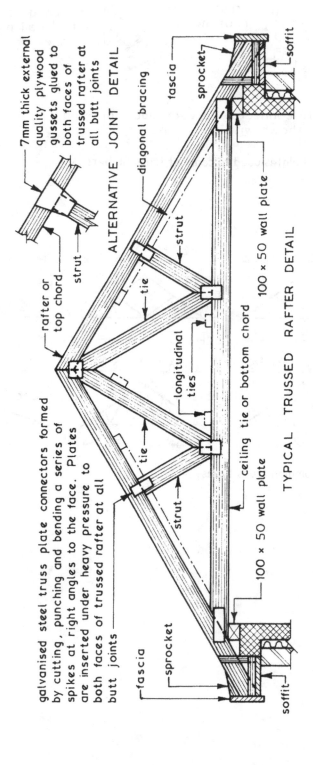

galvanised steel truss plate connectors formed by cutting, punching and bending a series of spikes at right angles to the face. Plates are inserted under heavy pressure to both faces of trussed rafter at all butt joints

7mm thick external quality plywood gussets glued to both faces of trussed rafter at all butt joints

ALTERNATIVE JOINT DETAIL

rafter or top chord

strut

diagonal bracing

fascia

sprocket

soffit

strut

tie

100 × 50 wall plate

longitudinal ties

tie

ceiling tie or bottom chord

strut

100 × 50 wall plate

TYPICAL TRUSSED RAFTER DETAIL

fascia

sprocket

soffit

Figure 12.8 Trussed rafter features

379

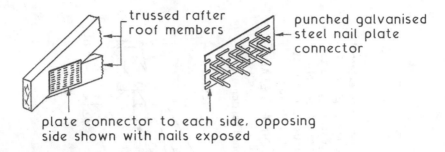

Figure 12.9 Nail plates used to connect truss timbers

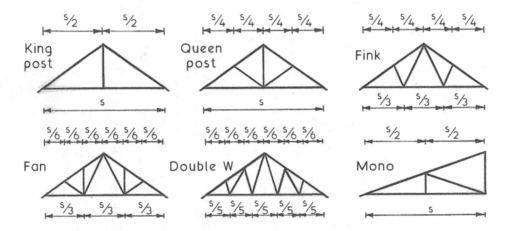

Figure 12.10 Common truss types

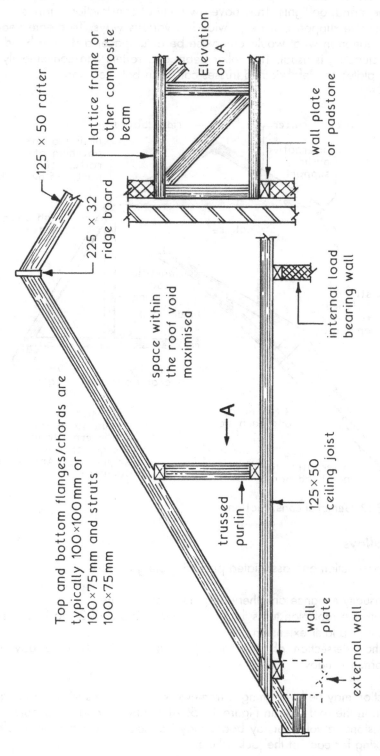

More economical for larger spans than with double rafters and ordinary purlins. Used where the size of purlins would be uneconomically large and/or an excessive number of struts, ties and hangers would be required

125 × 50 rafter

lattice frame or other composite beam

Elevation on A

wall plate or padstone

225 × 32 ridge board

space within the roof void maximised

internal load bearing wall

Top and bottom flanges/chords are typically 100×100mm or 100×75mm and struts 100×75mm

A

trussed purlin

125 × 50 ceiling joist

wall plate

external wall

The lattice frame can be faced with plasterboard or with plywood for additional strength.

Figure 12.11 Lattice or girder trusses

Structure

Gambrels are useful in providing more attic headroom and frequently incorporate dormers and rooflights. They have a variety of constructional forms.

Intermediate support can be provided in various ways. To create headroom for accommodation in what would otherwise be attic space, a double head plate and partition studding is usual. The collar beam and rafters can conveniently locate on the head plates or prefabricated trusses can span between partitions.

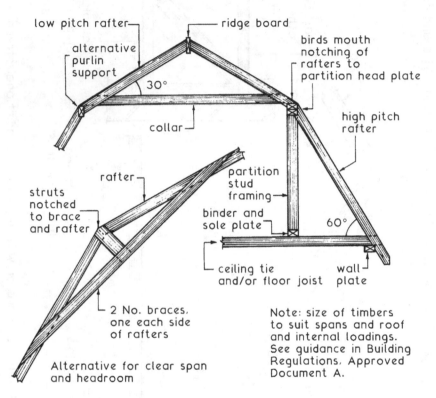

Figure 12.12 Gambrel construction

Roof valleys

Valley construction and associated pitched roofing is used:

- To visually enhance an otherwise plain roof structure.
- Where the roof plan turns through an angle (usually 90°) to follow the building layout or a later extension.
- At the intersection of main and projecting roofs above a bay window or a dormer window.

Construction may be by forming a framework of cut rafters trimmed to valley rafters as shown in the roof plan in Figure 12.5. Alternatively, and as favoured with building extensions, a valley or lay board may be located over the main rafters to provide a fixing for each of the jack rafters.

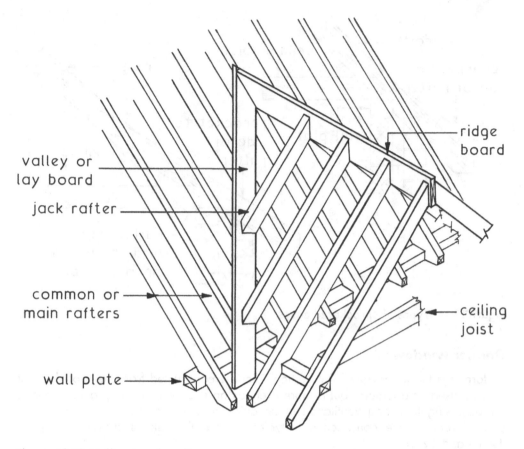

valley or
lay board

jack rafter

common or
main rafters

wall plate

ridge
board

ceiling
joist

Figure 12.13 Valley construction

Sprocketed eaves

Sprockets may be provided at the eaves to reduce the slope of a pitched roof. Sprockets are generally most suitable for use on wide, steeply pitched roofs to:

- Enhance the roof profile by creating a feature.
- To slow down the velocity of rainwater running off the roof and prevent it from overshooting the gutter.

Where the rafters overhang the external wall, taper cut timber sprockets can be attached to the top of the rafters. Alternatively, the ends of rafters can be birds-mouthed onto the wall plate and short lengths of timber the same size as the rafters secured to the rafter feet. In reducing the pitch angle, albeit for only a short distance, it should not be less than the minimum angle recommended for specific roof coverings.

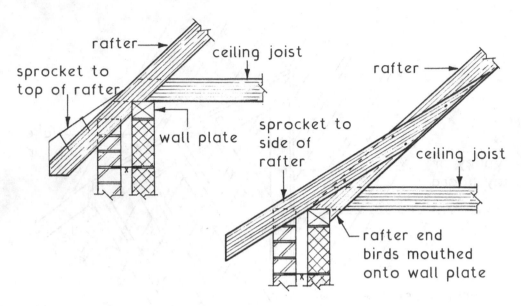

Figure 12.14 Types of sprocketed eaves

Dormer windows

A dormer is the framework for a vertical window constructed from the roof slope. It may be used as a feature, but is more likely as an economical and practical means for accessing light and ventilation in an attic room. Dormers can have a flat or pitched roof. Frame construction is typical of the illustrations shown in Figures 12.15 and 12.16.

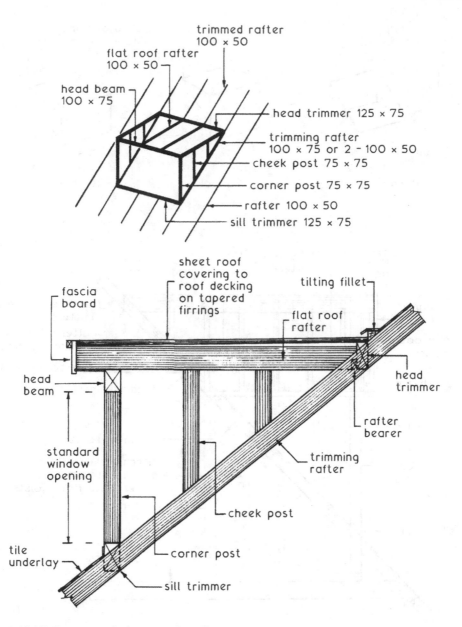

trimmed rafter
100 × 50

flat roof rafter
100 × 50

head beam
100 × 75

head trimmer 125 × 75

trimming rafter
100 × 75 or 2 – 100 × 50

cheek post 75 × 75

corner post 75 × 75

rafter 100 × 50

sill trimmer 125 × 75

sheet roof
covering to
roof decking
on tapered
firrings

tilting fillet

fascia
board

flat roof
rafter

head
beam

head
trimmer

rafter
bearer

standard
window
opening

trimming
rafter

cheek post

tile
underlay

corner post

sill trimmer

Figure 12.15 Dormer window construction

Structure

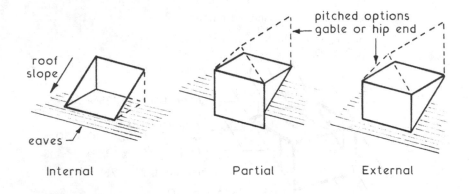

pitched options
gable or hip end

roof
slope

eaves

Internal   Partial   External

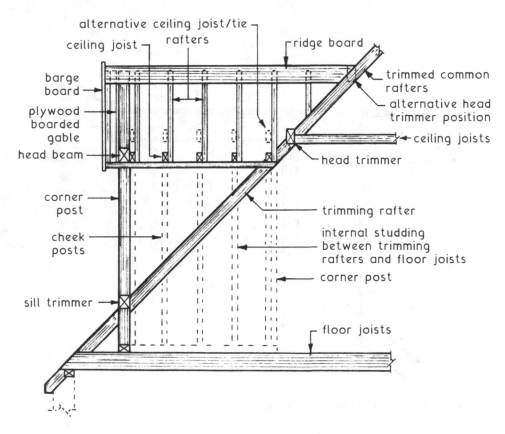

Figure 12.16 Section through gable ended external dormer

12.2 Stability

Stability of gable walls and construction at the eaves, plus integrity of the roof structure during excessive wind forces, requires lateral restraint. This can be achieved with 30 × 5mm cross-sectional area galvanised steel straps bracing the external walls to the roof structure.

Exceptions may occur if the following apply:

1. Exceeds 15° pitch.
2. Is tiled or slated.
3. Has the type of construction known locally to resist gusts.
4. Has ceiling joists and rafters bearing onto support walls at not more than 12m centres.

Application

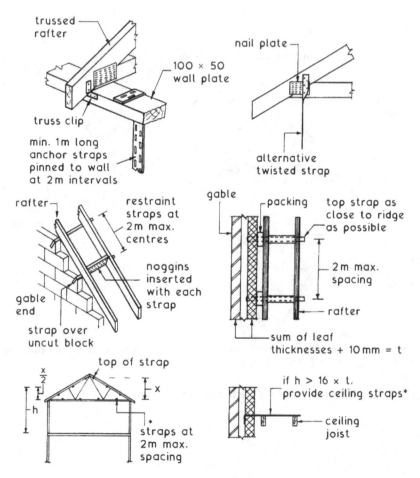

Figure 12.17 Lateral restraint methods to roofs

12.3 Underlays

Roof underlays or sarking felts under the roof tiles provide a barrier to the entry of snow, wind and rain blown between the tiles or slates. It also prevents the entry of water from capillary action.

Types

Bitumen fibre-based felts, supplied in rolls 1m wide, are traditionally used in house construction with a cold ventilated roof.

A breather membrane is an alternative to conventional bituminous felt as an under-tiling layer. It has the benefit of restricting liquid water penetration whilst allowing water vapour transfer from within the roof space. This permits air circulation without perforating the under-tiling layer. Underlay of this type should be installed taut with minimal sag across the rafters, with counter battens support to the tile battens. Where counter battens are not used, underlay should sag slightly between rafters to allow rain penetration to flow under tile battens.

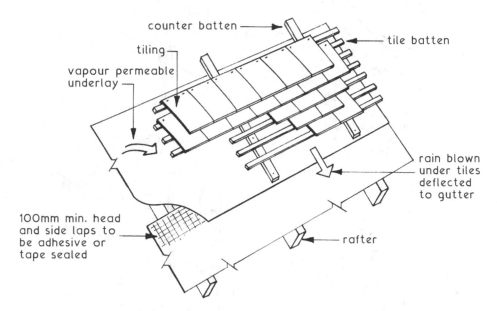

Underlays are fixed initially with galvanised clout nails or st/st staples but are finally secured with the tiling or slating batten fixings

Figure 12.18 Roof underlay details with counter battens

12.4 Double lap tiles

These are the traditional tile covering for pitched roofs made from clay, also known as plain tiles. Plain tiles have a slight camber in their length to ensure that the tail of the tile will bed and not ride on the tile below. There are always at least two layers of tiles covering any part of the roof. Each tile has at least two nibs on the underside of its head so that it can be hung on support battens nailed over the rafters. Two nail holes provide the means of fixing the tile to the batten. Traditionally, only every fourth course of tiles is nailed unless the roof exposure is high. Double lap tiles are laid to a bond so that the edge joints between the tiles are in the centre of the tiles immediately below and above the course under consideration.

Limitation for use of double lap tiles ~ minimum pitch 35° machine-made, 45° hand-made.

For other types, shapes and sections see BS EN 1304: *Clay Roofing Tiles and Fittings. Product Definitions and Specifications.*

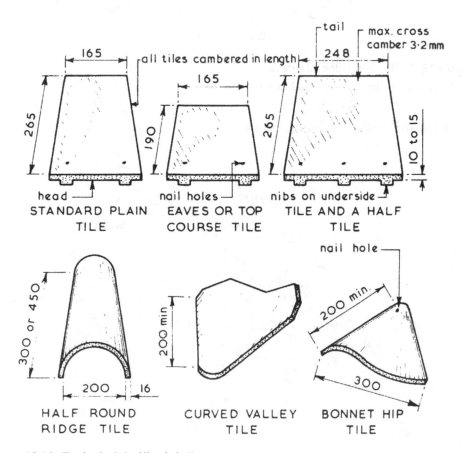

Figure 12.19 Typical plain tile details

Double lap tiles

Ridge details

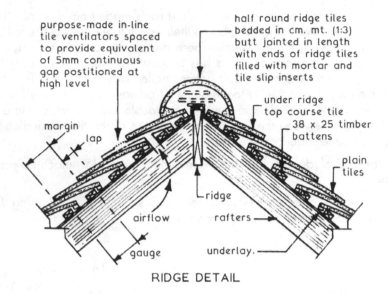

purpose-made in-line
tile ventilators spaced
to provide equivalent
of 5mm continuous
gap postitioned at
high level

half round ridge tiles
bedded in cm. mt. (1:3)
butt jointed in length
with ends of ridge tiles
filled with mortar and
tile slip inserts

under ridge
top course tile
38 x 25 timber
battens

plain
tiles

margin

lap

ridge

airflow

rafters

gauge

underlay.

RIDGE DETAIL

Figure 12.20 Typical ridge tile detail
Note: 38 x 25 timber battens (see Note 1)

Eaves detail with soffit ventilation

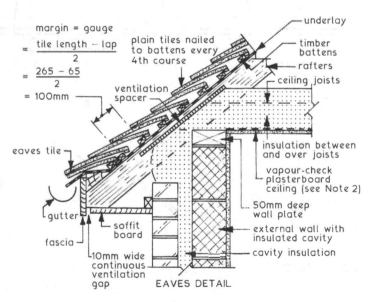

margin = gauge

$$= \frac{\text{tile length} - \text{lap}}{2}$$

$$= \frac{265 - 65}{2}$$

$$= 100mm$$

plain tiles nailed
to battens every
4th course

ventilation
spacer

eaves tile

gutter

fascia

soffit
board

10mm wide
continuous
ventilation
gap

underlay

timber
battens

rafters

ceiling joists

insulation between
and over joists

vapour-check
plasterboard
ceiling (see Note 2)

50mm deep
wall plate

external wall with
insulated cavity

cavity insulation

EAVES DETAIL

Figure 12.21 Typical eaves ventilation detail
Note 1: 50 x 25 where rafter spacing is 600mm.
Note 2: Through ventilation is necessary to prevent condensation from occurring in the
roof space. A vapour check can also help limit the amount of moisture entering the roof void.

Eaves detail with facia ventilation

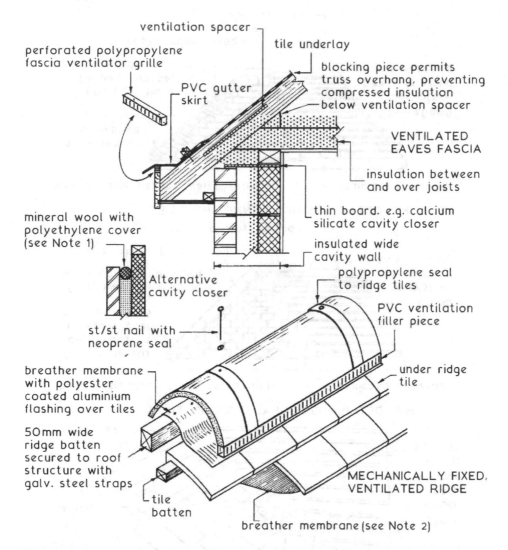

ventilation spacer

perforated polypropylene
fascia ventilator grille

tile underlay

blocking piece permits
truss overhang, preventing
compressed insulation
below ventilation spacer

PVC gutter
skirt

VENTILATED
EAVES FASCIA

insulation between
and over joists

mineral wool with
polyethylene cover
(see Note 1)

thin board. e.g. calcium
silicate cavity closer

insulated wide
cavity wall

Alternative
cavity closer

polypropylene seal
to ridge tiles

PVC ventilation
filler piece

st/st nail with
neoprene seal

breather membrane
with polyester
coated aluminium
flashing over tiles

under ridge
tile

50mm wide
ridge batten
secured to roof
structure with
galv. steel straps

MECHANICALLY FIXED,
VENTILATED RIDGE

tile
batten

breather membrane (see Note 2)

Figure 12.22 Typical over fascia and ridge ventilation details

Note 1: If a cavity closer is also required to function as a cavity barrier to prevent fire
spread, it should provide at least 30 minutes' fire resistance (B. Reg. A.D. B3 Section 6
[Vol. 1] and 9 [Vol. 2]).

Note 2: A breather membrane is an alternative to conventional bituminous felt as an
under-tiling layer. It has the benefit of restricting liquid water penetration whilst allowing
water vapour transfer from within the roof space. This permits air circulation without
perforating the under-tiling layer.

Double lap tiles

Hip and valley details

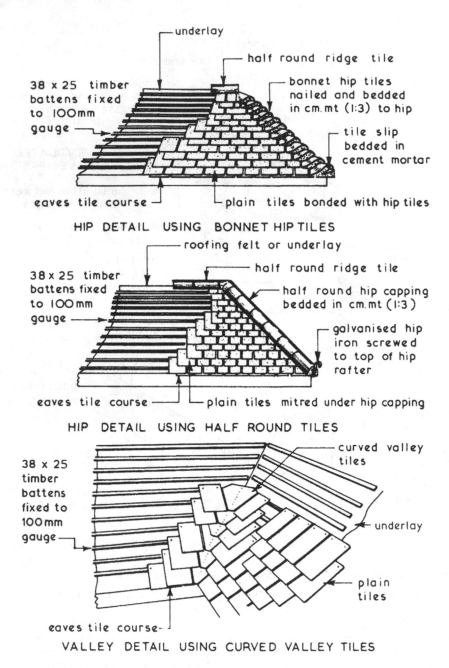

underlay

half round ridge tile

38 x 25 timber battens fixed to 100mm gauge

bonnet hip tiles nailed and bedded in cm. mt (1:3) to hip

tile slip bedded in cement mortar

eaves tile course

plain tiles bonded with hip tiles

HIP DETAIL USING BONNET HIP TILES

roofing felt or underlay

half round ridge tile

38 x 25 timber battens fixed to 100 mm gauge

half round hip capping bedded in cm. mt (1:3)

galvanised hip iron screwed to top of hip rafter

eaves tile course

plain tiles mitred under hip capping

HIP DETAIL USING HALF ROUND TILES

curved valley tiles

38 x 25 timber battens fixed to 100 mm gauge

underlay

plain tiles

eaves tile course

VALLEY DETAIL USING CURVED VALLEY TILES

Figure 12.23 Typical hip and valley details

392

Abutment and verge details

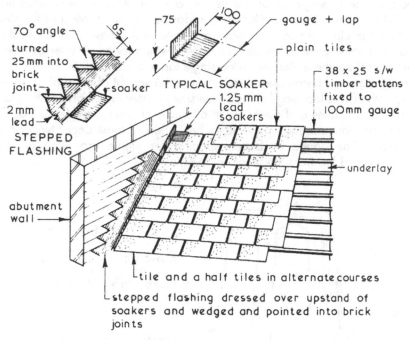

70° angle
turned
25mm into
brick
joint

65

soaker

2mm
lead

STEPPED
FLASHING

75

100

TYPICAL SOAKER

gauge + lap

plain tiles

38 x 25 s/w
timber battens
fixed to
100mm gauge

1.25 mm
lead
soakers

underlay

abutment
wall

tile and a half tiles in alternate courses

stepped flashing dressed over upstand of
soakers and wedged and pointed into brick
joints

ABUTMENT DETAIL

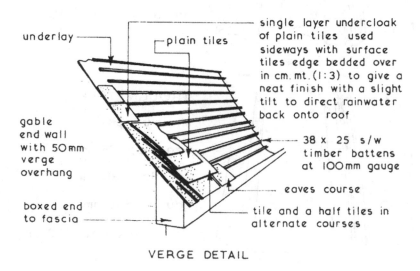

underlay

plain tiles

single layer undercloak
of plain tiles used
sideways with surface
tiles edge bedded over
in cm. mt. (1:3) to give a
neat finish with a slight
tilt to direct rainwater
back onto roof

gable
end wall
with 50mm
verge
overhang

38 x 25 s/w
timber battens
at 100mm gauge

eaves course

boxed end
to fascia

tile and a half tiles in
alternate courses

VERGE DETAIL

Figure 12.24 Typical verge and abutment details
Note: All verge course tiles are secured to battens and bedded in mortar.

12.5 Single lap tiles

So-called because the single lap of one tile over another provides the weather tightness, as opposed to the two layers of tiles used in double lap tiling. Most of the single lap tiles produced in clay and concrete have a tongue and groove joint along their side edges, and in some patterns on all four edges, which forms a series of interlocking joints and, therefore, these tiles are called single lap interlocking tiles.

Generally there will be an overall reduction in the weight of the roof covering when compared with double lap tiling but the batten size is larger than that used for plain tiles.

Commonly, a minimum of every tile in alternate courses is twice nailed. Contemporary specifications require every tile to be nailed at least once, twice in peripheral locations. The gauge or batten spacing for single lap tiling is found by subtracting the end lap from the length of the tile.

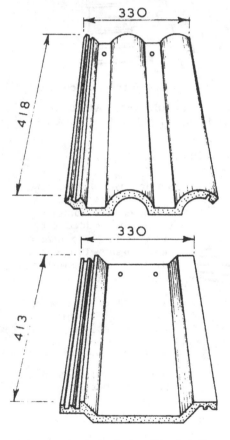

ROLL TYPE TILE

minimum pitch 30°

head lap 75 mm

side lap 30 mm

gauge 343 mm

linear coverage 300 mm

TROUGH TYPE TILE

minimum pitch 15°

head lap 75 mm

side lap 38 mm

gauge 338 mm

linear coverage 292 mm

Figure 12.25 Typical single lap tile details

Hips can be finished with a half round tile.

Valleys can be finished by using special valley trough tiles or with a lead lined gutter – see manufacturer's data.

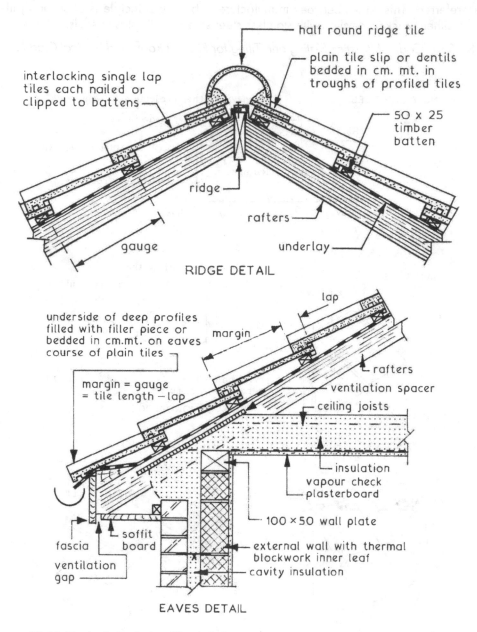

RIDGE DETAIL

half round ridge tile

plain tile slip or dentils bedded in cm. mt. in troughs of profiled tiles

interlocking single lap tiles each nailed or clipped to battens

50 x 25 timber batten

ridge

rafters

gauge

underlay

lap

margin

underside of deep profiles filled with filler piece or bedded in cm.mt. on eaves course of plain tiles

margin = gauge = tile length – lap

rafters

ventilation spacer

ceiling joists

insulation
vapour check
plasterboard

100 × 50 wall plate

external wall with thermal blockwork inner leaf
cavity insulation

fascia

soffit board

ventilation gap

EAVES DETAIL

Figure 12.26 Typical single lap tile details

12.6 Ridges and hips

The traditional method of laying and bedding of ridge and hip tiles, in cement and sand mortar alone, is not a secure means of fixing. A mechanical method of fixing is preferred. This varies between manufacturers, but the principle is of securing tiles with either brackets or clips with stainless steel screws with plastic seals.

BS 5534: *Code of Practice Slating and Tiling for Pitched Roofs and Vertical Cladding.*

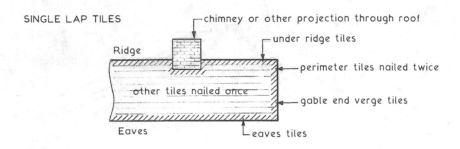

SINGLE LAP TILES

chimney or other projection through roof

under ridge tiles

Ridge

perimeter tiles nailed twice

other tiles nailed once

gable end verge tiles

Eaves

eaves tiles

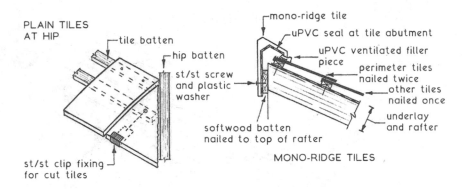

PLAIN TILES AT HIP

tile batten

hip batten

st/st screw and plastic washer

st/st clip fixing for cut tiles

mono-ridge tile

uPVC seal at tile abutment

uPVC ventilated filler piece

perimeter tiles nailed twice

other tiles nailed once

underlay and rafter

softwood batten nailed to top of rafter

MONO-RIDGE TILES

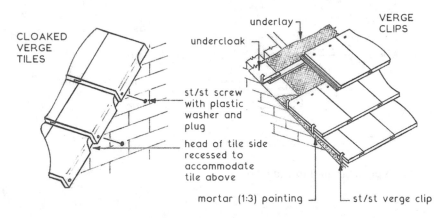

CLOAKED VERGE TILES

undercloak

underlay

VERGE CLIPS

st/st screw with plastic washer and plug

head of tile side recessed to accommodate tile above

mortar (1:3) pointing

st/st verge clip

Figure 12.27 Verge clips and cloaked verge tiles

12.7 Slates

Slate is a natural dense material which can be split into thin sheets and cut to form a small unit covering suitable for pitched roofs in excess of 25° pitch. Slates are graded according to thickness and texture, the thinnest being known as 'Bests'. These are of 4mm nominal thickness. Slates are laid to the same double lap principles as plain tiles. Ridges and hips are normally covered with half round or angular tiles whereas valley junctions are usually of mitred slates over soakers. Unlike plain tiles, every course is fixed to the battens by head or centre nailing.

The UK has been supplied with its own slate resources from quarries in Wales, Cornwall and Westmorland. Imported slate is also available from Spain, Argentina and parts of the Far East.

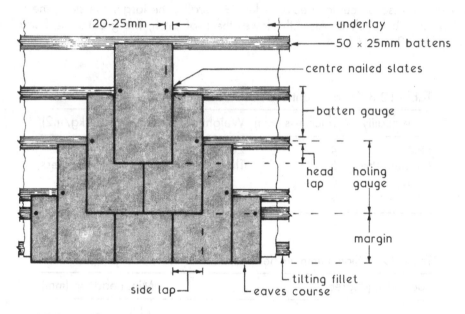

Figure 12.28   Slate lapping and coursing

Example

Countess slate, 510 × 255mm laid to a 30° pitch with 75mm head lap.
Batten gauge = (slate length – lap) ÷ 2 = (510–75) ÷ 2 = 218mm.
Holing gauge = batten gauge + head lap + 8 to 15mm,
= 218 + 75 + (8 to 15mm) = 301 to 308mm.
Side lap = 255 ÷ 2 = 127mm.
Margin = batten gauge of 218mm.
Eaves course length = head lap + margin = 293mm.

Traditional slate names and sizes (mm) are listed in Table 12.2.

Slates

Table 12.2 Slate types and dimensions

Empress	650 × 400	Wide Viscountess	460 × 255
Princess	610 × 355	Viscountess	460 × 230
Duchess	610 × 305	Wide Ladies	405 × 255
Small Duchess	560 × 305	Broad Ladies	405 × 230
Marchioness	560 × 280	Ladies	405 × 205
Wide Countess	510 × 305	Wide Headers	355 × 305
Countess	510 × 255	Headers	355 × 255
... ...	510 × 230	Small Ladies	355 × 203
... ...	460 × 305	Narrow Ladies	355 × 180

Sizes can also be cut to special order. Generally, the larger the slate, the lower the roof may be pitched. Also, the lower the roof pitch, the greater the head lap (see Tables 12.3 and 12.4).

Table 12.3 Slate weight and thickness

Slate quality	Thickness (mm)	Weight at 75mm head lap (kg/m2)
Best	4	26
Medium strong	5	Thereafter in proportion to thickness
Heavy	6	
Extra heavy	9	

Table 12.4 Slate minimum head laps for common roof pitches

Roof pitch (degrees)	Min. head lap (mm)
20	115
25	85
35	75
45	65

See also: BS EN 12326–1: *Slate and Stone for Discontinuous Roofing and External Cladding*.
BS 5534: *Slating and Tiling for Pitched Roofs and Vertical Cladding*.

Typical details

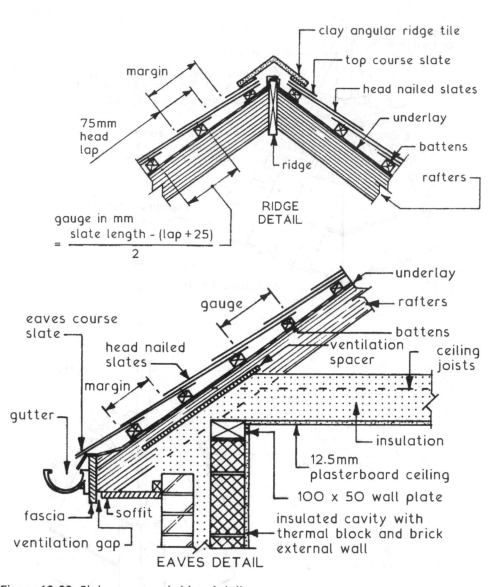

margin

75mm head lap

$$\text{gauge in mm} = \frac{\text{slate length} - (\text{lap} + 25)}{2}$$

clay angular ridge tile
top course slate
head nailed slates
underlay
battens
rafters

ridge

RIDGE DETAIL

eaves course slate

head nailed slates

margin

gutter

fascia

soffit

ventilation gap

EAVES DETAIL

underlay
rafters
battens
ventilation spacer
ceiling joists
insulation
12.5mm plasterboard ceiling
100 x 50 wall plate
insulated cavity with thermal block and brick external wall

gauge

Figure 12.29 Slate verge and ridge details
Note: Gauge for centre nailed slates = (slate length – lap) ÷ 2.

Slates

Roof hip examples

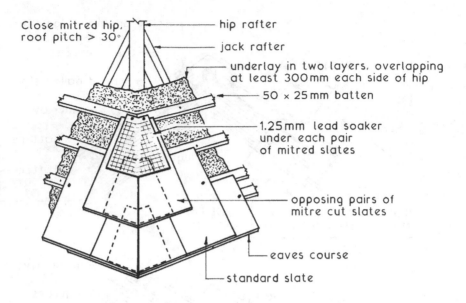

Close mitred hip, roof pitch > 30°

hip rafter

jack rafter

underlay in two layers, overlapping at least 300mm each side of hip

50 × 25mm batten

1.25mm lead soaker under each pair of mitred slates

opposing pairs of mitre cut slates

eaves course

standard slate

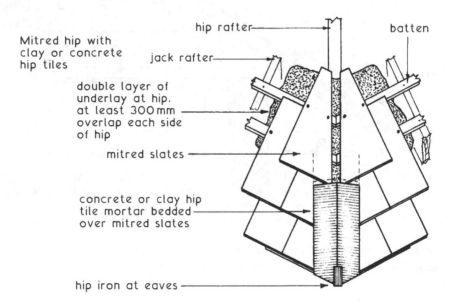

Mitred hip with clay or concrete hip tiles

hip rafter

batten

jack rafter

double layer of underlay at hip, at least 300mm overlap each side of hip

mitred slates

concrete or clay hip tile mortar bedded over mitred slates

hip iron at eaves

Figure 12.30 Cutting slates at hips

Roof valley examples

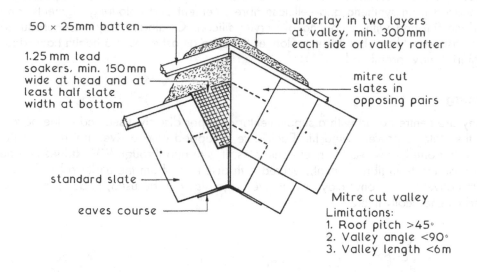

50 × 25mm batten

1.25 mm lead soakers, min. 150 mm wide at head and at least half slate width at bottom

underlay in two layers at valley, min. 300 mm each side of valley rafter

mitre cut slates in opposing pairs

standard slate

eaves course

Mitre cut valley
Limitations:
1. Roof pitch >45°
2. Valley angle <90°
3. Valley length <6 m

Alternatives

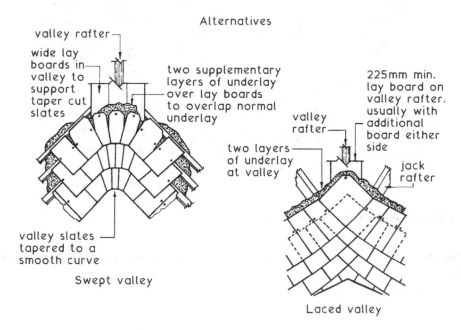

valley rafter

wide lay boards in valley to support taper cut slates

two supplementary layers of underlay over lay boards to overlap normal underlay

valley slates tapered to a smooth curve

Swept valley

valley rafter

two layers of underlay at valley

225 mm min. lay board on valley rafter, usually with additional board either side

jack rafter

Laced valley

Figure 12.31 Slate valley options
Note: In swept valleys, cut and tapered slates are interleaved with 1.25mm lead soakers.

12.8 Fibre cement slates

These are a manufactured alternative to natural slates, giving a similar appearance, produced from synthetic and cellulose fibres, cement and colouring pigments, and surface finished with a coating of acrylic silicon. Originally they were produced with asbestos fibres, but this is no longer the case as asbestos is a health hazardous material, now prohibited.

Fixing

They are centre nailed with a supplementary copper disc rivet secured to the bottom of the slate to prevent wind lift. Each slate is tapped onto a rivet that rests on the slate but one below. As the rivet penetrates, it is turned through 90°. Ridges can be purpose made in fibre cement, secured with disc rivets from the under ridge course. Alternatively, traditional clay or concrete ridge tiles can be used, bedded in mortar and mechanically fixed.

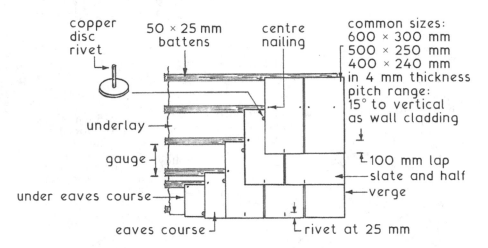

Figure 12.32 Fibre cement slate fixing

BS EN 492: *Fibre Cement Slates and Fittings. Production Specification and Test Methods.*

12.9 Roof ventilation

Air carries water vapour, the amount increasing proportionally with the air temperature. As the water vapour increases so does the pressure and this causes the vapour to migrate from warmer to cooler parts of a building. As the air temperature reduces, so does its ability to hold water and this manifests as condensation on cold surfaces. Insulation between living areas and roof spaces increases the temperature differential and potential for condensation in the roof void.

Condensation can be prevented by either of the following:

- Providing a vapour control layer on the warm side of any insulation.
- Removing the damp air by ventilating the colder area.

The most convenient form of vapour layer is vapour check plasterboard, which has a moisture resistant lining bonded to the back of the board. A typical patented product is a foil or metallised polyester backed plasterboard in 9.5 and 12.5mm standard thicknesses. This is most suitable where there are rooms in roofs and for cold deck flat roofs. Ventilation is appropriate to larger roof spaces.

Roof ventilation methods

Provision of eaves ventilation alone should allow adequate air circulation in most situations. However, in some climatic conditions and where the air movement is not directly at right angles to the building, moist air can be trapped in the roof apex. Therefore, supplementary ridge ventilation is recommended.

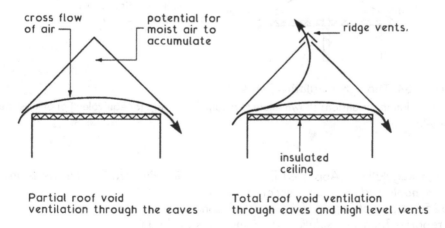

Figure 12.33 Ventilation paths through roofs

Roof ventilation

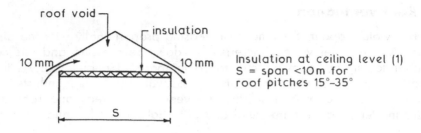

Insulation at ceiling level (1)
S = span <10m for
roof pitches 15°–35°

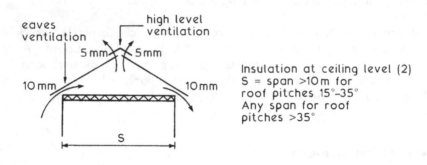

Insulation at ceiling level (2)
S = span >10m for
roof pitches 15°–35°
Any span for roof
pitches >35°

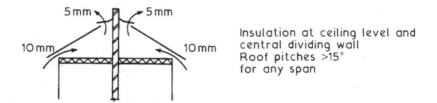

Insulation at ceiling level and
central dividing wall
Roof pitches >15°
for any span

Figure 12.34 Through ventilation paths

Note: Ventilation dimensions shown relate to a continuous strip (or equivalent) of at least the given gap.

Building Regulations, Approved Document C – *Site Preparation and Resistance to Contaminants and Moisture. Section 6 – Roofs.*
BS 5250: *Code of Practice for Control of Condensation in Buildings.*
BRE report – *Thermal Insulation: Avoiding Risks* (3rd edn).

Requirements

Permanent ventilation of a roof void can be achieved with:

- A breather or vapour permeable tile underlay.
- Open ventilation from continuous gaps at the eaves, with ventilation at the ridge in specific situations (see preceding pages). Purpose made tile vents may be used as an alternative, suitably spaced to provide equivalent ventilation.

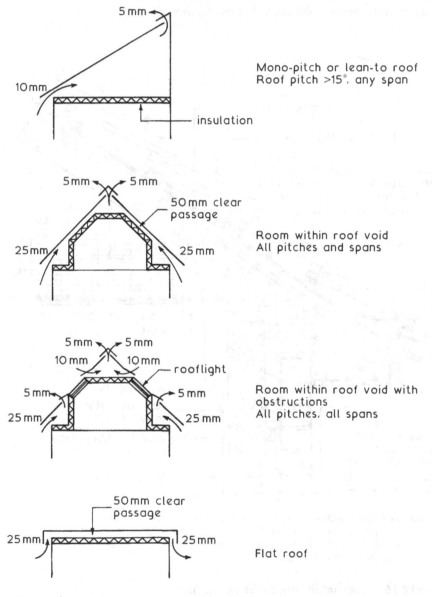

Figure 12.35 Cold and warm roof ventilation methods

12.10 Roof insulation

Insulating a roof will reduce heat loss and reduce the risk of condensation occurring.

Insulation may be placed between and over the ceiling joists, as shown in Figure 12.36, to produce a cold roof void. Where a roof space is used for habitable space, insulation must be provided within the roof slope, between or above the rafters – this is known as a 'warm roof'.

Insulation above the rafters eliminates the need for continuous ventilation. Insulation placed between the rafters requires a continuous 50mm ventilation void above the insulation to assist in the control of condensation.

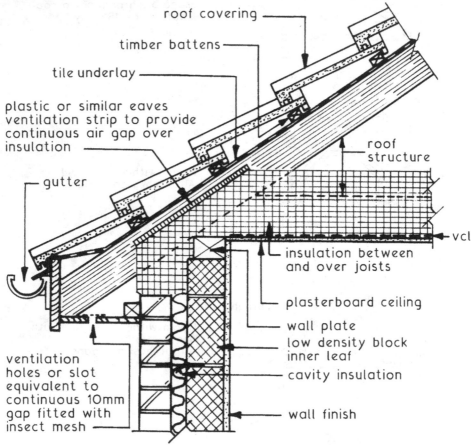

NB. All pipework in roof space should be insulated to prevent frost attack. The sides and top of cold water storage cisterns should be insulated to prevent freezing.

Figure 12.36 Eaves detail with soffit ventilation

Application of insulation with typical U-values

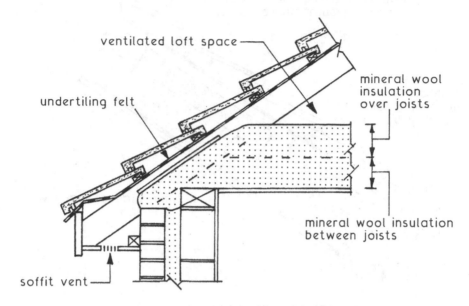

Figure 12.37 Cold roof

Table 12.5 Typical cold roof insulation U values

Insulation (mm) between/over joists		U-value (W/m²K)
100	80	0.22
100	100	0.20
100	150	0.16
100	200	0.13
100	250	0.11
100	300	0.10

Note: 200, 250 and 300mm over joist insulation in two layers.

Roof insulation

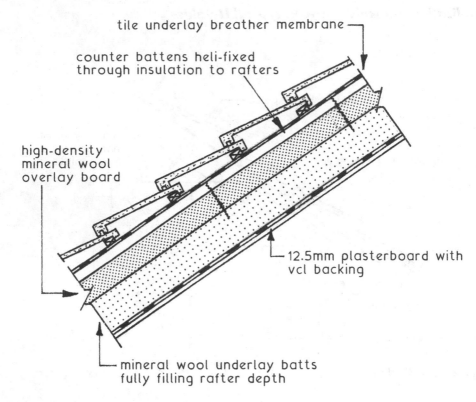

tile underlay breather membrane

counter battens heli-fixed
through insulation to rafters

high-density
mineral wool
overlay board

12.5mm plasterboard with
vcl backing

mineral wool underlay batts
fully filling rafter depth

Figure 12.38 Warm roof

Table 12.6 Typical warm roof insulation U values

Insulation (mm) underlay/overlay		U-value (W/m^{2}K)
100	50	0.25
100	80	0.21
100	100	0.19
150	50	0.20
150	80	0.17
100	100	0.16

Underlay insulation goes in between and fully fills the rafter depth. Overlay insulation board goes under counter battens secured to the top of rafters with helical skewers.

Suitable rigid insulants include low density polyisocyanurate (PIR) foam (both faces bonded to aluminium foil), or high density mineral wool slabs over rafters with less dense mineral wool between rafters.

An alternative location for the breather membrane is under the counter battens. This is often preferred as the insulation board will provide uniform support for the underlay. Otherwise, extra insulation could be provided between the counter battens, retaining sufficient space for the underlay to sag between rafter positions to permit any rainwater penetration to drain to eaves.

Insulation in between rafters is an alternative to placing it above (see Figure 12.39). The details given in Figure 12.40 show two possibilities where, if required, supplementary insulation can be secured to the underside of rafters.

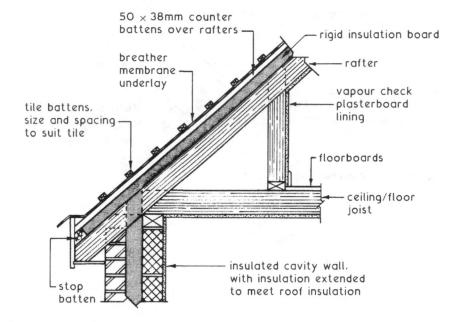

Figure 12.39 Insulation over the rafters

Roof insulation

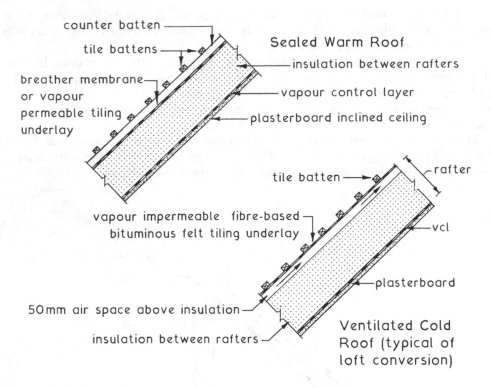

Figure 12.40 Insulation between rafters

12.11 Vapour control layer (VCL)

Condensation occurs where warm moist air contacts a cold layer. This could be in the roof space above inhabited rooms, where permeable insulation will not prevent the movement of moisture in the air and vapour from condensing on the underside of traditional tile underlay (sarking felt) and bituminous felt flat roof coverings. Venting of the roof space will control condensation.

Alternatively it can be controlled with a well-sealed vapour control layer (for instance, foil [metallised polyester] backed plasterboard) incorporated in the ceiling lining and used with a vapour permeable (breather membrane) underlay to the tiling. Joints and openings in the VCL ceiling (e.g. cable or pipe penetrations) should be sealed, but if this is impractical ventilation should be provided to the underside of the tile underlay.

12.12 Structural insulated panels (SIPs)

SIPs are prefabricated 'sandwich' panels that can be used as an alternative to traditional rafters or trusses for roof construction. They can also be used for structural wall panels. Surface layers of plywood or OSB are separated by a core of insulation. The outer face can be provided with counter battens for securing tile battens.

Properties

- High strength to weight ratio.
- Good thermal insulation.
- Continuity of insulation, no cold bridges.
- Good sound insulation.
- Fire-retardant core with improved resistance to fire by lining the inner face with plasterboard.
- Dimensionally coordinated.
- Factory cut to specification, including mitres and angles for ridge, valleys, hips, etc.
- Rapid and simple site assembly.
- Panel size is typically 1.2m wide, in lengths up to 8.0m.

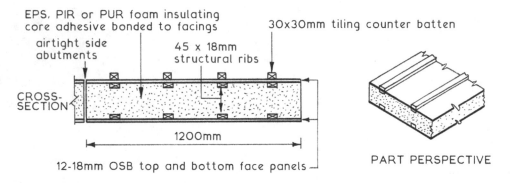

Figure 12.41 SIP panel details
Key: EPS – Expanded polystyrene. PIR – Polyisocyanurate, rigid foam. PUR – Rigid polyurethane. OSB – Oriented strand board.

Typical thermal insulation values are shown in Table 12.7.

Table 12.7 SIP panel thicknesses and U values

Depth/thickness, exc. battens (mm)	U-value (W/m^{2}K)
100	0.35
150	0.25
200	0.20
250	0.15

12.13 Thatched roofs

Materials

Water reed (Norfolk reed), wheat straw (Spring or Winter), Winter being the most suitable. Wheat for thatch is often known as wheat reed, long straw or Devon reed. Other thatches include rye and oat straws and sedge. Sedge is harvested every fourth year to provide long growth, making it most suitable as a ridging material.

There are various patterns and styles of thatching, relating to the skill of the thatcher and local traditions.

Typical details

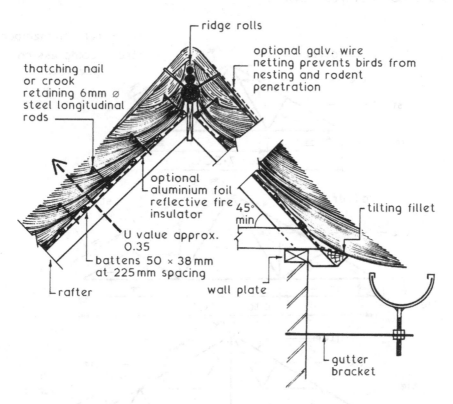

Figure 12.42 Thatched roof

The material composition of thatch with its natural voids and surface irregularities provides excellent insulation when dry and compact. However, when worn with possible accumulation of moss and rainwater, the U-value is less reliable. Thatch is also very vulnerable to fire. Therefore, in addition to imposing a premium, insurers may require application of a surface fire retardant and a fire insulant underlay.

12.14 Steel roof trusses

Used for industrial buildings up to the 1970s, these are triangulated plane frames that carry purlins to which the roof coverings can be fixed. Steel is stronger than timber and will not spread fire over its surface and for these reasons it is often preferred to timber for medium and long span roofs. The rafters are restrained from spreading by being connected securely at their feet by a tie member. Struts and ties are provided within the basic triangle to give adequate bracing. Steel angle sections are usually employed for steel truss members since they are economic and accept both tensile and compressive stresses. The members of a steel roof truss are connected together with rivets, bolts or by welding to shaped plates called gussets. Steel trusses are usually placed at 3m to 4.5m centres which give an economic purlin size at spans of up to 12m.

Typical steel roof truss format

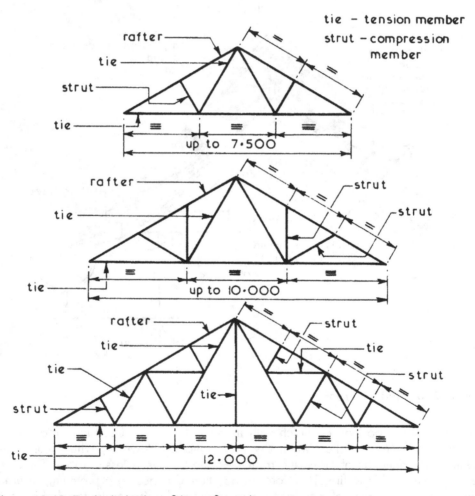

Figure 12.43 Typical steel roof truss formats

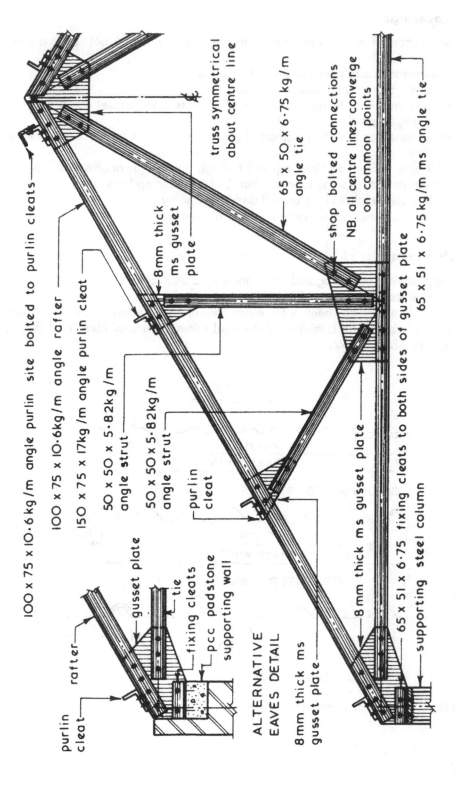

100 x 75 x 10·6 kg/m angle purlin site bolted to purlin cleats

truss symmetrical about centre line

65 x 50 x 6·75 kg/m angle tie

shop bolted connections
NB. all centre lines converge on common points

65 x 51 x 6·75 kg/m ms angle tie

100 x 75 x 10·6kg/m angle rafter

150 x 75 x 17kg/m angle purlin cleat

8mm thick ms gusset plate

50 x 50 x 5·82kg/m angle strut

50 x 50 x 5·82kg/m angle strut

purlin cleat

8mm thick ms gusset plate

65 x 51 x 6·75 fixing cleats to both sides of gusset plate

supporting steel column

ALTERNATIVE EAVES DETAIL

8mm thick ms gusset plate

purlin cleat

rafter

gusset plate

tie

fixing cleats

pcc padstone supporting wall

Figure 12.44 Typical steel roof truss details

415

Steel roof trusses

Sheet coverings

The basic functions of sheet coverings used in conjunction with steel roof trusses are to:

- Provide resistance to penetration by the elements.
- Provide restraint to wind and snow loads.
- Provide a degree of thermal insulation of not less than that set out in Part L of the Building Regulations.
- Provide resistance to surface spread of flame, as set out in Part B of the Building Regulations.
- Provide any natural daylight required through the roof in accordance with the maximum permitted areas set out in Part L of the Building Regulations.
- Be of low self-weight to give overall design economy.
- Be durable to keep maintenance needs to a minimum.

Suitable materials:

- Hot-dip galvanised corrugated steel sheets – BS 3083.
- Aluminium profiled sheets – BS 4868.
- Asbestos-free profiled sheets – there are various manufacturers whose products are usually based on a mixture of Portland cement, mineral fibres and density modifiers – BS EN 494.

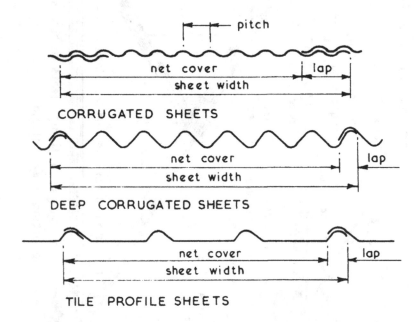

Figure 12.45 Profiles of typical roof sheets

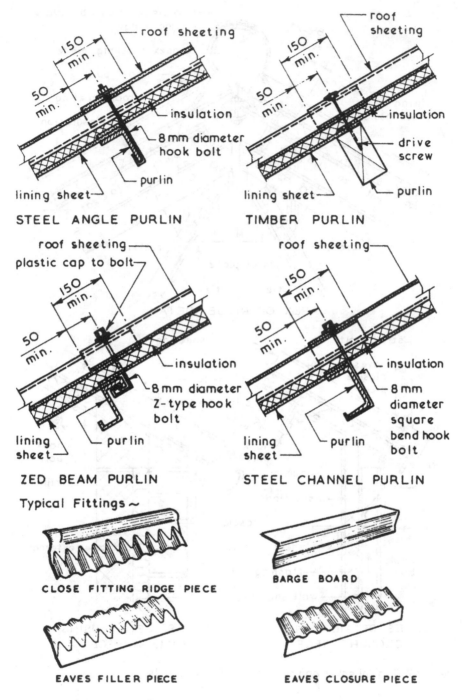

roof sheeting

150 min.

50 min.

insulation

8 mm diameter hook bolt

purlin

lining sheet

STEEL ANGLE PURLIN

roof sheeting

150 min.

50 min.

insulation

drive screw

purlin

lining sheet

TIMBER PURLIN

roof sheeting

plastic cap to bolt

150 min.

50 min.

insulation

8 mm diameter Z-type hook bolt

purlin

lining sheet

ZED BEAM PURLIN

roof sheeting

150 min.

50 min.

insulation

8 mm diameter square bend hook bolt

purlin

lining sheet

STEEL CHANNEL PURLIN

Typical Fittings ~

CLOSE FITTING RIDGE PIECE

BARGE BOARD

EAVES FILLER PIECE

EAVES CLOSURE PIECE

Figure 12.46 Typical purlin fixing details

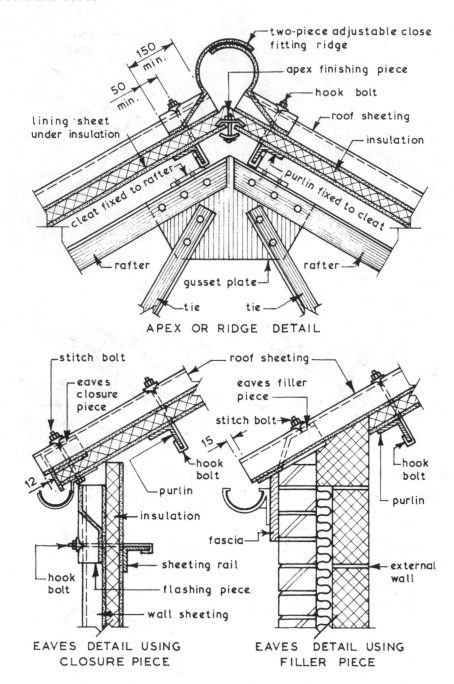

two-piece adjustable close fitting ridge

apex finishing piece

hook bolt

150 min.

50 min.

lining sheet under insulation

roof sheeting

insulation

cleat fixed to rafter

purlin fixed to cleat

rafter

rafter

gusset plate

tie

tie

APEX OR RIDGE DETAIL

stitch bolt

roof sheeting

eaves closure piece

eaves filler piece

stitch bolt

15

hook bolt

12

hook bolt

purlin

purlin

insulation

fascia

external wall

hook bolt

sheeting rail

flashing piece

wall sheeting

EAVES DETAIL USING CLOSURE PIECE

EAVES DETAIL USING FILLER PIECE

Figure 12.47 Typical eaves and ridge sheet fixing details

12.15 Double skin, energy roof systems

Use is limited to steel frame industrial and commercial use buildings, constructed to current thermal insulation standards. These systems can be specified to upgrade existing sheet profiled roofs with superimposed supplementary insulation and protective decking. Thermal performance with resin bonded mineral wool fibre of up to 250mm overall depth may provide U-values as low as 0.13W/m²K.

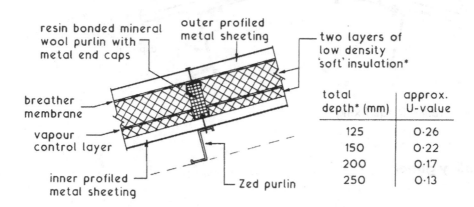

total depth* (mm)	approx. U-value
125	0·26
150	0·22
200	0·17
250	0·13

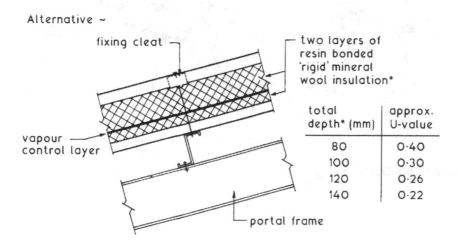

total depth* (mm)	approx. U-value
80	0·40
100	0·30
120	0·26
140	0·22

Figure 12.48 Zed purlin fixing of double skin panels

Further typical details are shown, using profiled galvanised steel or aluminium, colour coated if required.

419

Double skin, energy roof systems

Ridge

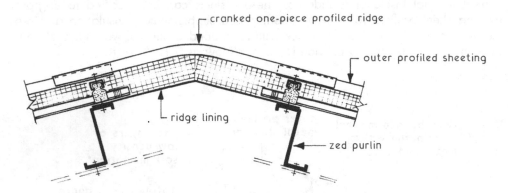

Figure 12.49 Ridge fixing

Valley gutter

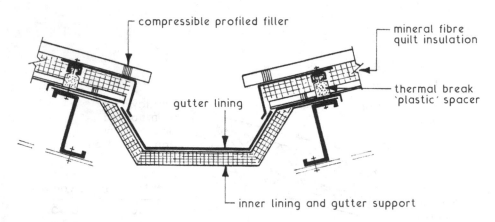

Figure 12.50 Gutter fixing

Eaves gutter

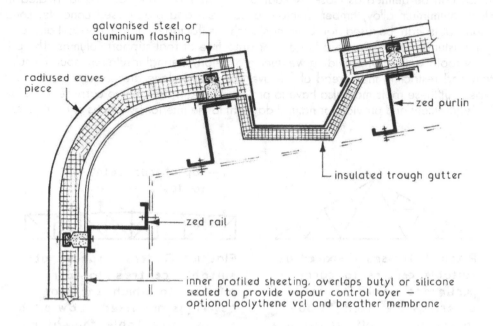

galvanised steel or aluminium flashing

radiused eaves piece

zed purlin

insulated trough gutter

zed rail

inner profiled sheeting, overlaps butyl or silicone sealed to provide vapour control layer — optional polythene vcl and breather membrane

Figure 12.51 Concealed gutter

12.16 Long span roofs

These can be defined as those exceeding 12m in span. They can be fabricated in steel, aluminium alloy, timber, reinforced concrete and pre-stressed concrete. Long span roofs can be used for buildings such as factories, large public halls and gymnasiums, which require a large floor area free of roof support columns. The primary roof functions of providing weather protection, thermal insulation, sound insulation and restricting the spread of fire over the roof surface are common to all roof types but these roofs may also have to provide strength sufficient to carry services lifting equipment and provide for natural daylight to the interior by means of rooflights.

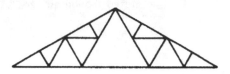

Pitched Trusses - spaced at suitable centres to carry purlins to which the roof coverings are fixed. Good rainwater run off - reasonable daylight spread from rooflights - high roof volume due to the triangulated format - on long spans roof volume can be reduced by using a series of short span trusses.

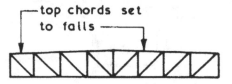

Flat Top Girders - spaced at suitable centres to carry purlins to which the roof coverings are fixed. Low pitch to give acceptable rainwater run off - reasonable daylight spread from rooflights - can be designed for very long spans but depth and hence roof volume increases with span.

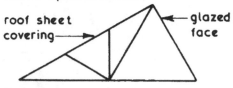

Northlight - spaced at suitable centres to carry purlins to which roof sheeting is fixed. Good rainwater run off - if correctly orientated solar glare is eliminated - long spans can be covered by a series of short span frames.

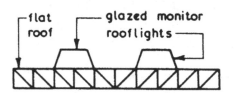

Monitor - girders or cranked beams at centres to suit low pitch decking used. Good even daylight spread from monitor lights which is not affected by orientation of building.

Figure 12.52 Basic roof forms

Pitched trusses

These can be constructed with a symmetrical outline or with an asymmetrical outline (north light – see Figure 12.53). They are usually made from standard steel sections with shop welded or bolted connections. Alternatively they can be fabricated using timber members joined together with bolts and timber connectors or formed as a precast concrete portal frame.

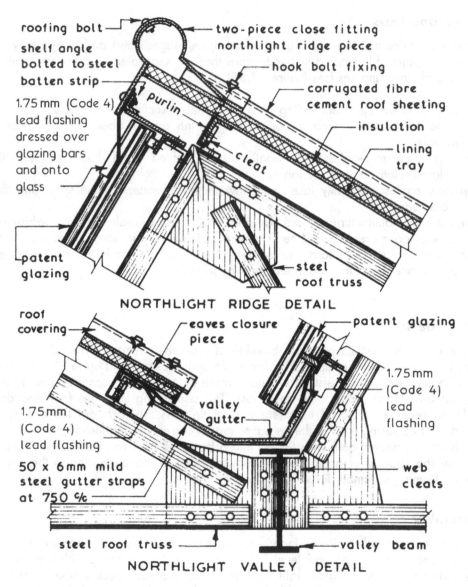

Figure 12.53 Typical multi-span north light roof details

Long span roofs

Monitor roofs

These are basically a flat roof with raised glazed portions called monitors, which forms a roof with a uniform distribution of daylight with no solar glare problems, irrespective of orientation, and with easy access for maintenance. These roofs can be constructed with light long span girders supporting the monitor frames, cranked welded beams following the profile of the roof, or they can be of a precast concrete portal frame format (see Figure 12.54).

Bowstring truss

A type of lattice truss formed with a curved upper edge. Bows and strings may be formed in pairs of laminated timber sections that are separated by solid web timber sections of struts and ties (see Figure 12.55).

Spacing: 4 to 6m apart depending on sizes of timber sections used and span.
Purlin: to coincide with web section meeting points and at about 1.000m interim intervals.
Decking: sheet material suitably weathered or profiled metal sheeting. Thermally insulated relative to application.
Top bow radius: generally taken as between three-quarters of the span and the whole span.
Application: manufacturing assembly areas, factories, aircraft hangers, exhibition centres, sports arenas and other situations requiring a very large open span with featured timbers. Standard steel sections may also be used in this profile where appearance is less important, e.g. railway termini.

Space deck

A roofing system based on a combination of repetitive inverted pyramidal units to give large clear spans of up to 22m for single span designs and up to 33m for two-way spans. Each unit consists of a square frame of steel angles connected with steel tubular bracing to a cast steel apex boss. The connecting boss has four threaded holes to receive horizontal high tensile steel tie bars. Threads are left and right handed at opposing ends of the tie bar to match corresponding threads in the boss. As the bar is rotated both ends tighten into a boss. Tie bars can be varied in length to alter the spacing of apex boss connectors. This will induce tension in the upper frame, forming a camber for surface water drainage.

Assembly

Usually transported to site as individual units for assembly into beams and complete space deck at ground level before a crane hoists into position onto the perimeter supports. A lightweight insulated structural roof deck system is appropriate in combination with rooflights mounted directly onto individual units if required (see Figure 12.56).

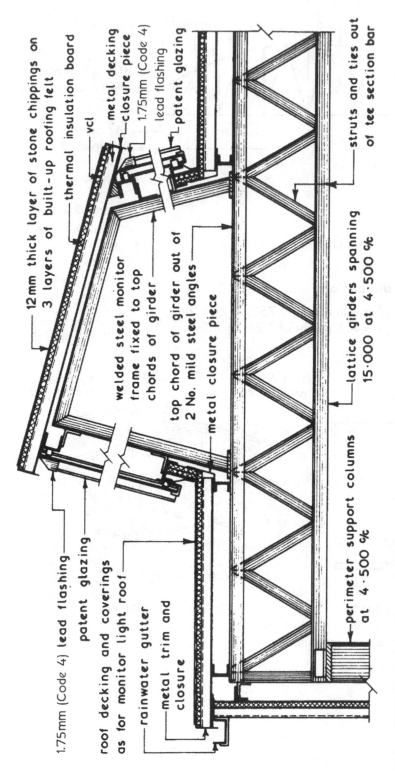

12mm thick layer of stone chippings on
3 layers of built-up roofing felt

thermal insulation board

vcl

metal decking
closure piece

1.75mm (Code 4)
lead flashing

patent glazing

metal decking
closure piece

welded steel monitor
frame fixed to top
chords of girder

top chord of girder out of
2 No. mild steel angles

metal closure piece

struts and ties out
of tee section bar

lattice girders spanning
15·000 at 4·500 ℀

perimeter support columns
at 4·500 ℀

1.75mm (Code 4) lead flashing
patent glazing

roof decking and coverings
as for monitor light roof

rainwater gutter

metal trim and
closure

Figure 12.54 Typical monitor roof details

Long span roofs

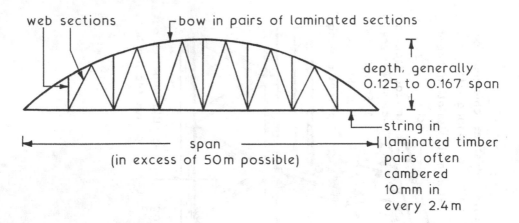

Figure 12.55 Bowstring roof truss

Standard unit dimensions

1.2m × 1.2m on plan with depths of 750mm or 1.2m.
1.5m × 1.5m on plan with depths of 1.2m or 1.5m.

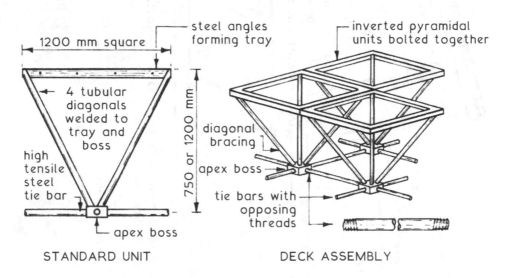

Figure 12.56 Space deck assemblies

Space deck frame support

Can be on columns or a peripheral ring beam through purpose made bearing joints. These joints should be designed with adequate strength to resist and transmit vertical and horizontal loads that combine compressive and tensile forces.

There are various possibilities for edge treatment and finishes, including vertical fascias, mansard and cantilever.

Space deck edge fixings and finishes are shown in Figure 12.57.

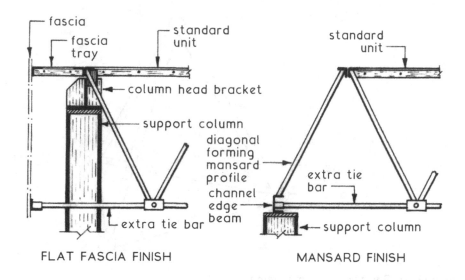

FLAT FASCIA FINISH   MANSARD FINISH

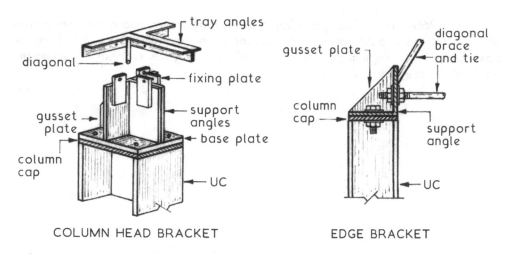

COLUMN HEAD BRACKET   EDGE BRACKET

Figure 12.57 Space deck structural frame connections

Long span roofs

Space frame

A concept based on a three-dimensional framework of structural members pinned at their ends, i.e. four joints or nodes that combine to produce a simple tetrahedron. As a roofing system it uses a series of connectors to join chords and bracing members. Single or double layer grids are possible; the former is usually employed for assembling small domes or curved roofs. Space frames are similar in concept to space decks but they have greater flexibility in design and layout possibilities. Steel or aluminium alloy tubes have been the favoured material for space frames. Fibre reinforced composites are also viable.

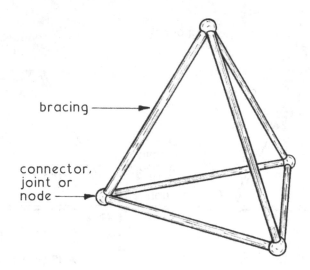

bracing

connector, joint or node

Figure 12.58 Basic space frame element

In the 1960s and 1970s space frame technology moved into a new generation of super-systems capable of very long spans often exceeding 50m. It is from the basis of the technological principles established during this era that contemporary applications are now designed.

Examples of space frames

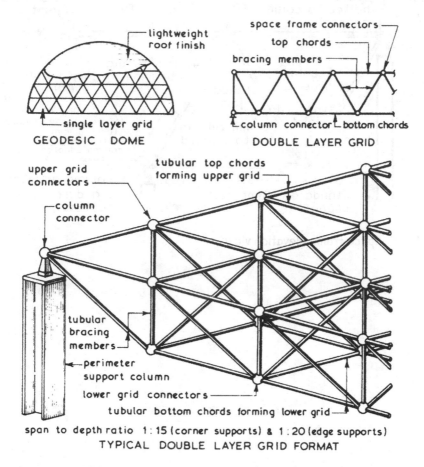

Figure 12.59 Space frame concept based on the 'Nodus' system of jointing

Development

During the 1960s and 1970s investment in public and private company research led to numerous systems of long span space frames being developed and introduced on a commercial scale. One of the better known is the 'Nodus' system created by British Steel Corporation (later Corus, now TATA). This concept formed the basis for later improvements and many contemporary systems use the principles of nodal jointing compatible with standard manufactured tubular metal sections.

• All factory fabrication, leaving simple assembly processes for site.
• Node steel castings in two parts with machined grooves and drilled holes. Machined steel teeth connector welded to chord ends.

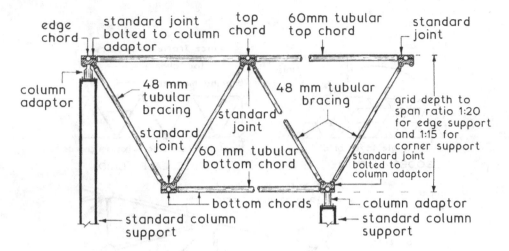

Figure 12.60 Space frame elevation view

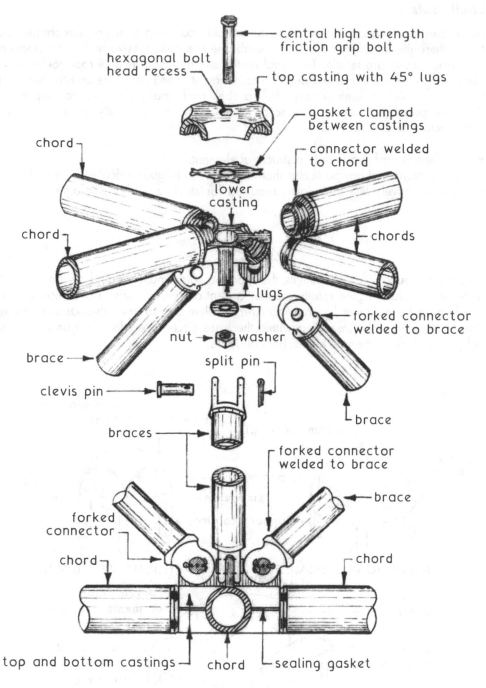

Figure 12.61 Tubular space frame connections

Long span roofs

Shell roofs

These can be defined as a structural curved skin covering a given plan shape and area where the forces in the shell or membrane are compressive and in the restraining edge beams are tensile. The usual materials employed in shell roof construction are in-situ reinforced concrete and timber. Concrete shell roofs are constructed over formwork, which in itself is very often a shell roof, making this format expensive since the principle of use and reuse of formwork cannot normally be applied. The main factors of shell roofs are:

* The entire roof is primarily a structural element.
* Basic strength of any particular shell is inherent in its geometrical shape and form.
* Comparatively less material is required for shell roofs than for other forms of roof construction.

Domes

These are double curvature shells that can be rotationally formed by any curved geometrical plane figure rotating about a central vertical axis. Translation domes are formed by a curved line moving over another curved line whereas pendentive domes are formed by inscribing within the base circle a regular polygon and vertical planes through the true hemispherical dome.

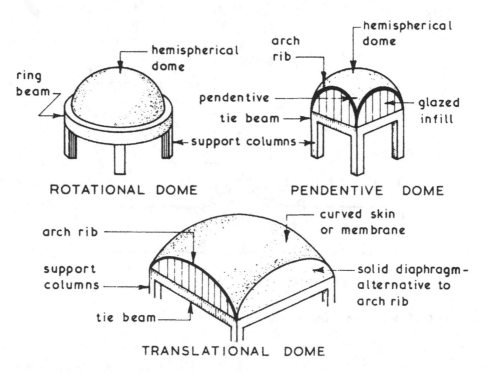

Figure 12.62 Dome roof examples

Barrel vaults

These are single curvature shells, which are essentially a cut cylinder which must be restrained at both ends to overcome the tendency to flatten. A barrel vault acts as a beam whose span is equal to the length of the roof. Long span barrel vaults are those whose span is longer than its width or chord length and, conversely, short barrel vaults are those whose span is shorter than its width or chord length. In every long span barrel vault thermal expansion joints will be required at 30m centres, which will create a series of abutting barrel vault roofs weather sealed together.

Typical single barrel vault principles

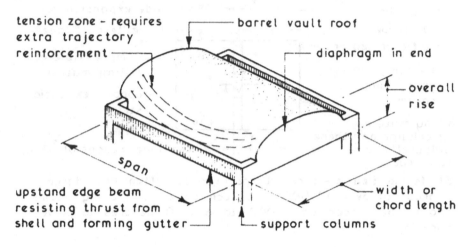

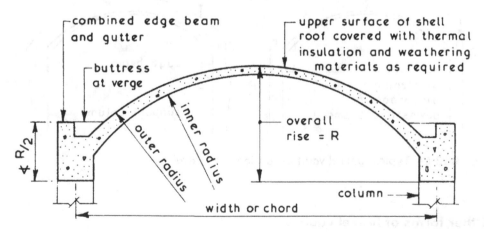

Figure 12.63 Single barrel vault roof

Note: Ribs not connected to support columns will set up extra stresses within the shell roof; therefore extra reinforcement will be required at the stiffening rib or beam positions.

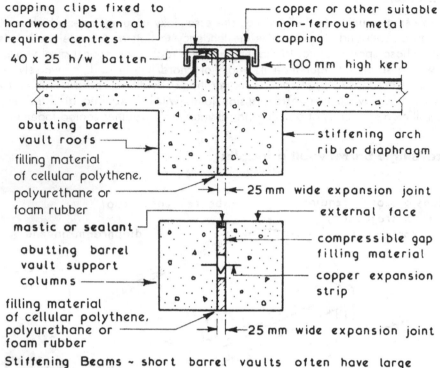

capping clips fixed to hardwood batten at required centres

40 x 25 h/w batten

copper or other suitable non-ferrous metal capping

100 mm high kerb

abutting barrel vault roofs

filling material of cellular polythene, polyurethane or foam rubber

mastic or sealant

abutting barrel vault support columns

filling material of cellular polythene, polyurethane or foam rubber

stiffening arch rib or diaphragm

25 mm wide expansion joint

external face

compressible gap filling material

copper expansion strip

25 mm wide expansion joint

Stiffening Beams ~ short barrel vaults often have large chords of over 12·000. In such cases stiffening beams should be placed at 3·000 to 6·000 centres to prevent buckling.

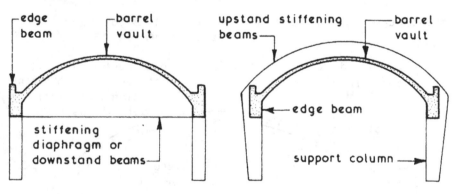

edge beam

barrel vault

stiffening diaphragm or downstand beams

upstand stiffening beams

barrel vault

edge beam

support column

Figure 12.64 Typical barrel vault expansion joint details

Other forms of barrel vault

By cutting intersecting and placing at different levels the basic barrel vault roof can be formed into a groin or north light barrel vault roof.

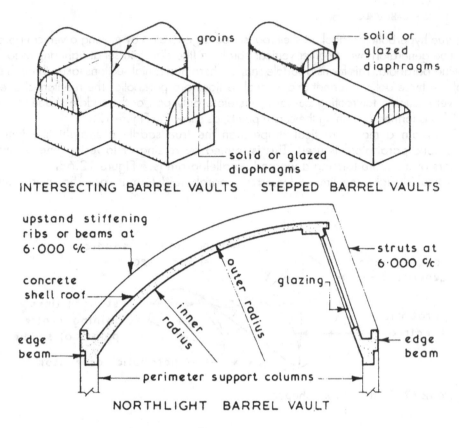

Figure 12.65 Examples of barrel roofs

CONOIDS

These are double curvative shell roofs which can be considered as an alternative to barrel vaults. Spans up to 12m with chord lengths up to 24m are possible. Typical chord to span ratio 2:1.

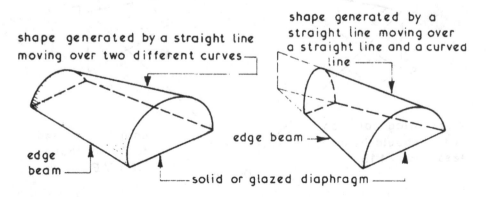

Figure 12.66 Conoid shapes

Long span roofs

The true hyperbolic paraboloid shell roof shape is generated by moving a vertical parabola (the generator) over another vertical parabola (the directrix) set at right angles to the moving parabola. This forms a saddle shape where horizontal sections taken through the roof are hyperbolic in format and vertical sections are parabolic. The resultant shape is not very suitable for roofing purposes; therefore only part of the saddle shape is used and this is formed by joining the centre points, as shown in Figure 12.67.

To obtain a more practical shape than the true saddle a straight line limited hyperbolic paraboloid is used. This is formed by raising or lowering one or more corners of a square forming a warped parallelogram (see Figure 12.68).

Typical straight line limited hyperbolic paraboloid formats are shown in Figure 12.69.

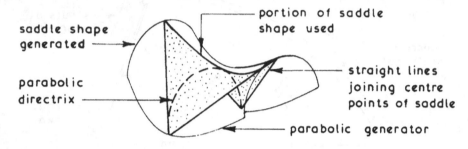

Figure 12.67 Parabolic roof shapes

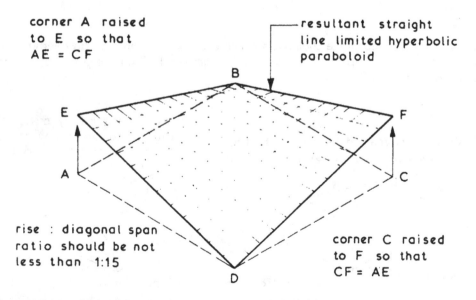

Figure 12.68 Straight line limited hyperbolic paraboloid shape

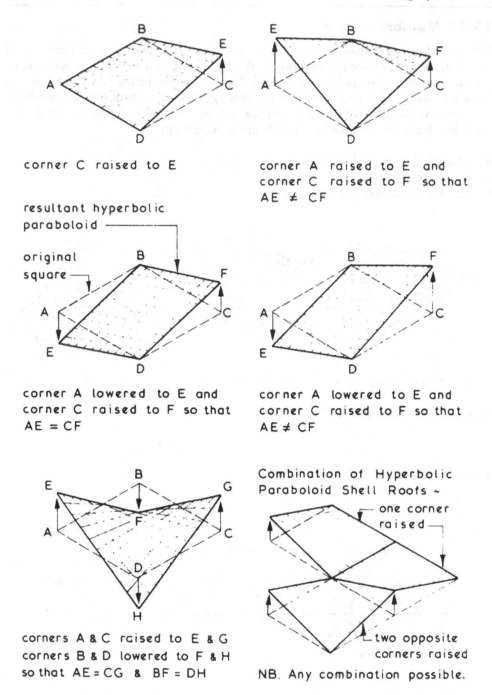

corner C raised to E

corner A raised to E and
corner C raised to F so that
AE ≠ CF

resultant hyperbolic
paraboloid

original
square

corner A lowered to E and
corner C raised to F so that
AE = CF

corner A lowered to E and
corner C raised to F so that
AE ≠ CF

corners A & C raised to E & G
corners B & D lowered to F & H
so that AE = CG & BF = DH

Combination of Hyperbolic
Paraboloid Shell Roofs ~
one corner
raised

two opposite
corners raised

NB. Any combination possible.

Figure 12.69 Straight line limited hyperbolic paraboloid formats

12.17 Membrane roof

This is a form of tensioned cable structural support system with a covering of stretched fabric. In principle and origin, this compares to a tent with poles as compression members secured to the ground. The fabric membrane is attached to peripheral stressing cables suspended in a catenary between vertical support members.

There are limitless three-dimensional possibilities. The following geometric shapes provide a basis for imagination and elegance in design:

- Hyperbolic paraboloid (Hypar).
- Barrel vault.
- Conical or double conical.

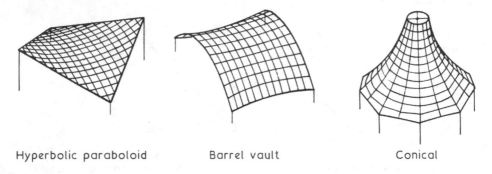

Hyperbolic paraboloid   Barrel vault   Conical

Figure 12.70 Membrane roof structures

Double conical

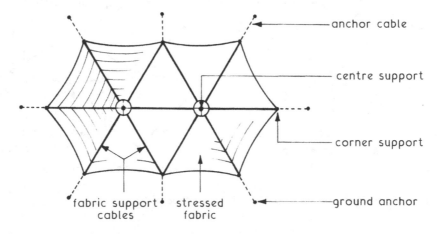

Figure 12.71 Plan view double conical

Simple support structure as viewed from the underside

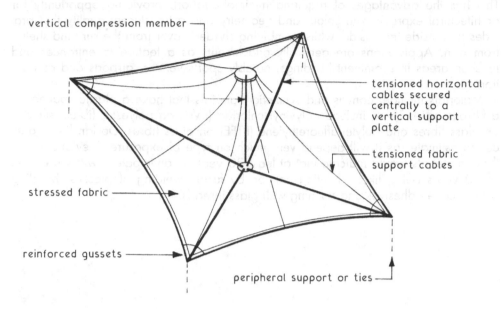

vertical compression member

tensioned horizontal cables secured centrally to a vertical support plate

tensioned fabric support cables

stressed fabric

reinforced gussets

peripheral support or ties

Figure 12.72 View from underside

12.18 Fabric

This has the advantages of requiring minimal support, providing opportunity for architectural expression in colour and geometry and a translucent quality that provides an outside feel inside, whilst combining shaded cover from the sun and shelter from rain. Applications are generally attachments as a feature to entrances and function areas in prominent buildings, notably sports venues, airports and convention centres.

Materials include: canvas and synthetic materials that have a plastic coating on a fibrous base. These include polyvinyl chloride (PVC) on polyester fibres, silicone on glass fibres and polytetrafluorethylene (PTFE) on glass fibres. Design life is difficult to estimate, as it will depend very much on type of exposure. Previous use of these materials would indicate that at least 20 years is anticipated, with an excess of 30 years being likely. Jointing can be by fusion welding of plastics, bonding with silicone adhesives and stitching with glass threads.

12.19 Rooflights

The useful penetration of daylight through the windows in external walls of buildings is from 6m to 9m depending on the height and size of the window. In buildings with spans over 18m, side wall daylighting needs to be supplemented by artificial lighting or, in the case of top floors or single-storey buildings, by rooflights. The total maximum area of wall window openings and rooflights for the various purpose groups is set out in the Building Regulations, with allowances for increased areas if double or triple glazing is used. In pitched roofs, such as north light and monitor roofs, the rooflights are usually in the form of patent glazing (see 'Long span roofs').

Patent glazing ~ systems of steel or aluminium alloy glazing bars that span the distance to be glazed whilst giving continuous edge support to the glass. They can be used in the roof forms noted previously as well as in pitched roofs with profiled coverings where the patent glazing bars are fixed above and below the profiled sheets.

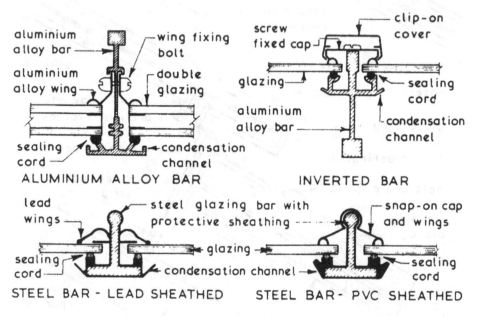

Figure 12.73 Typical patent glazing sections

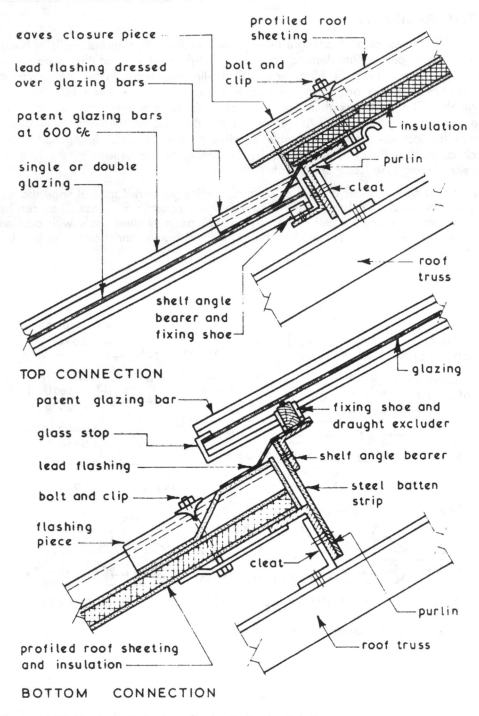

eaves closure piece

profiled roof sheeting

lead flashing dressed over glazing bars

bolt and clip

patent glazing bars at 600 c/c

insulation

single or double glazing

purlin

cleat

shelf angle bearer and fixing shoe

roof truss

glazing

TOP CONNECTION

patent glazing bar

fixing shoe and draught excluder

glass stop

shelf angle bearer

lead flashing

steel batten strip

bolt and clip

flashing piece

cleat

purlin

profiled roof sheeting and insulation

roof truss

BOTTOM   CONNECTION

Figure 12.74 Typical pitched roof patent glazing details

13 FLAT ROOFS

Opening remarks

A roof is classified as flat if the pitch/fall does not exceed 10°. Flat roofs are classified as either: cold deck, warm deck or inverted warm deck according to where the insulation is in the construction.

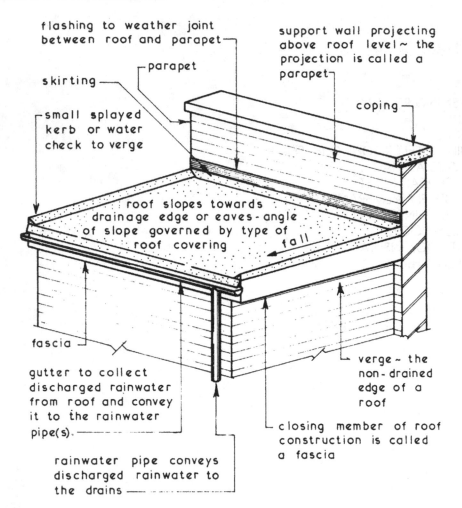

Figure 13.1 Flat roof elements

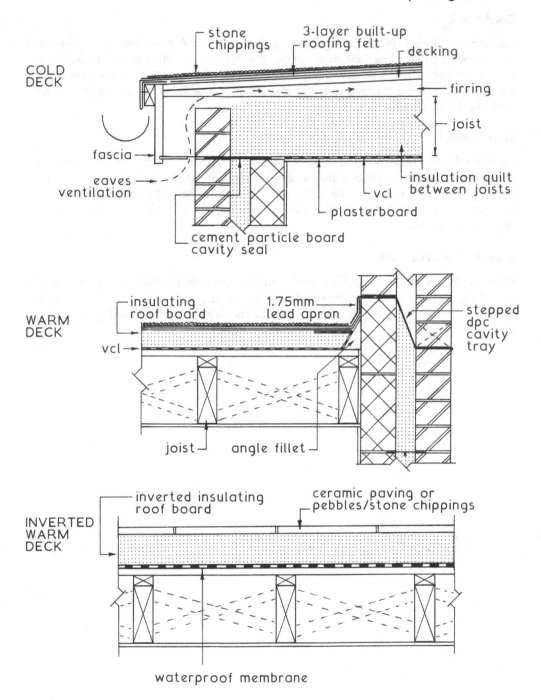

Figure 13.2 Types of flat roof
Note: Insulation type and thickness varies with application and situation.

Opening remarks

Cold deck

Insulation is placed on the ceiling lining, between joists. A metallised polyester lined plasterboard ceiling functions as a vapour control layer, with a minimum 50mm air circulation space between insulation and decking. The air space corresponds with eaves vents and both provisions will prevent moisture buildup, condensation and possible decay of timber.

Warm deck

Rigid insulation (resin bonded mineral fibre roof boards, expanded polystyrene or polyurethane slabs) is placed below the waterproof covering and above the roof decking. The insulation must be sufficient to maintain the vapour control layer and roof members at a temperature above dewpoint, as this type of roof does not require ventilation.

Inverted warm deck

Rigid (resin bonded mineral fibre roof boards, expanded polystyrene or polyurethane slabs) insulation is positioned above the waterproof covering. The insulation must be unaffected by water and capable of receiving a stone dressing or ceramic pavings.

13.1 Falls

The pitch chosen can be governed by the roof covering selected and/or by the required rate of rainwater discharge off the roof. As a general rule, the minimum pitch for smooth surfaces such as asphalt should be 1:80 or 0°–43° and for sheet coverings with laps 1:60 or 0°–57°.

Methods of obtaining falls

Wherever possible joists should span the shortest distance of the roof plan.

1. Joists cut to falls

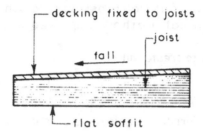

Simple to fix but could be wasteful in terms of timber unless two joists are cut from one piece of timber

Figure 13.3 Joists cut to fall

2. Joists laid to falls

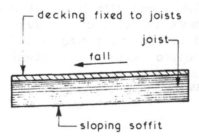

Economic and simple but sloping soffit may not be acceptable, but this could be hidden by a flat suspended ceiling

Figure 13.4 Joist laid to fall

3. Firrings with joist run

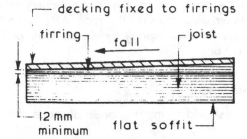

decking fixed to firrings

firring — fall — joist

12 mm minimum — flat soffit

Simple and effective but does not provide a means of natural cross ventilation. Usual method employed.

Figure 13.5 Firrings used to create fall

4. Firrings against joist run

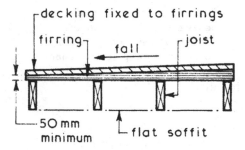

decking fixed to firrings

firring — fall — joist

50 mm minimum — flat soffit

Simple and effective but uses more timber than 3 but does provide a means of natural cross ventilation

Figure 13.6 Firrings across joists

Typical eaves details

BS EN 13707: *Flexible Sheets for Waterproofing. Reinforced Bitumen Sheets for Roof Waterproofing.*

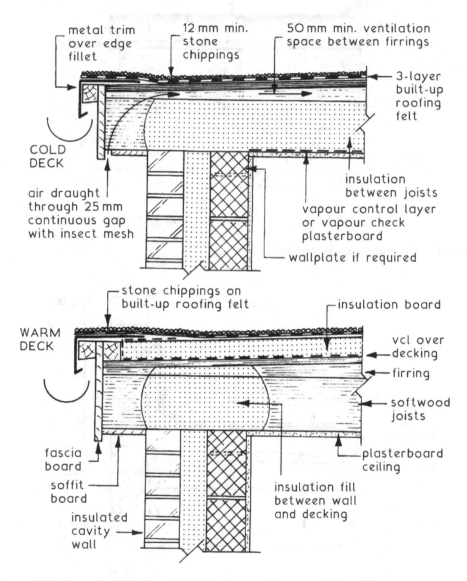

Figure 13.7 Eaves details for cold and warm decks

BS 8217: *Reinforced Bitumen Membranes for Roofing.*

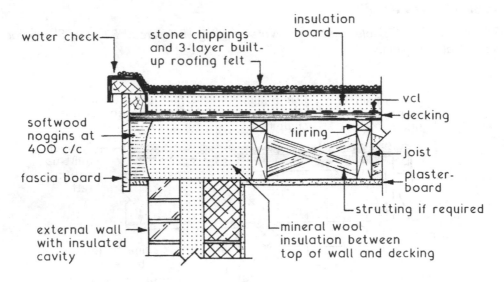

water check

stone chippings and 3-layer built-up roofing felt

insulation board

vcl

decking

firring

joist

plasterboard

softwood noggins at 400 c/c

fascia board

strutting if required

external wall with insulated cavity

mineral wool insulation between top of wall and decking

TYPICAL VERGE DETAILS – WARM DECK

Figure 13.8 Warm deck verge detail

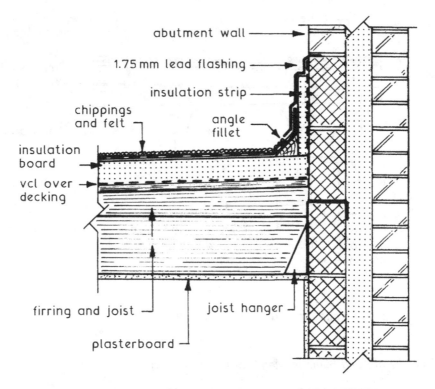

abutment wall

1.75 mm lead flashing

insulation strip

chippings and felt

angle fillet

insulation board

vcl over decking

firring and joist

joist hanger

plasterboard

TYPICAL ABUTMENT DETAILS – WARM DECK

Figure 13.9 Warm deck wall abutment details

450

13.2 Decks

Plywood ~ exterior grade boarding often specified as WBP (water boiled proof), a reference to the bonding quality of the adhesive securing the ply veneers. Sheets are fixed on all four edges, requiring noggins/struts between the joists. BS EN 636.

Oriented strand board (OSB) ~ Bitumen coated OSB is a common and cheaper favoured alternative to plywood chipboard as it is more stable in moist conditions. BS EN 300 types 3 and 4.

Cement bonded particle board ~ another alternative that has a high density and greater moisture resistance than other particle boards. BS EN 634–2.
Woodwool slabs ~ composed of wood fibre shreds bonded together with cement. Produced in a variety of thicknesses, but for roof decking the minimum thickness is 50mm to satisfy strength requirements when accessed. Widths of 600mm are produced in lengths up to 4m. BS EN 13168.

In-situ concrete or precast concrete slabs are standard for framed buildings.

13.3 Built up felt roofing (BUFR)

The traditional method of providing a waterproof covering for flat roofs with a timber structure is built up felt roofing (BUFR)

Standard rolls of 1m width, modern bituminous felts have a matrix of glass fibre or polyester matting as reinforcement to a stabilising bitumen coating on both sides. These materials provide an effective binding and have superseded rags as a more robust base. The upper surface is lightly coated with sand or fine mineral granules to prevent the sheet from sticking to itself when rolled.

BUFR application

The first layer is laid at right angles to the fall commencing at the eaves. If the decking is plywood or a wood composite, the first layer can be secured with large flat head nails, with subsequent layers bonded together with hot molten bitumen. Side laps are at least 50mm with 75mm minimum laps at upper and lower ends.

A variation known as torch-on is for use with specially made sheet. This is heated with a blowtorch roller to the underside to produce a wave of molten bitumen as the sheet is unrolled. Timber product decking is not suitable for torch-on applications due to the fire risk, unless the surface is pre-felted and taped.

Finish

Limestone, light-coloured shingle or granite chippings of 10 to 12mm are suitable as weatherproofing, protection from solar radiation and resistance to fire. These are bonded to the surface with a cold or hot molten bitumen solution.

BS EN 13707: *Flexible Sheets for Waterproofing. Reinforced Bitumen Sheets for Roof Waterproofing. Definitions and Characteristics.*
BS 8217: *Reinforced Bitumen Membranes for Roofing. Code of Practice.*

13.4 Mastic asphalt

Mastic asphalt ~ a dense mixture of coarse and fine aggregate with a bitumen binder plus additives (polymers, waxes). The mixture is designed to be of low void content. Mastic asphalt is pourable and able to be spread in its working temperature condition. It requires no compaction on site.

On site, blocks of mastic asphalt are heated in a cauldron to about 200°C. The then molten asphalt is transferred by bucket and spread on the deck manually with a wooden trowel. Application consists of two layers of mastic asphalt laid on breaking joints and built up to a minimum thickness of 20mm. It should be laid to the recommendations of BS 8218. The mastic asphalt is laid over an isolating membrane of black sheathing felt which should be laid loose with 50mm minimum overlaps.

Typical details

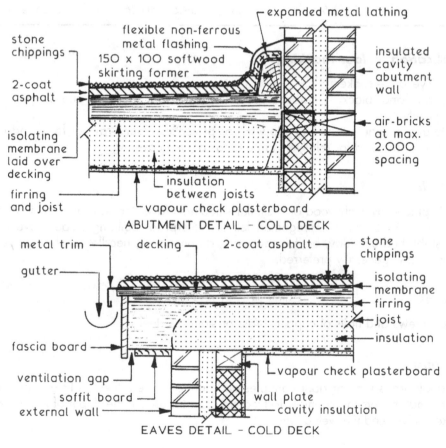

Figure 13.10 Mastic asphalt roof details

BS 8218: *Code of Practice for Mastic Asphalt Roofing.*

13.5 Milled lead sheet

Lead is a soft and malleable material used traditionally for flat and pitched roofs.

Available in a variety of weights/thicknesses and widths of sheets for different applications.

Table 13.1 Lead sheet classifications and standards

BS 1178 code no.	BS EN 12588/standard milled thickness (mm)	Weight (kg/m^2) BS EN/milled	Colour marking
3	1.25/1.32	14.17/14.97	Green
–	1.50/1.59	17.00/18.03	Yellow
4	1.75/1.80	19.84/20.41	Blue
5	2.00/2.24	22.67/25.40	Red
6	2.50/2.65	28.34/30.05	Black
7	3.00/3.15	34.02/35.72	White
8	3.50/3.55	39.69/40.26	Orange

Application (colour marking)

Green/yellow – soakers.

Blue, red and black – flat roof covering in small, medium and large areas respectively.

White and orange – lead lining to walls as protection from X-rays or for sound insulation. Can be used for relatively large areas of roof covering.

Underlay

This is placed over plywood or a similar smooth surface decking, or over rigid insulation boards. Bitumen impregnated felt or waterproof building paper have been the established underlay, but for new work, a non-woven, needle punched polyester textile is now generally preferred.

Fixings

Clips, screws and nails of copper, brass or stainless steel.

Jointing

For small areas such as door canopies and dormers where there is little opportunity for thermal movement, a simply formed welt can be used if the depth of rainwater is unlikely to exceed the welt depth.

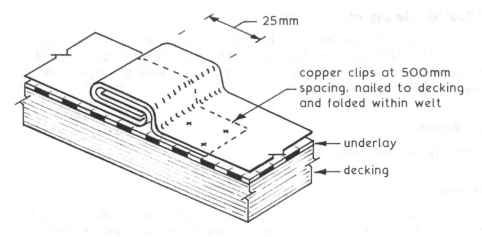

25mm

copper clips at 500mm
spacing, nailed to decking
and folded within welt

underlay

decking

Figure 13.11 Welted joint

Jointing to absorb movement

- Wood cored rolls in the direction of the roof slope.
- Drips at right angles to and across the roof slope.

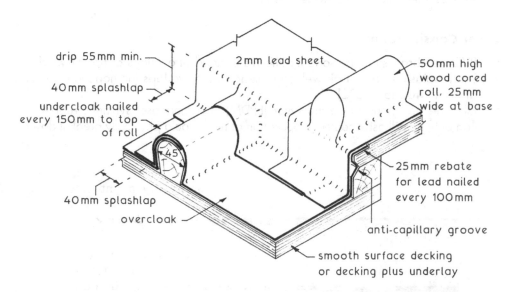

drip 55mm min.

40mm splashlap

undercloak nailed
every 150mm to top
of roll

2mm lead sheet

50mm high
wood cored
roll, 25mm
wide at base

45

40mm splashlap

overcloak

25mm rebate
for lead nailed
every 100mm

anti-capillary groove

smooth surface decking
or decking plus underlay

Figure 13.12 Typical provision of wood cored rolls and drips – detail at junction

13.6 Single ply membranes

These are durable, resilient, flexible and lightweight sheet materials composed mainly of synthetic polymers. Some are reinforced with glass fibres depending on application and coverage area. A backing of glass fibre or polyester matting is often provided as a bonding interface.

Thickness

Generally between 1 and 2mm.

Fixing

Product manufacturer's recommended adhesive applied to the sub-surface. Purpose-made mechanical fixing devices are an alternative in situations that may be exposed to wind lift.

Materials

- Polyvinyl chloride (PVC).
- Thermoplastics: Thermoplastic polyolefin (TPO).
- Chlorinated polyethylene (CPE).
- Ethylene interpolymer (EIP).
- Copolymer alloy (CPA).
- Acrylonitrile butadiene polymer (NBP).

Other Considerations

- PVC membranes can be solvent adhesive bonded at overlaps or hot air welded.
- Thermoplastics are hot air welded at seams and overlaps (homogenous jointing at about 400 to 500°C).
- Ethylene propylene diene monomer (EPDM) is a thermoset synthetic rubber that can only be adhesive sealed. Application of heat would physically break it down.

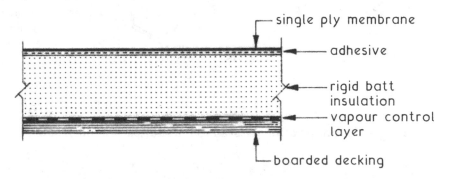

Figure 13.13 Single ply membrane over rigid insulation

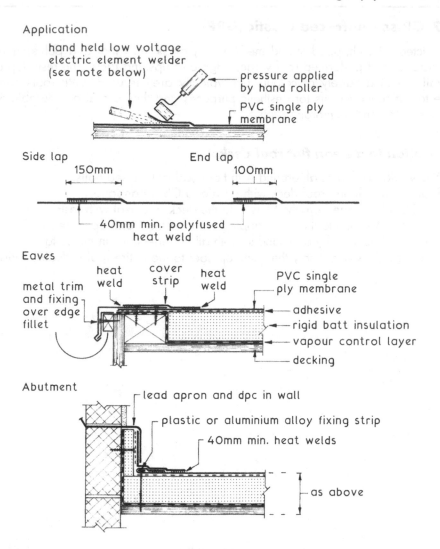

Application

hand held low voltage
electric element welder
(see note below)

pressure applied
by hand roller

PVC single ply
membrane

Side lap

150mm

End lap

100mm

40mm min. polyfused
heat weld

Eaves

metal trim
and fixing
over edge
fillet

heat
weld

cover
strip

heat
weld

PVC single
ply membrane

adhesive

rigid batt insulation

vapour control layer

decking

Abutment

lead apron and dpc in wall

plastic or aluminium alloy fixing strip

40mm min. heat welds

as above

Figure 13.14 Single ply membrane details

Note: Hand-held welder and roller are used mainly for small detail areas, automatic (manually directed) welders/rollers are more effective for continuous seaming.

13.7 Glass reinforced plastic (GRP)

Constructed with chopped strand mesh (CSM) glass fibre mat and polyester resin with catalyst and preformed roof verge, edge and upstand trims. The resin top coat is available in a variety of colours but typically grey is used. GRP roofs can be made to mimic lead roofs for aesthetic purposes and they are also a durable alternative to BUFR and membrane roofs.

Application to a clean flat roof deck

1. Apply GRP roof trims all around with flat head galvanised nails.
2. Seal all joints in the roof deck with resin and CSM bandage.
3. Lap CSM bandage from the trims onto the deck and seal with resin.
4. Apply CSM sheets to the deck lapping onto the trims and apply resin.
5. Allow 24 hours to dry and sand smooth all imperfections in the surface.
6. Clean the deck and apply the resin top coat to the entire roof including trims.

13.8 Green roofs

Application is mainly on flat roofs, being a composite structure of insulation, drainage layers soils and vegetation to create an environmentally and ecologically friendly building element.

Green roof advantages

- Absorbs and controls water run-off.
- Integral thermal insulation.
- Integral sound insulation.
- Absorbs air pollutants, dust and CO_2.
- Passive heat storage potential.

Green roof disadvantages

- Weight.
- Maintenance.

Construction

The following buildup will be necessary to fulfil the objectives and to create stability:

- Vapour control layer above the roof structure.
- Rigid slab insulation.
- Root resilient waterproof under-layer.
- Drainage layer.
- Filter.
- Growing medium (soil).
- Vegetation (grass, etc.).

Categories

Typical extensive roof buildup

Formed with a relatively shallow soil base (typically 50mm) and of lightweight construction, maximum roof pitch is 40° and slopes greater than 20° will require a system of baffles to prevent the soil moving. Plant life is limited by the shallow soil base to grasses, mosses, herbs and sedum (succulents, generally with fleshy leaves producing pink or white flowers).

Green roofs

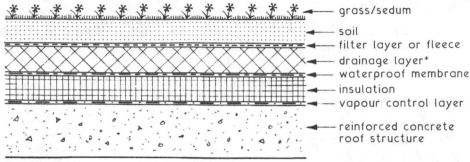

* typically, expanded polystyrene with slots

Figure 13.15 Green roof layers

Typical intensive roof buildup

Otherwise known as a roof garden, this type has a deeper soil base (typically 400mm) that will provide for landscaping features, small ponds, occasional shrubs and small trees. A substantial building structure is required for support and it is only feasible to use a flat roof.

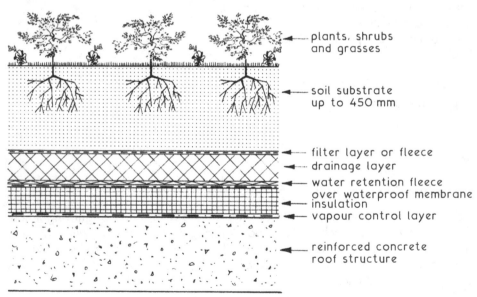

Depth to vcl, approximately 560 mm at about 750 kg/m² saturated weight. 750 kg/m² × 9.81 = 7358 N/m² or 7.36 kN/m².

Figure 13.16 Intensive roof layers

13.9 Flat roof insulation

Timber warm flat roof

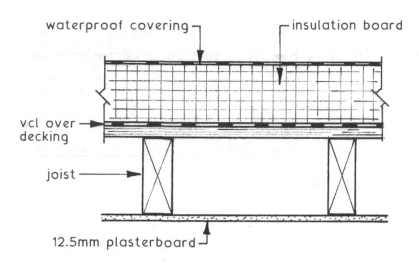

Figure 13.17 Insulation above timber deck

Table 13.2 Insulation above timber deck typical U values

Insulation thickness (mm)	U-value W/m2K
130	0.25
165	0.22
190	0.18
220	0.16
270	0.13

Concrete warm roof

Note: Insulation is mineral wool roofing board, with a thermal conductivity (λ) of 0.038W/mK.

Flat roof insulation

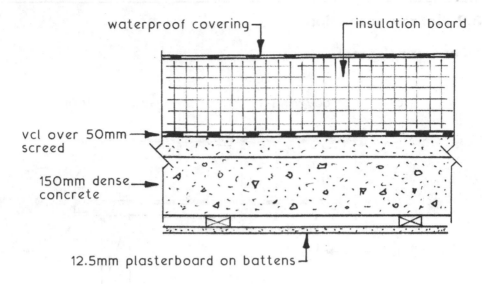

waterproof covering┐   ┌insulation board

vcl over 50mm →
screed

150mm dense →
concrete

12.5mm plasterboard on battens ┘

Figure 13.18 Insulation above concrete deck

Table 13.3 Insulation above concrete deck typical
U values

Insulation thickness (mm)	U-value W/m2K
130	0.25
165	0.22
190	0.18
220	0.16
270	0.13

13.10 Rooflights

Lantern lights

These are a form of rooflight used in conjunction with flat roofs. They consist of glazed vertical sides and fully glazed pitched roof, which is usually hipped at both ends. Part of the glazed upstand sides is usually formed as an opening light or alternatively glazed with louvres to provide a degree of controllable ventilation. They can be constructed of timber, metal or a combination of these two materials. Lantern lights in the context of new buildings have been generally superseded by the various forms of dome light.

Lens lights

These are small square or round blocks of translucent toughened glass especially designed for casting into concrete and are suitable for use in flat roofs and curved roofs such as barrel vaults. They can also be incorporated into precast concrete frames for inclusion in a cast in-situ roof.

Dome, pyramid and similar rooflights

These are used in conjunction with flat roofs and may be framed or unframed. The glazing can be of glass or plastics such as polycarbonate, acrylic, PVC and glass fibre reinforced polyester resin (GRP). The whole component is fixed to a kerb and may have a raising piece containing hit and miss ventilators, louvres or flaps for controllable ventilation purposes.

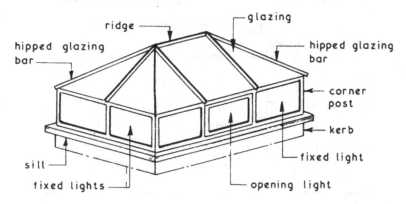

Figure 13.19 Typical lantern light details

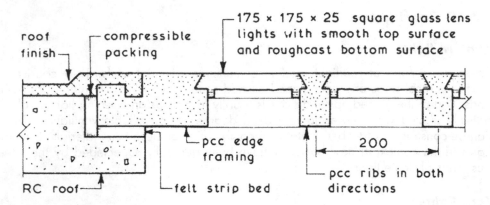

Figure 13.20 Lens lights

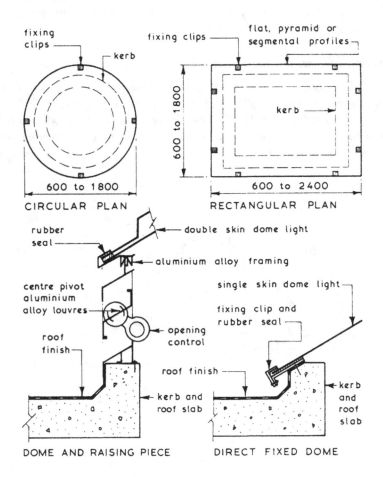

Figure 13.21 Glazing edge details

14 STEEL FRAMED BUILDINGS

DEVELOPMENT
STEEL FRAME STRUCTURE
HOT ROLLED STEEL SECTIONS
STEELWORK CONNECTIONS
FIRE PROTECTION

14.1 Development

Load-bearing wall structures transmit the loads of the roof and all floors through the walls into the foundations and the ground. The tallest solid load-bearing wall building ever built is the Monadnock building in Chicago (1893), which reached 17 storeys, approximately 66m high. The disadvantages of this form of construction is the thickness of the walls at the ground floor (about 2m thick), which takes up a large amount of the building footprint, and the internal floor areas of upper floors vary with the reducing in thickness of the external walls with height. The building is also very heavy, needing very substantial foundations. All of these factors make high rise load-bearing wall construction uneconomical.

Also in Chicago in the late 19th century the first high rise steel framed buildings were being constructed. The benefits of this type of construction were soon realised and rapidly developed. These frames of hot rolled steel columns and beams were hot riveted together, forming very strong rigid joints. The external envelope or curtain walling is supported by the frame. This form of structure is far lighter than a solid wall building, reducing the load that needs to be carried by the foundation.

The late 19th century also saw the development of reinforced concrete as a structural material for engineering and building.

Today, the choice between reinforced concrete or steel for the framing is largely an economic decision dependent on the availability and cost of materials and appropriately qualified and experienced workforce.

Design constraints

The relationship between the superstructure, its foundations and the bearing capacity of the subsoil is the most important consideration – in particular, loading from vertical or gravitational forces in addition to exposure to horizontal or lateral wind forces and possibly ground movement from earthquakes.

Other significant design factors:

- Occupancy type, building function and purpose will influence internal layout and structural floor load.
- Internal circulation: up to 20% of the floor area may be required to provide for vertical access in the form of stairs, lifts and escalators.
- Building services: up to 15% of the volume of a building may be required to accommodate vertical and horizontal distribution of services.
- Site constraints, accessibility for plant, deliveries and site establishment.

14.2 Steel frame structure

A structural frame consists of:

- Beams (horizontal).
- Columns (vertical).
- Bracing (diagonal).

Often a rigid core housing lifts, stairs and services is also required. Connections between steel beams and columns can either be: simple, semi-rigid or rigid.

Rigid steel frame

The structure is designed with rigidly connected welded and bolted joints between columns and beams. These joints sustain bending and provide resistance to lateral/horizontal forces. Columns are positioned quite close together and internal columns may impose to some extent on the design of the interior layout.

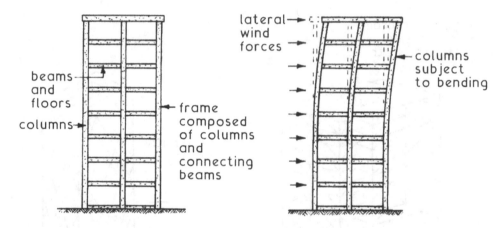

Figure 14.1 Rigid steel frame

Core structures

These are designed with a rigid structural core extending the full height of the building. The core is located centrally and the void within it is used to contain lifts and stairs. Lateral wind forces are transferred from the external wall cladding through the floor structure and into the core.

Hull or tube core

In a variation of the central core structure, the perimeter wall and its structural bracing contain all the floor area. The outer framing, bracing and cladding is known as the hull and it functions structurally with the internal core through the intermediary

Steel frame structure

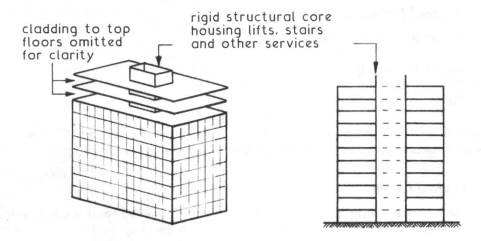

Figure 14.2 Core structures

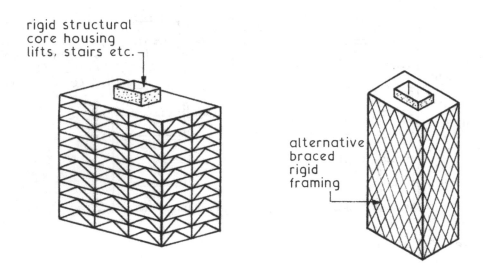

Figure 14.3 Core structure with braced hull

floors. The result is a rigid form of construction with a design potential for very tall buildings.

Shear wall

Lateral loading is resisted by rigid reinforced concrete external and internal walls. Floors act as diaphragms that transmit the lateral wind loads from façade cladding to shear walls. Long rectangular plan shaped buildings are suited to this design.

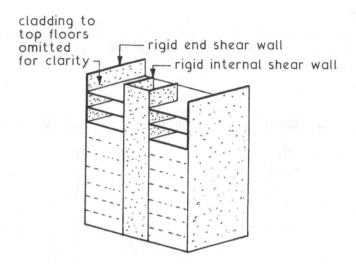

Figure 14.4 Shear wall construction

Diagonal braced

A steel variation of the concrete shear wall. Structural walls are substituted with steel bracing, which is, in effect, a series of vertical steel trusses. Columns are designed solely as compression members.

Suspended, propped and cantilevered are variations on a design theme that reduce the ground space occupied by the structural components. This facilitates unimpeded and sheltered pedestrian and vehicular access.

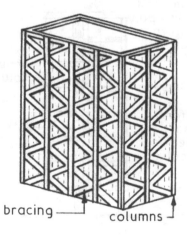

Figure 14.5 Diagonal braced frame

Steel frame structure

Suspended

Uses a central services structural core to resist lateral forces and to transmit vertical loads to the foundations. Floors are supported on beams secured to peripheral columns that hang from a roof level truss.

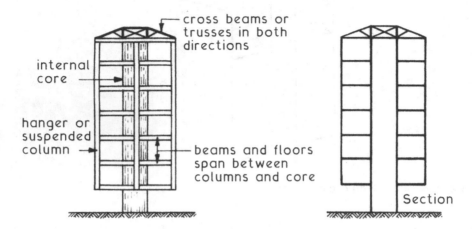

Figure 14.6 Suspended structure

Propped

A first-floor cantilevered reinforced concrete slab or a steel structure that forms a base for peripheral columns. These columns carry the floor beams and structural floor slabs.

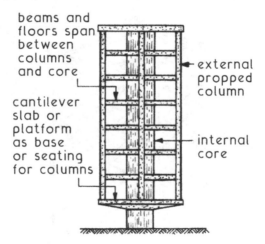

Figure 14.7 Externally propped structure

Cantilevered

Each floor slab is cantilevered from a structural inner core. Peripheral columns are not required and external cladding can be lightweight infill panelling. The infilling must have sufficient structural resistance to lateral wind loads.

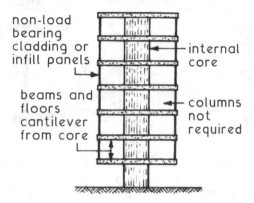

Figure 14.8 Cantilever floor structure

14.3 Hot rolled steel sections

Structural steelwork standard sections

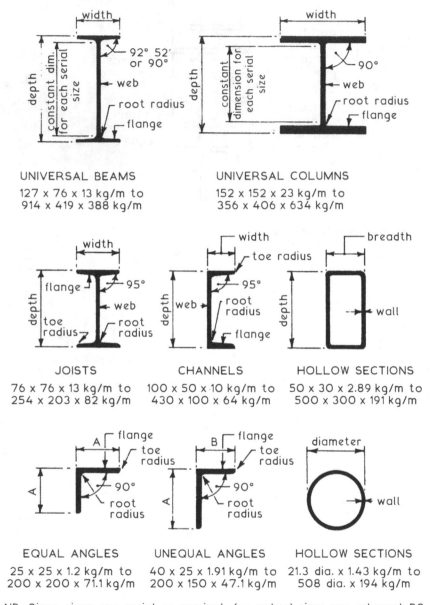

UNIVERSAL BEAMS
127 x 76 x 13 kg/m to
914 x 419 x 388 kg/m

UNIVERSAL COLUMNS
152 x 152 x 23 kg/m to
356 x 406 x 634 kg/m

JOISTS
76 x 76 x 13 kg/m to
254 x 203 x 82 kg/m

CHANNELS
100 x 50 x 10 kg/m to
430 x 100 x 64 kg/m

HOLLOW SECTIONS
50 x 30 x 2.89 kg/m to
500 x 300 x 191 kg/m

EQUAL ANGLES
25 x 25 x 1.2 kg/m to
200 x 200 x 71.1 kg/m

UNEQUAL ANGLES
40 x 25 x 1.91 kg/m to
200 x 150 x 47.1 kg/m

HOLLOW SECTIONS
21.3 dia. x 1.43 kg/m to
508 dia. x 194 kg/m

NB. Sizes given are serial or nominal, for actual sizes see relevant BS.

Figure 14.9 Standard steel sections

BS EN 10365: *Hot rolled steel channels, I and H sections.*
BS EN 10056: *Specification for Structural Steel Equal and Unequal Angles.*
BS EN 10210: *Hot Finished Structural Hollow Sections of Non-Alloy and Fine Grain Steels.*

Compound sections

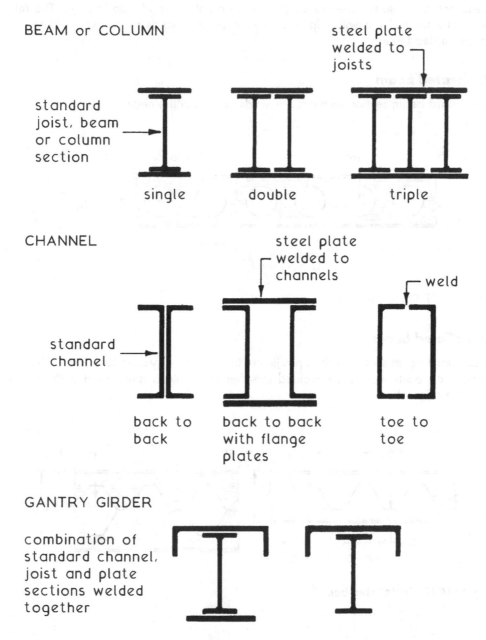

Figure 14.10 Compound steel elements

Hot rolled steel sections

These are produced by welding together standard sections. Various profiles are possible, which can be designed specifically for extreme situations such as very high loads and long spans, where standard sections alone would be insufficient. Some popular combinations of standard sections are shown in Figure 14.12.

Open web beams

These are particularly suited to long spans with light to moderate loading. The relative increase in depth will help resist deflection and voids in the web will reduce structural dead load.

Perforated beam

A standard beam section with circular voids cut about the neutral axis.

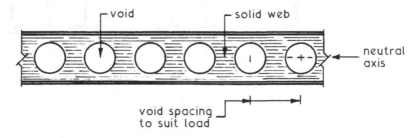

Figure 14.11 Perforated beam

Castellated beam

A standard beam section web is profile cut into two by oxy-acetylene torch. The projections on each section are welded together to create a new beam 50% deeper than the original.

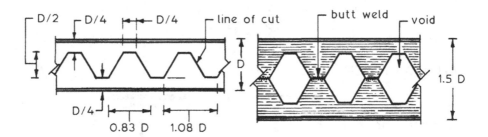

Figure 14.12 Castellated beam

Litzka beam

A standard beam cut as the castellated beam but with overall depth increased further by using spacer plates welded to the projections, giving a minimal increase in weight for greater depth.

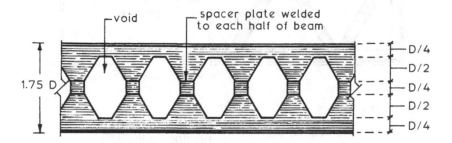

Figure 14.13 Litzka castellated beam
Note: Voids at the end of open web beams should be filled with a welded steel plate, as this is the area of maximum shear stress in a beam.

Lattice beams

These are an alternative type of open web beam, using standard steel sections to fabricate high depth to weight ratio units capable of spans up to about 15m. The range of possible components is extensive, and some examples are shown in Figure 14.14.

Hot rolled steel sections

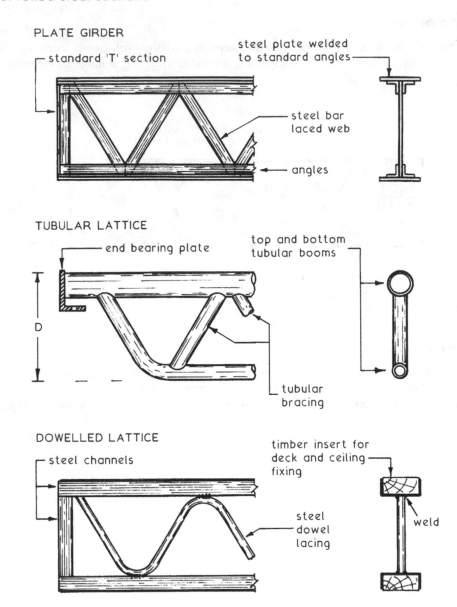

PLATE GIRDER

standard 'T' section

steel plate welded to standard angles

steel bar laced web

angles

TUBULAR LATTICE

end bearing plate

top and bottom tubular booms

D

tubular bracing

DOWELLED LATTICE

steel channels

timber insert for deck and ceiling fixing

steel dowel lacing

weld

Figure 14.14 Lattice beam forms
Note: Span potential for lattice beams is approximately 24 × D.

14.4 Steelwork connections

These are either workshop or site connections according to where the fabrication takes place. Most site connections are bolted whereas workshop connections are very often carried out by welding. The design of structural steelwork members and their connections is the province of the structural engineer who selects the type and number of bolts or the size and length of weld to be used according to the connection strength to be achieved.

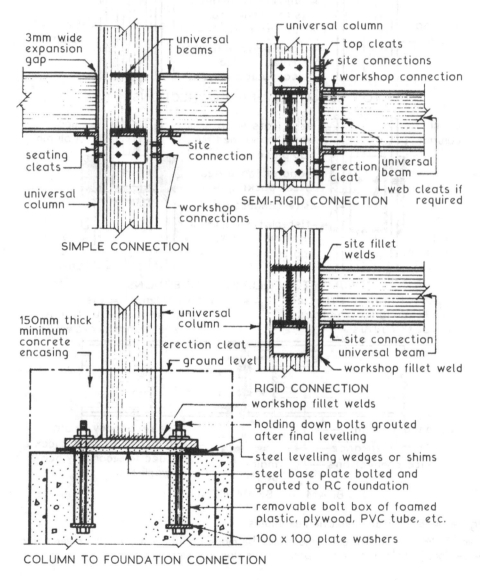

Figure 14.15 Types of steel frame connections

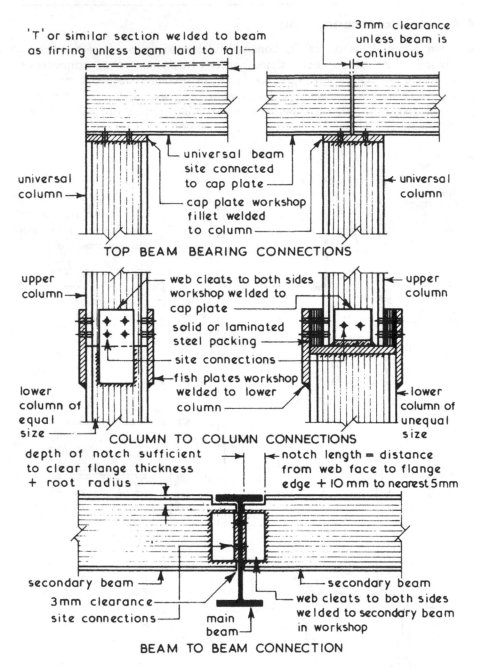

'T' or similar section welded to beam as firring unless beam laid to fall

3 mm clearance unless beam is continuous

universal column →

universal beam site connected to cap plate

cap plate workshop fillet welded to column

→ universal column

TOP BEAM BEARING CONNECTIONS

upper column →

web cleats to both sides workshop welded to cap plate

solid or laminated steel packing

site connections

fish plates workshop welded to lower column

lower column of equal size

← upper column

←lower column of unequal size

COLUMN TO COLUMN CONNECTIONS

depth of notch sufficient to clear flange thickness + root radius

notch length = distance from web face to flange edge + 10 mm to nearest 5 mm

secondary beam

3 mm clearance

site connections

main beam

secondary beam

web cleats to both sides welded to secondary beam in workshop

BEAM TO BEAM CONNECTION

Figure 14.16 Typical connection examples

Note: All holes for bolted connections must be made from backmarking the outer surface of the section(s) involved. For actual positions see structural steelwork manuals.

Column base connections

The type selected will depend on the load carried by the column and the distribution area of the base plate. The cross-sectional area of a UC concentrates the load into a relatively small part of the base plate. Therefore, to resist bending and shear, the base must be designed to resist the column loads and transfer them onto the pad foundation below. Types: slab or bloom base, gusset base.

Bolt boxes

These are used to accurately locate columns holding down bolts into wet concrete, with plastic tubes providing space around the bolts when the concrete has set. The bolts can then be moved slightly to aid alignment with the column base.

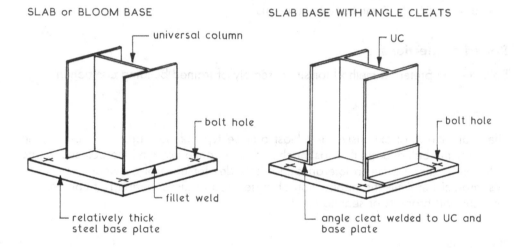

Figure 14.17 Steel column base connections

Steelwork connections

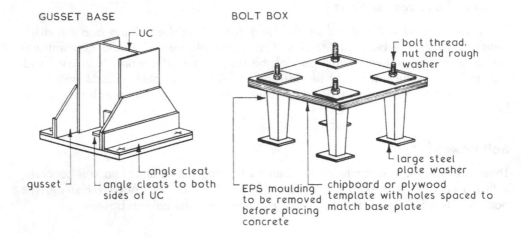

Figure 14.18 Steel column base connections

Bolted connections

Bolts are the preferred method for site assembly of framed building components

BLACK BOLTS

These are the least expensive and least precise type of bolt, produced by forging with only the bolt and nut threads machined. Clearance between the bolt shank and bolt hole is about 2mm, a tolerance that provides for ease of assembly. However, this imprecision limits the application of these bolts to direct bearing of components onto support brackets or seating cleats.

BRIGHT BOLTS

Also known as turned and fitted bolts. These are machined under the bolt head and along the shank to produce a close fit of 0.5mm hole clearance. They are specified where accuracy is paramount.

HIGH STRENGTH FRICTION GRIP BOLTS

Also known as torque bolts as they are tightened to a predetermined shank tension by a torque-controlled wrench. This procedure produces a clamping force that transfers the connection by friction between components and not by shear or bearing on the bolts. These bolts are manufactured from high-yield steel. The number of bolts used to make a connection is less than otherwise required.

BS 4190: *ISO Metric Black Hexagon Bolts, Screws and Nuts. Specification.*
BS 3692: *ISO Metric Precision Hexagon Bolts, Screws and Nuts. Specification.*
BS EN 14399 (10 parts): *High Strength Structural Bolting Assemblies for Preloading.*

14.5 Fire protection

Although steel is a noncombustible material with negligible surface spread of flame properties it does not behave very well under fire conditions. During the initial stages of a fire the steel will actually gain in strength, but this reduces to normal at a steel temperature range of 250 to 400°C and continues to decrease until the steel temperature reaches 550°C, when it has lost most of its strength. Since the temperature rise during a fire is rapid, most structural steelwork will need protection to give it a specific degree of fire resistance in terms of time. Approved Document B of the Building Regulations sets out the minimum requirements related to building usage and size, BRE Report 128 'Guidelines for the Construction of Fire Resisting Structural Elements' gives acceptable methods.

Typical examples for 120 minutes fire resistance

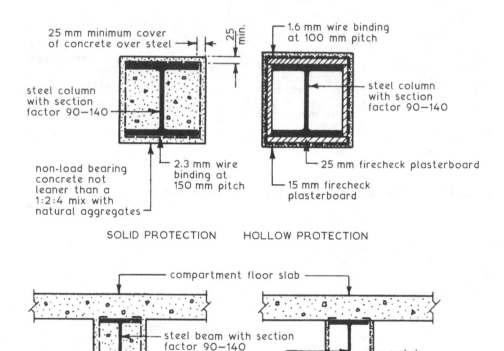

Figure 14.19 Solid and hollow fire protection examples to columns and beams

Fire protection

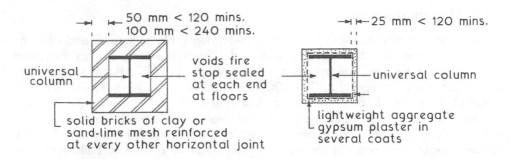

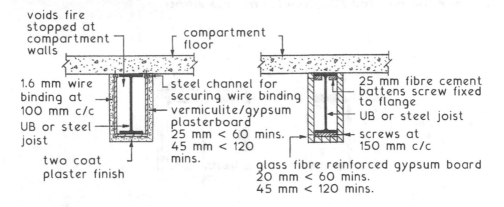

Figure 14.20 Further example of hollow protection to columns

Figure 14.21 Further example of hollow protection to beams

15 CONCRETE FRAMED BUILDINGS

Opening remarks

There are two principal methodologies: reinforced in-situ concrete frames or precast concrete frames.

In the cast in-situ method formworks are erected to contain the wet concrete and form the element, steel reinforcement is installed within the form work or the formwork is erected around the steel reinforcement, concrete is then pumped into the form. A vibrating poker is inserted into the wet concrete to remove air and consolidate the concrete. After the concrete is set the formwork is removed.

Construction sequence:

1. Assemble and erect formwork.
2. Prepare and place reinforcement.
3. Pour and compact or vibrate concrete.
4. Strike and remove formwork in stages as curing proceeds.

15.1 Formwork

When first mixed, concrete is a fluid and, therefore, to form any concrete member the wet concrete must be placed in a mould or formwork to retain its shape, size and position as it sets. Falsework is the term used for the temporary structure which supports the formwork.

Formwork is either fabricated on site from plywood and timber or proprietary aluminium, steel or composite systems are used.

Beam formwork

Constructed with plywood and timber bracings, this is basically a three-sided box supported and propped in the correct position and to the desired level. The beam formwork sides have to retain the wet concrete in the required shape and be able to withstand the initial hydrostatic pressure of the wet concrete, whereas the formwork soffit, apart from retaining the concrete, has to support the initial load of the wet concrete and finally the set concrete until it has gained sufficient strength to be self-supporting. It is essential that all joints in the formwork are constructed to prevent the escape of grout, which could result in honeycombing and/or feather

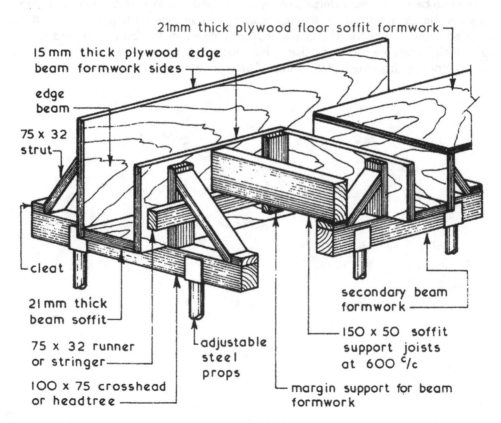

Figure 15.1 Plywood and timber beam formwork

Formwork

edging in the cast beam. The removal time for the formwork will vary with air temperature, humidity and consequent curing rate. A releasing agent is applied to the formwork to prevent the concrete bonding to the formwork.

Typical formwork striking times

Table 15.1 In-situ concrete typical formwork striking times

Beam sides – 9 to 12 hours	Using OPC air temp
Beam soffits – 8 to 14 days (props left under)	7 to 16°C
Beam props – 15 to 21 days	

Column formwork

Vertical formwork is commonly erected around the reinforcement on all sides and can be plywood and timber or purpose-made aluminium forms clamped together. The head of the column formwork can be used to support the incoming beam formwork, which gives good top lateral restraint but results in complex formwork. Alternatively the column can be cast to the underside of the beams and at a later stage a collar of formwork can be clamped around the cast column to complete casting and support the incoming beam formwork. Column forms are located at the bottom around a 75–100mm high concrete plinth or kicker, which has the dual function of location and preventing grout loss from the bottom of the column formwork.

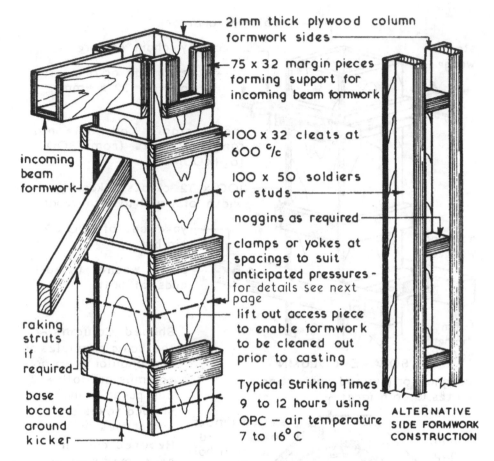

21mm thick plywood column formwork sides

75 x 32 margin pieces forming support for incoming beam formwork

incoming beam formwork

100 x 32 cleats at 600 °/c

100 x 50 soldiers or studs

noggins as required

raking struts if required

clamps or yokes at spacings to suit anticipated pressures - for details see next page

lift out access piece to enable formwork to be cleaned out prior to casting

base located around kicker

Typical Striking Times 9 to 12 hours using OPC – air temperature 7 to 16°C

ALTERNATIVE SIDE FORMWORK CONSTRUCTION

Figure 15.2 Plywood and timber column formwork

Wall forms

Bespoke wall forms are made of plywood sheeting and timber framing with repetitive forms commonly of steel or aluminium proprietary systems. The forms may be plastic or wood faced for specific concrete finishes. Stability is provided by vertical studs and horizontal walings retained in place by adjustable props. Base location is by a kicker of 50 of 75mm height of width to suit the wall thickness.

To keep the wall forms apart at the correct distance tube spacers are placed over the bolts between the forms. For greater load applications, variations include purpose-made high tensile steel bolts or dowels. These too are sleeved with plastic tubes and have removable spacer cones inside the forms. Surface voids from the spacers can be made good with strong mortar. Some examples are shown in Figure 15.4 with the alternative coil tie system.

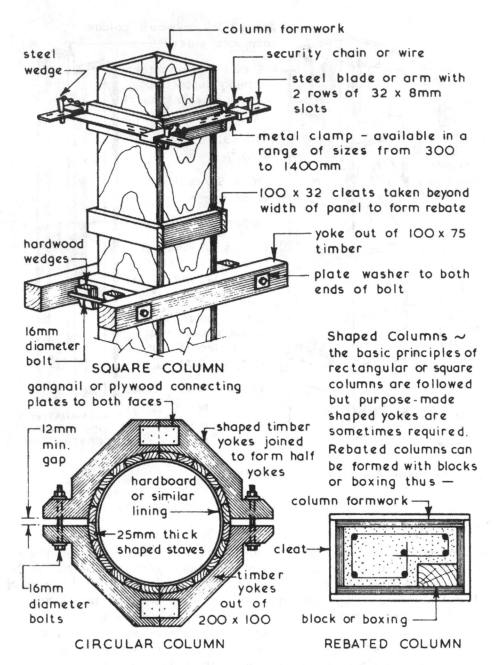

column formwork

security chain or wire

steel blade or arm with 2 rows of 32 x 8mm slots

steel wedge

metal clamp - available in a range of sizes from 300 to 1400mm

100 x 32 cleats taken beyond width of panel to form rebate

hardwood wedges

yoke out of 100 x 75 timber

plate washer to both ends of bolt

16mm diameter bolt

SQUARE COLUMN

gangnail or plywood connecting plates to both faces

Shaped Columns ~ the basic principles of rectangular or square columns are followed but purpose-made shaped yokes are sometimes required. Rebated columns can be formed with blocks or boxing thus —

12mm min. gap

shaped timber yokes joined to form half yokes

hardboard or similar lining

25mm thick shaped staves

column formwork

16mm diameter bolts

timber yokes out of 200 x 100

cleat

block or boxing

CIRCULAR COLUMN

REBATED COLUMN

Figure 15.3 Column formwork clamping

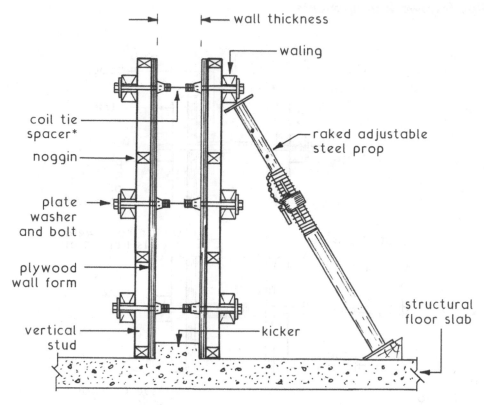

Figure 15.4 Wall formwork principles
Note: Reinforcement omitted for clarity.

Floor formwork

Comprises plywood sheet forms with timber framing or proprietary aluminium or steel sheet forms to provide a level deck for steel reinforcement bars on spacers and as a soffit form for the subsequent pour of wet concrete. The deck is supported on falsework and steel props.

BS 6100–9: *Building and Civil Engineering. Vocabulary. Work with Concrete and Plaster.*
BS 8000–2.2: *Workmanship on Building Sites. Code of Practice for Concrete Work. Sitework with In-Situ and Precast Concrete.*
BRE Report 495: *Concrete Frame Buildings, Modular Formwork.*

Formwork

Slab formwork components

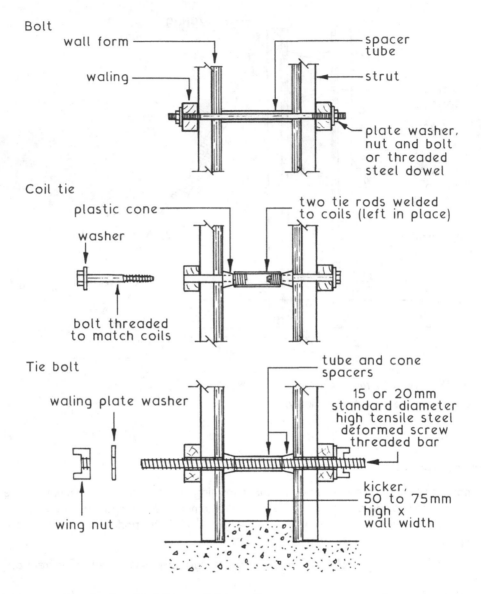

Figure 15.5 Methods of holding wall forms apart
Note: Reinforcement omitted for clarity.

Falsework

This term relates to the temporary support props and frames that support the form-work in position.

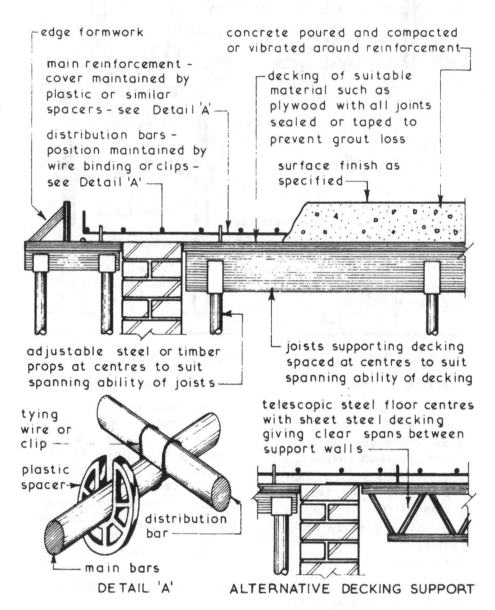

Figure 15.6 Concrete deck falsework

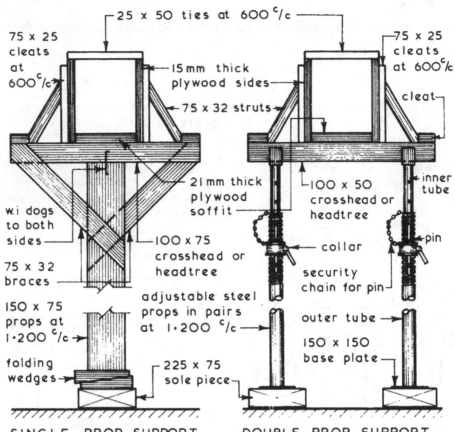

SINGLE PROP SUPPORT   **DOUBLE PROP SUPPORT**

Erecting Formwork

1. Props positioned and levelled through.
2. Soffit placed, levelled and position checked.
3. Side forms placed, their position checked before being fixed.
4. Strutting position and fixed.
5. Final check before casting.

Suitable Formwork Materials~ timber, steel and special plastics.

Striking or Removing Formwork

1. Side forms as soon as practicable usually within hours of casting this allows drying air movements to take place around the setting concrete.
2. Soffit formwork as soon as practicable usually within days but as a precaution some props are left in position until concrete member is self-supporting.

Figure 15.7 Typical simple beam formwork details

15.2 Reinforcement

Concrete is a material that is strong in compression and weak in tension, and if the member is overloaded its tensile resistance may be exceeded, leading to structural failure.

Steel bars and mesh are used to provide the tensile strength that plain concrete lacks. The number, diameter, spacing, shape and type of bars to be used have to be designed. Reinforcement is placed as near to the outside as practicable, with sufficient cover of concrete over the reinforcement to protect the steel bars from corrosion and to provide a degree of fire resistance. Slabs that are square in plan are considered to be spanning in two directions and therefore main reinforcing bars are used both ways. Slabs that are rectangular in plan, however, are considered to span across the shortest distance and main bars are used in this direction only, with smaller diameter distribution bars placed at right angles, forming a mat or grid.

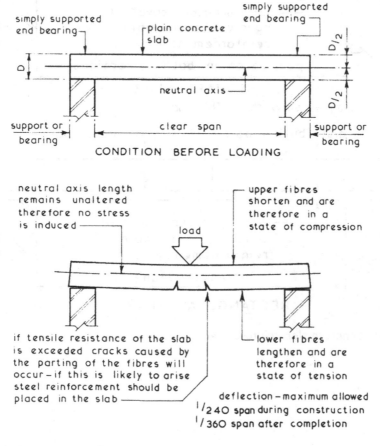

Figure 15.8 Beam loading forces

Reinforcement

Rebar

Mild steel or high yield steel both contain about 99% iron. The remaining constituents are manganese, carbon, sulphur and phosphorus. The proportion of carbon determines the quality and grade of steel; mild steel has 0.25% carbon, high yield steel 0.40%. High yield steel may also be produced by cold working or deforming mild steel until it is strain hardened. Mild steel has the letter R preceding the bar diameter in mm, e.g. R20, and high yield steel the letter T or Y.

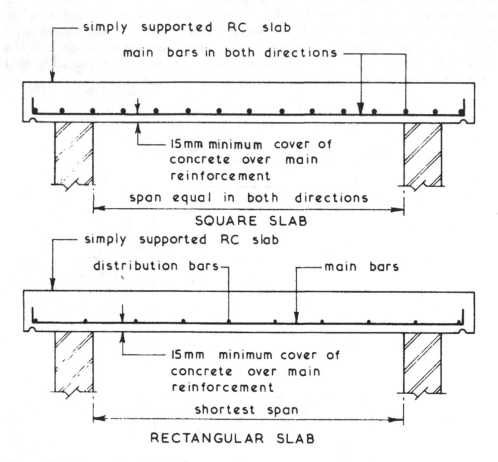

Figure 15.9 Concrete slab reinforcement

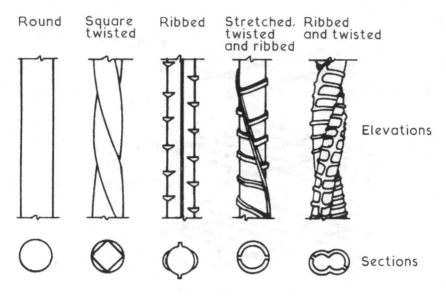

Round Square   Ribbed   Stretched,   Ribbed
       twisted   twisted   and twisted
                         and ribbed

Elevations

Sections

Figure 15.10 Examples of steel reinforcement

Standard bar diameters = 6, 8, 10, 12, 16, 20, 25, 32 and 40mm.

Steel mesh

Steel reinforcement mesh or fabric is produced in four different formats for different applications:

Standard sheet size = 4.8m long × 2.4m wide.

Standard roll size = 48 and 72m long × 2.4m wide.

Specification ~ format letter plus a reference number. This number equates to the cross sectional area in mm^2 of the main bars per metre width of mesh.

E.g. B385 is rectangular mesh with 7mm dia. main bars, i.e. 10 bars of 7mm dia. @ 100mm spacing = 385mm^2.

BS 4483: *Steel Fabric for the Reinforcement of Concrete. Specification.*

BS 4482: *Steel Wire for the Reinforcement of Concrete Products. Specification.*

Table 15.2 Steel reinforcement mesh types and applications

Format	Type	Typical application
A	Square mesh	Floor slabs
B	Rectangular mesh	Floor slabs
C	Long mesh	Roads and pavements
D	Wrapping mesh	Binding wire with concrete fire protection to structural steelwork

Reinforcement

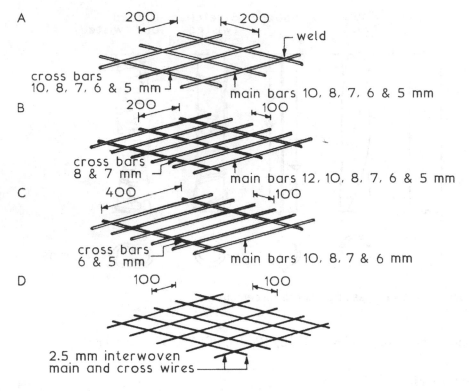

Figure 15.11 Steel mesh reinforcement types

Reinforcement concrete cover

Cover to reinforcement in columns, beams, foundations, etc. is required for the following reasons:

- To protect the steel against corrosion.
- To provide sufficient bond or adhesion between steel and concrete.
- To ensure sufficient protection of the steel in a fire.

If the cover is insufficient, concrete will spall away from the steel.

Minimum cover ~ never less than the maximum size of aggregate in the concrete, or the largest reinforcement bar size (take greater value).

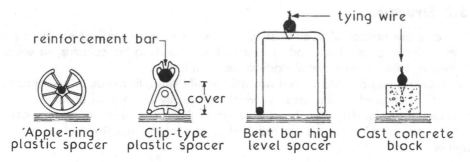

Figure 15.12 Concrete cover spacers

Guidance on minimum cover for particular locations

Below ground:

- Foundations, retaining walls, basements, etc., 40mm, binders 25mm.
- Marine structures, 65mm, binders 50mm.
- Uneven earth and fill 75mm, blinding 40mm.

Above ground:

- Ends of reinforcing bars, not less than 25mm nor less than 2 × bar diameter.
- Column longitudinal reinforcement, 40mm, binders 20mm.
- Columns <190mm min. dimension with bars <12mm diameter, 25mm.
- Beams, 25mm, binders 20mm.
- Slabs, 20mm (15mm where max. aggregate size is <15mm).

Note: Minimum cover for corrosion protection and bond may not be sufficient for fire protection and severe exposure situations.

15.3 Structure

RC beams are horizontal, load-bearing members that are classified as either main beams, which transmit floor and secondary beam loads to the columns, or secondary beams, which transmit floor loads to the main beams.

Concrete being a material that has little tensile strength needs to be reinforced to resist the induced tensile stresses, which can be in the form of ordinary tension or diagonal tension (shear). The calculation of the area, diameter, type, position and number of reinforcing bars required is one of the functions of a structural engineer.

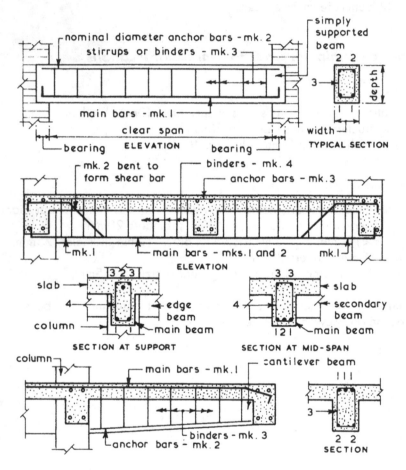

Figure 15.13 Typical RC beam details

RC columns

Columns are the vertical, load-bearing members of the structural frame, which transmit the beam loads down to the foundations. They are usually constructed in storey heights and, therefore, the reinforcement must be lapped to provide structural continuity.

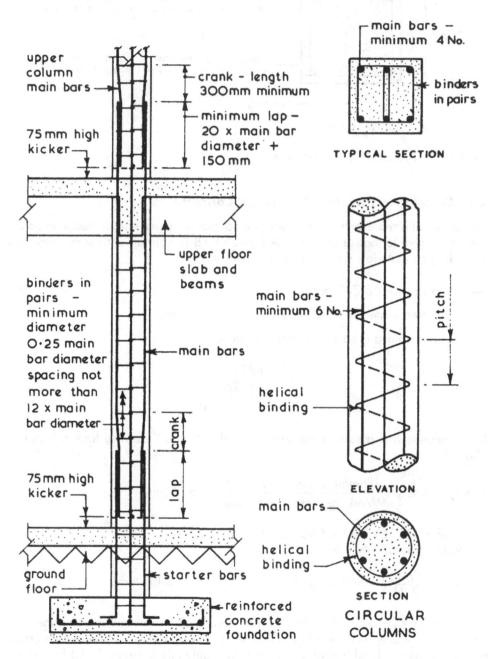

Figure 15.14 Typical RC column details

Structure

e.g.

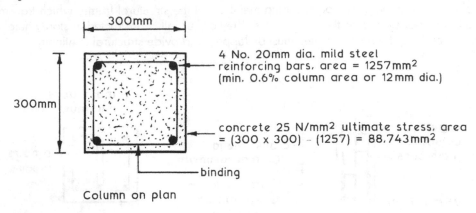

Figure 15.15 Typical RC column details

Load Calculation for Steel Reinforced Concrete

A modular ratio represents the amount of load that a square unit of steel can safely transmit relative to that of concrete. A figure of 18 is normal, with some variation depending on materials specification and quality.

Area of concrete = 88.743 mm^2

Equivalent area of steel = 18 x 1257 mm^2 = 22626 mm^2

Equivalent combined area of concrete and steel:

$$\begin{array}{r} 88743 \\ +22626 \\ \hline 111369 \text{mm}^2 \end{array}$$

Using concrete with a safe or working stress of 5 N/mm^2, derived from a factor of safety of 5, i.e.

$$\text{Factory of safety} = \frac{\text{Ultimate stress}}{\text{Working stress}} = \frac{25 \text{ N/mm}^2}{5 \text{ N/mm}^2} = 5 \text{ N/mm}^2$$

5 N/mm^2 x 111369 mm^2 = 556845 Newtons
kg x 9.81 (gravity) = Newtons

Therefore: $\frac{556845}{9.81}$ = 56763kg or 56·76 tonnes permissible load

NB. This is the safe load calculation for a reinforced concrete column where the load is axial and bending is minimal or non-existent, due to a very low slenderness ratio (effective length to least lateral dimension). In reality this is unusual and factors for buckling will usually be incorporated into the calculation.

15.4 Fire protection

Concrete is an inherently fire resistant material. The following are examples of fire resistance.

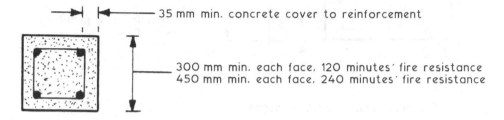

35 mm min. concrete cover to reinforcement

300 mm min. each face, 120 minutes' fire resistance
450 mm min. each face, 240 minutes' fire resistance

Figure 15.16 Column fully exposed

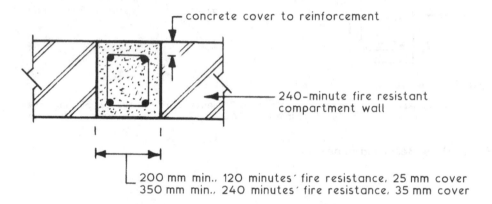

concrete cover to reinforcement

240-minute fire resistant compartment wall

200 mm min., 120 minutes' fire resistance, 25 mm cover
350 mm min., 240 minutes' fire resistance, 35 mm cover

Figure 15.17 Column, maximum 50% exposed

Fire protection

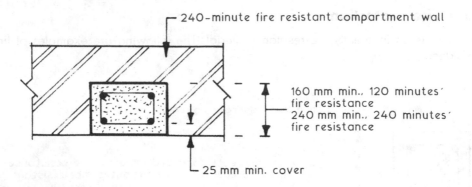

240-minute fire resistant compartment wall

160 mm min., 120 minutes' fire resistance
240 mm min., 240 minutes' fire resistance

25 mm min. cover

Figure 15.18 Column, one face only exposed

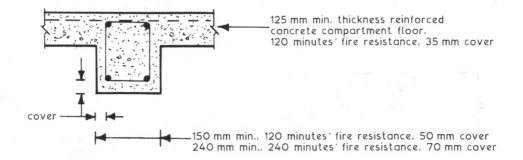

125 mm min. thickness reinforced concrete compartment floor, 120 minutes' fire resistance, 35 mm cover

cover

150 mm min., 120 minutes' fire resistance, 50 mm cover
240 mm min., 240 minutes' fire resistance, 70 mm cover

Figure 15.19 Beam and floor slab

15.5 Precast concrete frames

Applications

Precast concrete (PCC) frames are suitable for single-storey and low rise applications. They provide the skeleton for the building and clad externally. The frames are usually produced as part of a manufacturer's standard range of designs and are therefore seldom purpose made, due mainly to the high cost of the moulds.

Advantages

* Factory-controlled conditions for quality and accuracy.
* Repetitive casting lowers the cost of individual members.
* Off-site production releases site space for other activities.
* Frames can be assembled in cold weather and generally by semi-skilled labour.

Disadvantages

* Lack the design flexibility of cast in-situ, purpose-made frames.
* Site planning can be limited by manufacturer's delivery programme.
* Require lifting plant of a type and size not normally required by traditional construction methods.

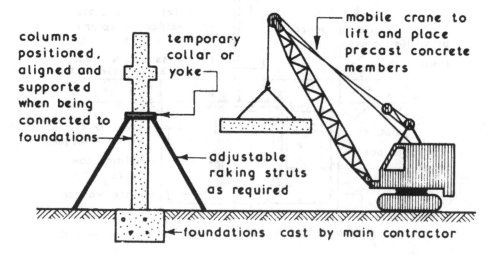

columns positioned, aligned and supported when being connected to foundations

temporary collar or yoke

mobile crane to lift and place precast concrete members

adjustable raking struts as required

foundations cast by main contractor

Figure 15.20 Typical site activities

15.6 Connections

Foundation connections

The preferred method of connection is to set the column into a pocket cast into a reinforced concrete pad foundation and is suitable for light to medium loadings.

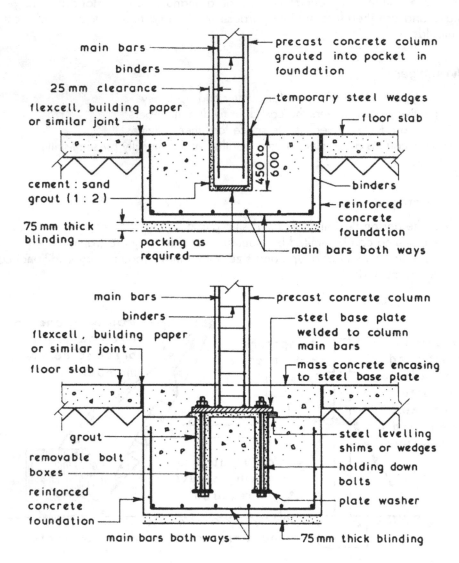

Figure 15.21 Pre-cast column to foundation connections

Where heavy column loadings are encountered it may be necessary to use a steel base plate secured to the reinforced concrete pad foundation with holding down bolts.

Column to column connection

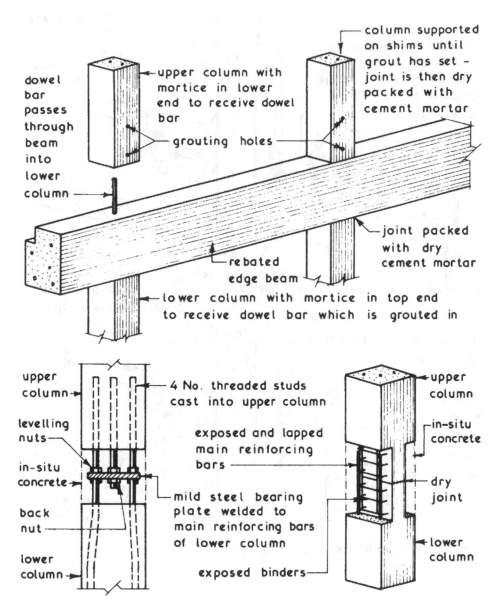

dowel bar passes through beam into lower column

upper column with mortice in lower end to receive dowel bar

grouting holes

column supported on shims until grout has set – joint is then dry packed with cement mortar

joint packed with dry cement mortar

rebated edge beam

lower column with mortice in top end to receive dowel bar which is grouted in

upper column

levelling nuts

in-situ concrete

back nut

lower column

4 No. threaded studs cast into upper column

exposed and lapped main reinforcing bars

mild steel bearing plate welded to main reinforcing bars of lower column

exposed binders

upper column

in-situ concrete

dry joint

lower column

Figure 15.22 Pre-cast frame connections

Beam to column connections

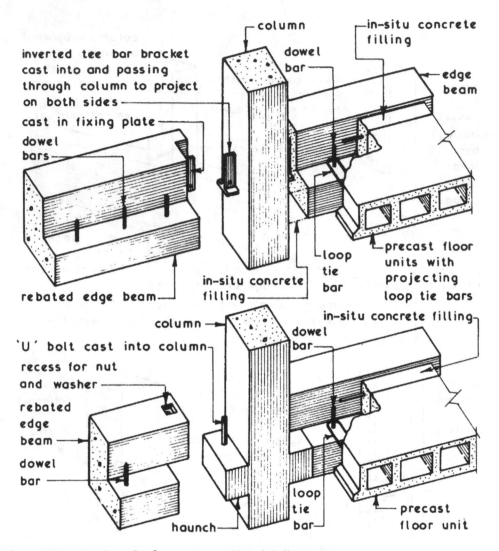

inverted tee bar bracket
cast into and passing
through column to project
on both sides

cast in fixing plate

dowel
bars

rebated edge beam

column

dowel
bar

in-situ concrete
filling

edge
beam

in-situ concrete
filling

loop
tie
bar

precast floor
units with
projecting
loop tie bars

'U' bolt cast into column

recess for nut
and washer

rebated
edge
beam

dowel
bar

column

dowel
bar

in-situ concrete filling

haunch

loop
tie
bar

precast
floor unit

Figure 15.23 Further RC frame connection details

15.7 Pre-stressed concrete

Concrete has high compressive strength and low tensile strength. Adding steel reinforcement in a predetermined pattern gives the concrete the required tensile strength.

In pre-stressed concrete a pre-compression is induced into the member to make full use of its own inherent compressive strength when loaded. The design aim is to achieve a balance of tensile and compressive forces so that the end result is a concrete member that is resisting only stresses that are compressive. In practice a small amount of tension may be present but, providing this does not exceed the tensile strength of the concrete being used, tensile failure will not occur (see Figure 15.24).

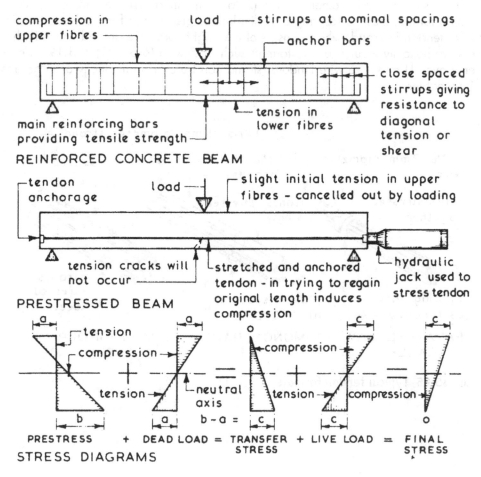

Figure 15.24 Comparison of reinforced and pre-stressed concrete

Pre-stressed concrete

Concrete will shrink whilst curing and it can also suffer sectional losses due to creep when subjected to pressure. The amount of shrinkage and creep likely to occur can be controlled by designing the strength and workability of the concrete, high strength and low workability giving the greatest reduction in both shrinkage and creep.

Mild steel will suffer from relaxation losses – where the stresses in steel under load decrease to a minimum value after a period of time – and this can be overcome by increasing the initial stress in the steel. If mild steel is used for pre-stressing the summation of shrinkage, creep and relaxation losses will cancel out any induced compression, therefore special alloy steels must be used to form tendons for pre-stressed work.

Tendons

These consist of small diameter wires (2 to 7mm) in a plain round, crimped or indented format; these wires may be individual or grouped to form cables. Another form of tendon is strand, which consists of a straight core wire around which further wires are helically wound to give formats such as 7 wire (6 over 1) and 19 wire (9 over 9 over 1) and, like wire tendons, strand can be used individually or in groups to form cables.

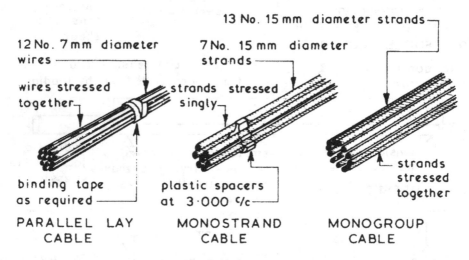

Figure 15.25 Typical tendon formats

15.8 Pre-tensioning

In pre-tensioning the wires are stressed within the mould before the concrete is placed around them. Steam curing is often used to accelerate this process to achieve a 24-hour characteristic strength of $28N/mm^2$, with a typical 28-day cube strength of $40N/mm^2$. Stressing of the wires is carried out by using hydraulic jacks operating from one or both ends of the mould to achieve an initial 10% overstress to counteract expected losses. After curing, the wires are released or cut and the bond between the stressed wires and the concrete prevents the tendons from regaining their original length, thus maintaining the pre-compression or pre-stress.

At the extreme ends of the members the bond between the stressed wires and concrete is not fully developed due to low frictional resistance. This results in a small contraction and swelling at the ends of the wire, forming, in effect, a cone shape anchorage. The distance over which this contraction occurs is called the transfer length and is equal to 80 to 120 times the wire diameter. To achieve a greater total surface contact area it is common practice to use a larger number of small diameter wires rather than a smaller number of large diameter wires giving the same total cross-sectional area.

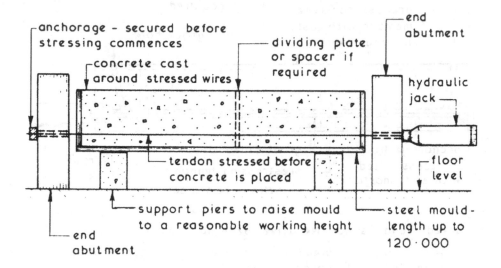

Figure 15.26 Typical pre-tensioning arrangement

15.9 Post-tensioning

This method is usually employed where stressing is to be carried out on site after casting an in-situ component or where a series of precast concrete units are to be joined together to form the required member. It can also be used where curved tendons are to be used to overcome negative bending moments. In post-tensioning, the concrete is cast around ducts or sheathing in which the tendons are to be housed. Stressing is carried out after the concrete has cured by means of hydraulic jacks operating from one or both ends of the member. The anchorages, which form part of the complete component, prevent the stressed tendon from regaining its original length, thus maintaining the pre-compression or pre-stress.

After stressing the annular space in the tendon ducts should be filled with grout to prevent corrosion of the tendons due to any entrapped moisture and to assist in stress distribution. Due to the high local stresses at the anchorage positions it is usual for a reinforcing spiral to be included in the design.

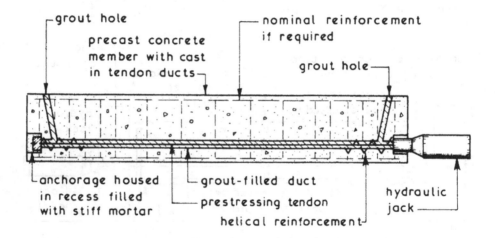

Curved Tendons for Negative Bending Moments~

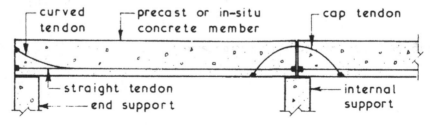

Figure 15.27 Post–tensioning details

Anchorages

The anchorages used in conjunction with post-tensioned, pre-stressed concrete works depends mainly on whether the tendons are to be stressed individually or as a group, but most systems use a form of split cone wedges or jaws acting against a form of bearing or pressure plate.

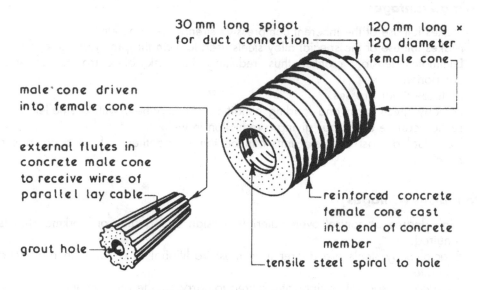

30 mm long spigot for duct connection

120 mm long × 120 diameter female cone

male cone driven into female cone

external flutes in concrete male cone to receive wires of parallel lay cable

grout hole

reinforced concrete female cone cast into end of concrete member

tensile steel spiral to hole

FREYSSINET ANCHORAGE

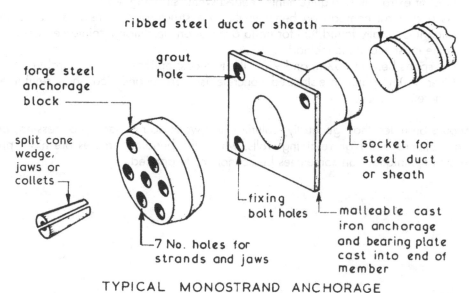

ribbed steel duct or sheath

forge steel anchorage block

grout hole

split cone wedge, jaws or collets

7 No. holes for strands and jaws

fixing bolt holes

socket for steel duct or sheath

malleable cast iron anchorage and bearing plate cast into end of member

TYPICAL MONOSTRAND ANCHORAGE

Figure 15.28 Typical anchorage details

15.10 Comparison of pre-stressed and reinforced concrete

When comparing pre-stressed concrete with conventional reinforced concrete the main advantages and disadvantages can be enumerated but in the final analysis each structure and/or component must be decided on its own merit.

Main advantages

• Makes full use of the inherent compressive strength of concrete.
• Makes full use of the special alloy steels used to form the pre-stressing tendons.
• Eliminates tension cracks, thus reducing the risk of corrosion of steel components.
• Reduces shear stresses.
• For any given span and loading condition a component with a smaller cross section can be used, thus giving a reduction in weight.
• Individual precast concrete units can be joined together to form a composite member.

Main disadvantages

• High degree of control over materials, design and quality of workmanship is required.
• Special alloy steels are dearer than most traditional steels used in reinforced concrete.
• Extra cost of special equipment required to carry out the pre-stressing activities.
• Cost of extra safety requirements needed whilst stressing tendons.
• As a general comparison between the two structural options under consideration it is usually found that for up to 6.0m spans traditional reinforced concrete is the most economic method.
• For spans between 6.0m and 9.0m the two cost options are comparable.
• Over 9.0m spans, pre-stressed concrete is more economical than reinforced concrete.

It should be noted that, generally, columns and walls do not need pre-stressing, but in tall columns and high retaining walls, where the bending stresses are high, pre-stressing techniques can sometimes be economically applied.

15.11 Portal frames

Portal frames are available in precast concrete from a range of standard sections.

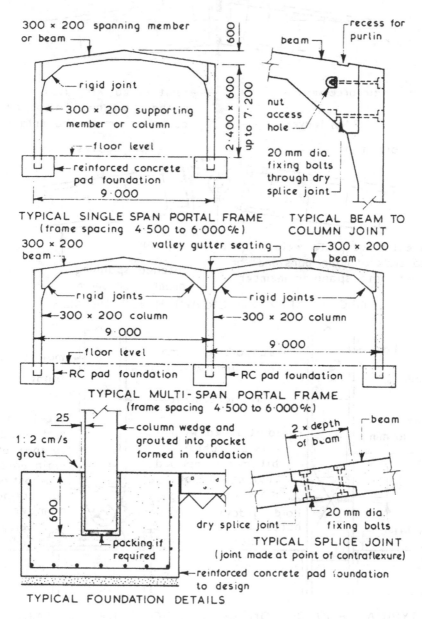

Figure 15.29 Typical precast concrete portal frame details

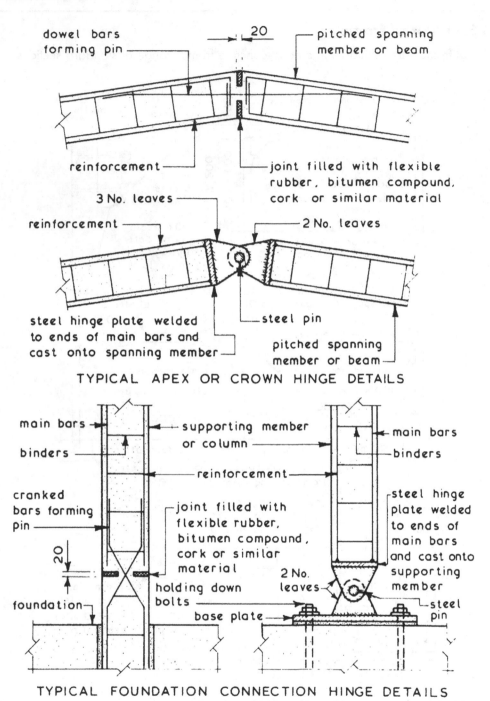

TYPICAL APEX OR CROWN HINGE DETAILS

TYPICAL FOUNDATION CONNECTION HINGE DETAILS

Figure 15.30 Typical precast concrete portal frame hinge details

16 CLADDING OF FRAMED BUILDINGS

Opening remarks

Framed buildings are designed to support the external envelope on the frame, in-between the frame or a combination of both.

16.1 Non-load-bearing brick panels

These are used in conjunction with framed structures with an inner leaf infill between the beams and columns and outer facing leaf supported on steel angles fixed to floor slabs or perimeter beams. They are constructed in the same manner as ordinary brick walls with the openings being formed by traditional methods.

Basic requirements

- To be adequately supported by and tied to the structural frame.
- Have sufficient strength to support own self-weight plus any attached finishes and imposed loads such as wind pressures.
- Provide the necessary resistance to penetration by the natural elements.
- Provide the required degree of thermal insulation, sound insulation and fire resistance.
- Have sufficient durability to reduce maintenance costs to a minimum.
- Provide for movements due to moisture and thermal expansion of the panel and for contraction of the frame.

These are used for multi-storey framed buildings, where a traditional brick façade is required.

Brickwork movement

To allow for climatic changes and differential movement between the cladding and main structure, a 'soft' joint (cellular polyethylene, cellular polyurethane, expanded rubber or sponge rubber with polysulphide or silicon pointing) should be located below the support angle. Vertical movement joints may also be required at a maximum of 12m spacing.

Lateral restraint

This is provided by normal wall ties between the inner and outer leaf of masonry, plus sliding brick anchors below the support angle.

Infill panel walls ~ these can be used between the framing members of a building to provide the cladding and division between the internal and external environments and are distinct from claddings and facing.

Functional requirements

All forms of infill panel should be designed and constructed to fulfil the following functional requirements:

- Be self-supporting between structural framing members.
- Provide resistance to the penetration of the elements.
- Provide resistance to positive and negative wind pressures.
- Give the required degree of thermal insulation.

Non-load-bearing brick panels

- Give the required degree of sound insulation.
- Give the required degree of fire resistance.
- Have sufficient openings to provide the required amount of natural ventilation.
- Have sufficient glazed area to fulfil the natural daylight and vision out requirements.
- Be economical in the context of construction and maintenance.
- Provide for any differential movements between panel and structural frame.

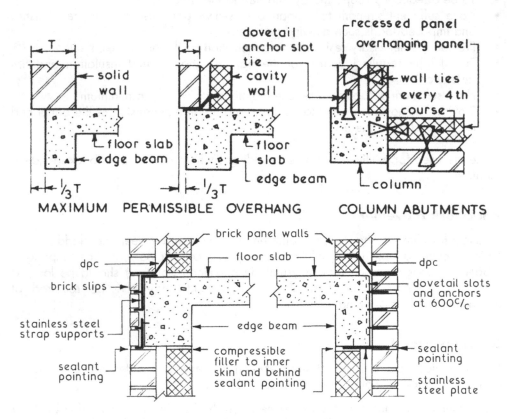

Figure 16.1 Non-load-bearing panel connection details

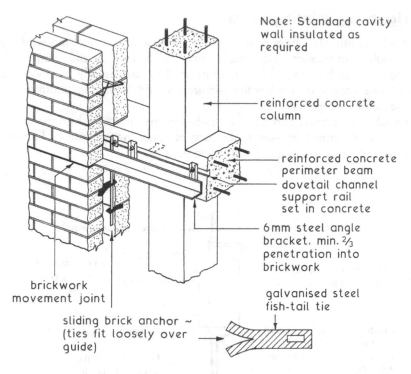

Note: Standard cavity wall insulated as required

reinforced concrete column

reinforced concrete perimeter beam

dovetail channel support rail set in concrete

6mm steel angle bracket, min. ⅔ penetration into brickwork

brickwork movement joint

galvanised steel fish-tail tie

sliding brick anchor ~ (ties fit loosely over guide)

Figure 16.2 Brickwork cladding support system

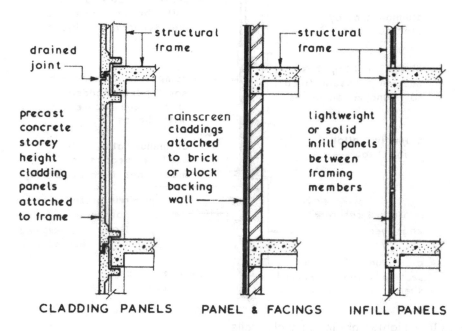

drained joint

structural frame

structural frame

precast concrete storey height cladding panels attached to frame

rainscreen claddings attached to brick or block backing wall

lightweight or solid infill panels between framing members

CLADDING PANELS   PANEL & FACINGS   INFILL PANELS

Figure 16.3 Infill panel walls

16.2 Lightweight infill panels

These can be constructed from a wide variety or combination of materials, such as timber, metals and plastics, into which single or double glazing can be fitted. If solid panels are to be used below a transom they are usually of a composite or sandwich construction to provide the required sound insulation, thermal insulation and fire resistance properties.

Lightweight infill panels can be fixed between the structural horizontal and vertical members of the frame or fixed to the face of either the columns or beams to give a grid, horizontal or vertical emphasis to the façade.

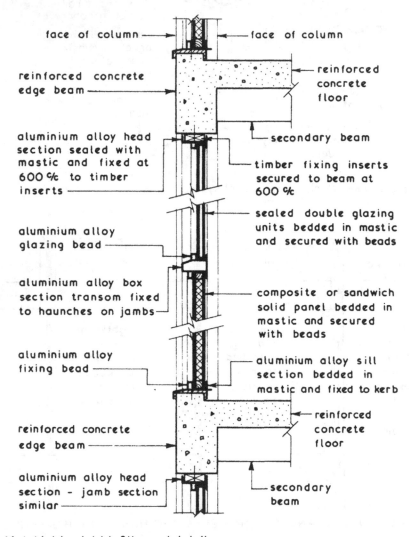

Figure 16.4 Lightweight infill panel details

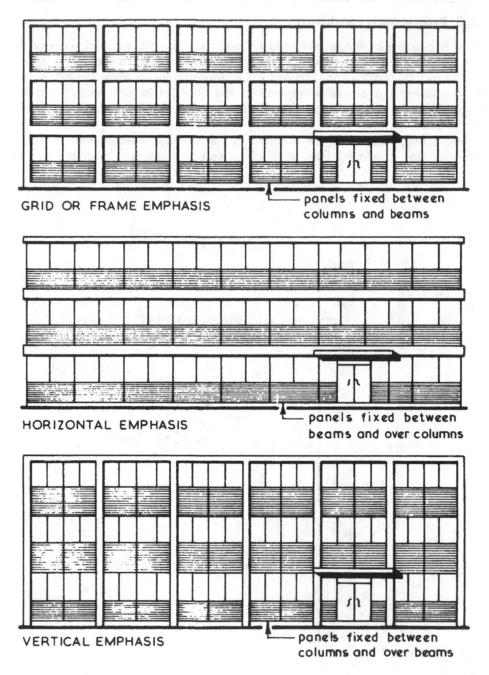

GRID OR FRAME EMPHASIS

panels fixed between
columns and beams

HORIZONTAL EMPHASIS

panels fixed between
beams and over columns

VERTICAL EMPHASIS

panels fixed between
columns and over beams

Figure 16.5 Infill panel examples

16.3 Rainscreen cladding

This is a building envelope produced by over cladding the external walls of new construction, or as a decorative façade and insulation enhancement to the external walls of existing construction. This concept provides an inexpensive *loose-fit* weather resistant layer. It is simple to replace to suit changes in occupancy, corporate image, client tastes, new material innovations and design changes in the appearance of buildings.

Sustainability objectives are satisfied by reuse and refurbishment instead of demolition and rebuilding. Existing buildings can be seamlessly extended with uniformity of finish.

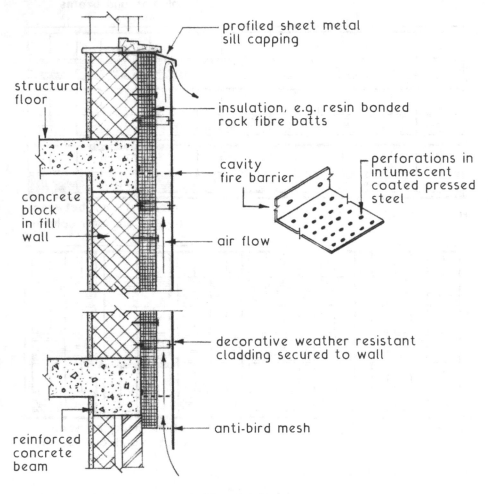

Figure 16.6 Typical rainscreen cladding details

Principal features

- Weather-proof outer layer includes plastic laminates, fibre-cement, ceramics, aluminium, enamelled steel and various stone effects.
- Decorative finish.
- Support frame attached to structural wall.
- Ventilated and drained cavity behind cladding.
- Facility for sound and thermal insulation.
- *Loose-fit* – simple and economic for ease of replacement.

Rainscreen cladding support framework is aluminium alloy standard profile sections with extendable support brackets.

Note: RSC materials should satisfy tests for fire propagation and surface spread of flame (BS 476-6 and 7 respectively).

Rainscreen cladding

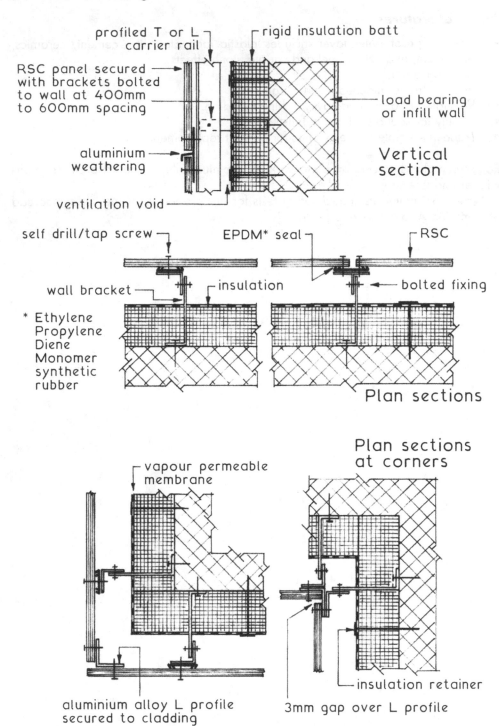

profiled T or L carrier rail

rigid insulation batt

RSC panel secured with brackets bolted to wall at 400mm to 600mm spacing

load bearing or infill wall

aluminium weathering

Vertical section

ventilation void

self drill/tap screw

EPDM* seal

RSC

wall bracket

insulation

bolted fixing

* Ethylene Propylene Diene Monomer synthetic rubber

Plan sections

Plan sections at corners

vapour permeable membrane

aluminium alloy L profile secured to cladding

insulation retainer

3mm gap over L profile

Figure 16.7 Metal support systems

16.4 Glazing

Systems – two edge and four edge

The two edge system relies on conventional glazing beads/fixings to the head and sill parts of a frame, with sides silicone-bonded to mullions and styles.

The four edge system relies entirely on structural adhesion, using silicone bonding between glazing and support frame.

Structural glazing, as shown in Figures 16.8 and 16.9, is in principle a type of curtain walling.

Due to its unique appearance it is usual to consider full glazing of the building façade as a separate design and construction concept.

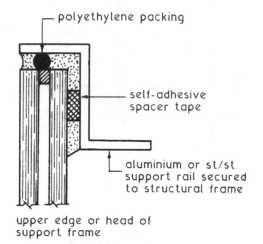

Figure 16.8 Glazing head detail

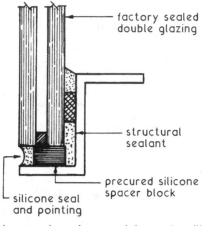

Figure 16.9 Glazing bottom detail

Note: Sides of frame as head.

BS EN 13830: *Curtain Walling. Product Standard* defines curtain walling as an external vertical building enclosure produced by elements mainly of metal, timber or plastic. Glass as a primary material is excluded.

16.5 **Structural glazing**

Structural glazing is otherwise known as frameless glazing. It is a system of toughened glass cladding without the visual impact of surface fixings and supporting components. Unlike curtain walling, the self-weight of the glass and wind loads are carried by the glass itself and transferred to a subsidiary lightweight support structure behind the glazing.

Assembly principles

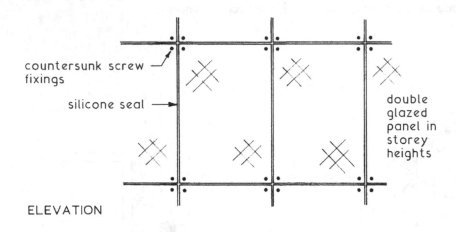

countersunk screw fixings

silicone seal

double glazed panel in storey heights

ELEVATION

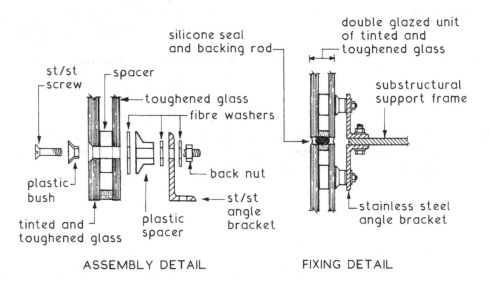

st/st screw

spacer

silicone seal and backing rod

double glazed unit of tinted and toughened glass

toughened glass

fibre washers

substructural support frame

back nut

plastic bush

plastic spacer

st/st angle bracket

tinted and toughened glass

stainless steel angle bracket

ASSEMBLY DETAIL

FIXING DETAIL

Figure 16.10 Frameless structural glazing details

16.6 Curtain walling

This is a form of lightweight, non-load-bearing, external cladding that forms a complete envelope or sheath around the structural frame. In low rise structures the curtain wall framing could be of timber or patent glazing, but in the usual high rise context, box or solid members of steel or aluminium alloy are normally employed.

Basic requirements for curtain walls

- Provide the necessary resistance to penetration by the elements.
- Have sufficient strength to carry own self-weight and provide resistance to both positive and negative wind pressures.
- Provide required degree of fire resistance – glazed areas are classified in the Building Regulations as unprotected areas; therefore any required fire resistance must be obtained from the infill or under sill panels and any backing wall or beam.
- Be easy to assemble, fix and maintain.
- Provide the required degree of sound and thermal insulation.
- Provide for thermal and structural movements.

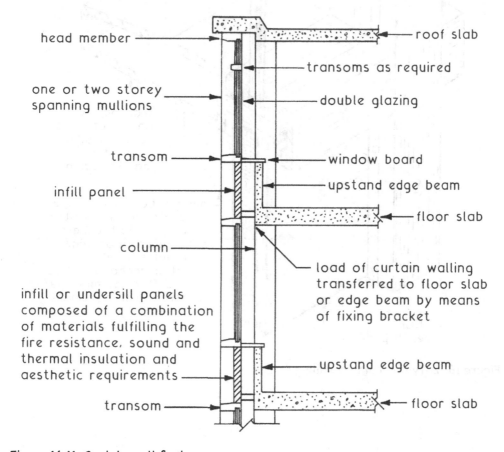

Figure 16.11 Curtain wall features

Curtain walling

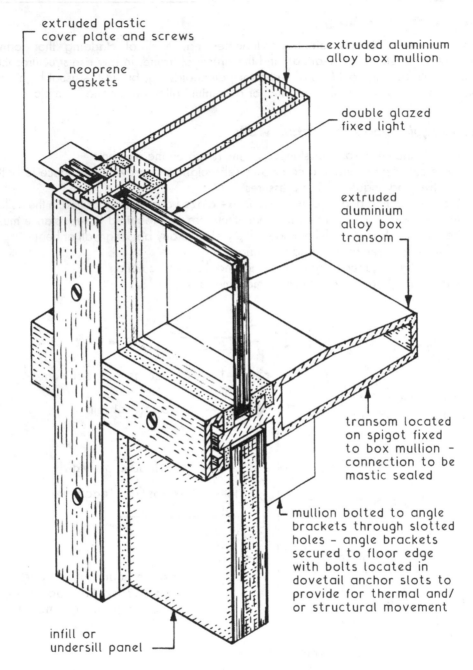

extruded plastic
cover plate and screws

neoprene
gaskets

extruded aluminium
alloy box mullion

double glazed
fixed light

extruded
aluminium
alloy box
transom

transom located
on spigot fixed
to box mullion –
connection to be
mastic sealed

mullion bolted to angle
brackets through slotted
holes – angle brackets
secured to floor edge
with bolts located in
dovetail anchor slots to
provide for thermal and/
or structural movement

infill or
undersill panel

Figure 16.12 Typical curtain walling details

528

Fixing curtain walling to the structure

In curtain walling systems it is the main vertical component or mullion which carries the loads and transfers them to the structural frame at every, or alternate floor levels, depending on the spanning ability of the mullion. At each fixing point the load must be transferred and an allowance made for thermal expansion and differential movement between the structural frame and curtain walling. The usual method employed is slotted bolt fixings.

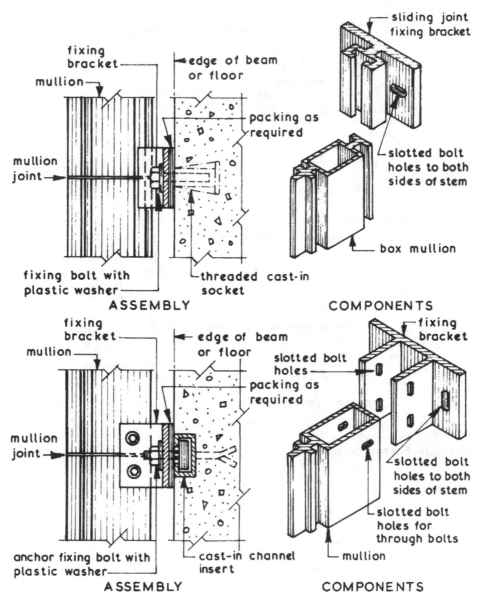

Figure 16.13 Curtain wall fixing to the structure

Curtain walling

Curtain wall over cladding

Recladding existing framed buildings has become an economical alternative to complete demolition and rebuilding. This may be justified when a building has a change of use or it is in need of an image upgrade. Current energy conservation measures can also be achieved by the redressing of older buildings.

Typical section through an existing structural floor slab with a replacement system attached:

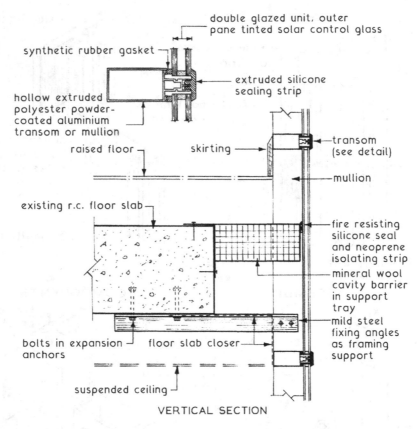

Figure 16.14 Framing detail

16.7 Concrete cladding panels

These are usually of reinforced, precast concrete to an undersill or storey height format, the former sometimes being called apron panels. All precast concrete cladding panels should be designed and installed to fulfil the following functions:

- Be self-supporting between framing members.
- Provide resistance to penetration by the natural elements.
- Resist both positive and negative wind pressures.
- Provide required degree of fire resistance.
- Provide required degree of thermal insulation by having the insulating material incorporated within the body of the cladding or, alternatively, allow the cladding to act as the outer leaf of cavity wall panel.
- Provide required degree of sound insulation.

Undersill or apron cladding panels

These are designed to span from column to column and provide a seating for the windows located above. Levelling is usually carried out by wedging and packing from the lower edge before being fixed with grouted dowels.

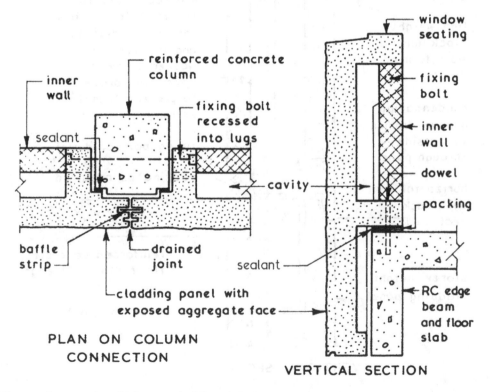

Figure 16.15 Undersill/apron cladding details

Concrete cladding panels

Storey height cladding panels

These are designed to span vertically from beam to beam and can be fenestrated if required. Levelling is usually carried out by wedging and packing from floor level before being fixed.

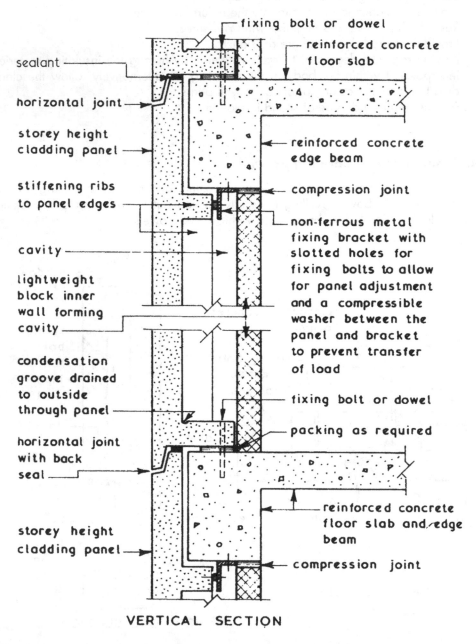

fixing bolt or dowel

reinforced concrete floor slab

sealant

horizontal joint

storey height cladding panel

reinforced concrete edge beam

stiffening ribs to panel edges

compression joint

non-ferrous metal fixing bracket with slotted holes for fixing bolts to allow for panel adjustment and a compressible washer between the panel and bracket to prevent transfer of load

cavity

lightweight block inner wall forming cavity

condensation groove drained to outside through panel

fixing bolt or dowel

packing as required

horizontal joint with back seal

storey height cladding panel

reinforced concrete floor slab and edge beam

compression joint

VERTICAL SECTION

Figure 16.16 Storey height cladding panels

Concrete cladding joints

Single stage

The application of a compressible filling material and a weatherproofing sealant between adjacent cladding panels. This may be adequate for relatively small areas and where exposure to thermal or structural movement is limited. Elsewhere, in order to accommodate extremes of thermal movement between exposed claddings, the use of only a sealant and filler would require an over-frequency of joints or over-wide joints that could slump or fracture.

Two stage

These are otherwise known as open drained joints. This is the preferred choice as there is a greater facility to absorb movement. Drained joints to cladding panels comprise a sealant to the inside or back of the joint and a baffle to the front, both separated by an air seal.

Comparison of single- and two-stage jointing principles

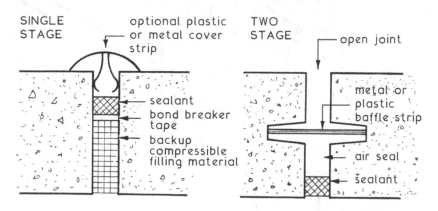

Figure 16.17 Single and two stage gaskets

Typical coefficients of linear thermal expansion $(10^{-6}\text{m/mK}) \sim$ Dense concrete aggregate 14, lightweight concrete aggregate 10.

BS 6093: *Design of Joints and Jointing in Building Construction.*

Concrete cladding panels

Drained joint intersection

Where the horizontal lapped joint between upper and lower cladding panels coincides with the vertical open drained joint, a baffle panel and stepped apron flashing of reinforced synthetic rubber is required to weather the intersection.

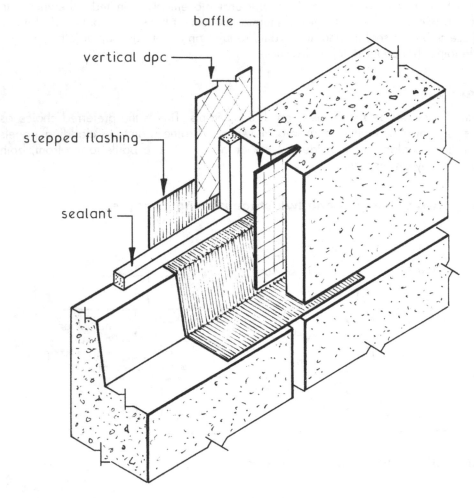

Figure 16.18 Typical weathering detail

Gasket joints

They are used specifically where movements or joint widths are greater than could be accommodated by sealants. For this purpose a gasket is defined in BS 6093 as 'flexible, generally elastic, preformed material that constitutes a seal when compressed'.

Profiles are solid or hollow extrusions in a variety of shapes. Generally non-structural but vulcanised polychloroprene rubber can be used if a structural specification is required.

Materials (non-structural)

Synthetic rubber including neoprene, silicone, ethylene propylene diene monomer (EPDM) and thermoplastic rubber (TPR). These materials are very durable with excellent resistance to compression, heat, water, ultra-violet light, ozone, ageing, abrasion and chemical cleaning agents such as formaldehyde. They also have exceptional elastic memory, i.e. will resume original shape after stressing. Polyvinyl chloride (PVC) and similar plastics can also be used but they will need protection from the effects of direct sunlight.

BS 4255-1: *Rubber Used in Preformed Gaskets for Weather Exclusion from Buildings. Specification for Non-Cellular Gaskets.*
BS 6093: *Design of Joints and Jointing in Building Construction. Guide.*

Non-structural gasket types (shown plan view)

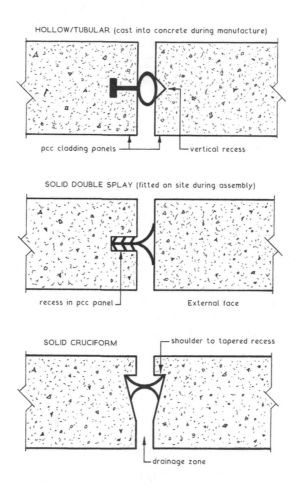

Figure 16.19 Non structural gaskets

Concrete cladding panels

Concrete surface finishes

Apart from a plain surface concrete the other main options are:

- Textured and profiled surfaces.
- Tooled finishes.
- Exposed aggregate finishes.
- Textured and profiled surfaces.
- Granolithic.
- Terrazzo.

All of these are produced in the factory.

17 WINDOWS AND DOORS

Opening remarks

A window must be aesthetically acceptable in the context of building design and the surrounding environment.

Windows should be selected or designed to resist wind loadings, be easy to clean and provide for safety and security. They should be sited to provide visual contact with the outside.

Habitable upper floor rooms should have a window for emergency escape. Min. opening area, 0.330m². Min. height and width, 0.450m. Max height of opening, 1.100m above floor.

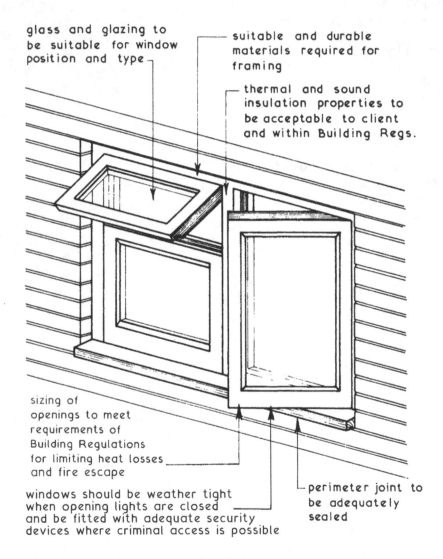

glass and glazing to be suitable for window position and type

suitable and durable materials required for framing

thermal and sound insulation properties to be acceptable to client and within Building Regs.

sizing of openings to meet requirements of Building Regulations for limiting heat losses and fire escape

windows should be weather tight when opening lights are closed and be fitted with adequate security devices where criminal access is possible

perimeter joint to be adequately sealed

Figure 17.1 Window performance requirements

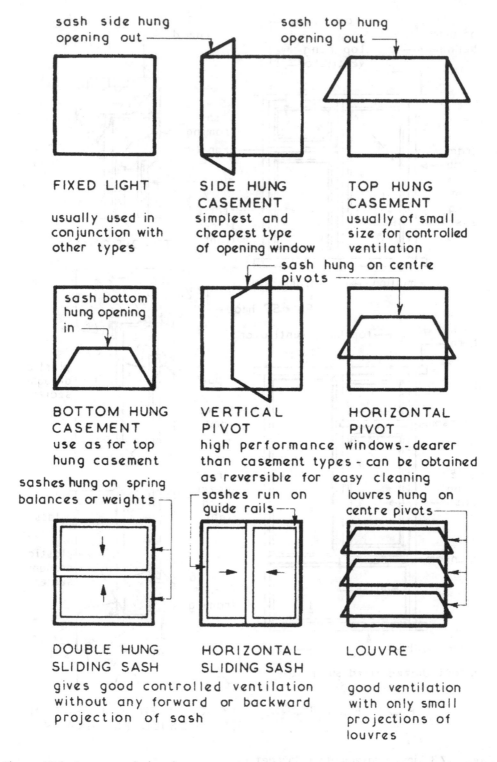

sash side hung
opening out

sash top hung
opening out

FIXED LIGHT

usually used in
conjunction with
other types

**SIDE HUNG
CASEMENT**

simplest and
cheapest type
of opening window

**TOP HUNG
CASEMENT**

usually of small
size for controlled
ventilation

sash bottom
hung opening
in

sash hung on centre
pivots

**BOTTOM HUNG
CASEMENT**

use as for top
hung casement

**VERTICAL
PIVOT**

high performance windows-dearer
than casement types-can be obtained
as reversible for easy cleaning

**HORIZONTAL
PIVOT**

sashes hung on spring
balances or weights

sashes run on
guide rails

louvres hung on
centre pivots

**DOUBLE HUNG
SLIDING SASH**

gives good controlled ventilation
without any forward or backward
projection of sash

**HORIZONTAL
SLIDING SASH**

LOUVRE

good ventilation
with only small
projections of
louvres

Figure 17.2 Common window types

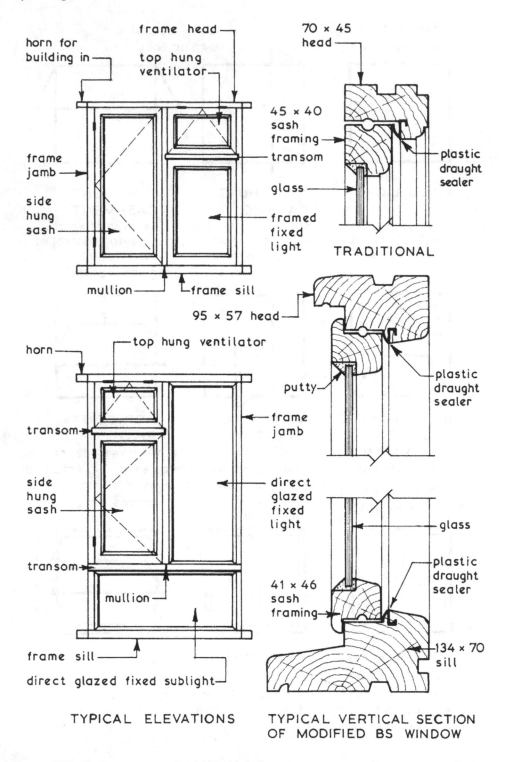

TYPICAL ELEVATIONS

TYPICAL VERTICAL SECTION OF MODIFIED BS WINDOW

Figure 17.3 Timber casement window details

17.1 Casement windows

The standard range of casement windows used in the UK was derived from the English Joinery Manufacturer's Association (EJMA) designs of some 50 years ago. These became adopted in BS 644: *Timber Windows and Doorsets. Fully Finished Factory Assembled Windows of Various Types.*

Contemporary building standards require higher levels of performance in terms of thermal and sound insulation (Bldg. Regs. Pt. L and E), air permeability, watertightness and wind resistance (BS ENs 1026, 1027 and 12211, respectively). This has been achieved by adapting Scandinavian designs with double and triple glazing to attain U-values as low as 1.2W/m²K and a sound reduction of 50dB.

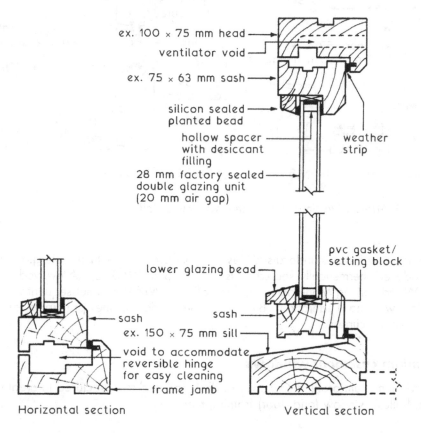

ex. 100 × 75 mm head
ventilator void
ex. 75 × 63 mm sash
silicon sealed planted bead
hollow spacer with desiccant filling
28 mm factory sealed double glazing unit (20 mm air gap)
weather strip
pvc gasket/ setting block
lower glazing bead
sash
sash
ex. 150 × 75 mm sill
void to accommodate reversible hinge for easy cleaning
frame jamb

Horizontal section   Vertical section

Figure 17.4 **Double glazed timber casement window details**

Further references:
BS 6375 series: Performance of windows and doors.
BS 6375-1: Classification for weather tightness.
BS 6375-2: Classification for operation and strength characteristics.
PAS 24: Enhanced security performance requirements for doorsets and windows in the UK.

Casement windows

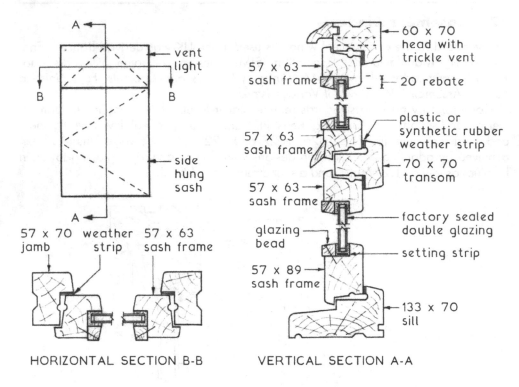

Figure 17.5 Alternative timber casement window details

Details show EJMA standardised designs for casements with double glazed, factory produced, hermetically sealed units. In the early 1960s EJMA evolved into the British Woodworking Manufacturers Association (BWMA) and subsequently the British Woodworking Federation (BWF). Although dated, the principles of these designs remain current.

Aluminium casement windows

Extruded aluminium profiled sections are designed and manufactured to create lightweight, hollow window (and door) framing members.

Finish

Untreated aluminium is prone to surface oxidisation. This can be controlled by paint application, but most manufacturers provide a variable colour range of polyester coatings finished in gloss, satin or matt.

Thermal insulation

Poor insulation and high conductivity are characteristics of solid profile metal windows. This is much less apparent with hollow profile outer members, as they can be considerably enhanced by a thermal infilling of closed cell foam.

Condensation

A high strength, two-part polyurethane resin thermal break between internal and external profiles inhibits cold bridging. This reduces the opportunity for condensation to form on the surface. An indicative U-value of $1.2W/m^2K$ relates to a thermal break of 4mm. If this is increased to 16mm the values can be reduced by up to $0.2W/m^2K$.

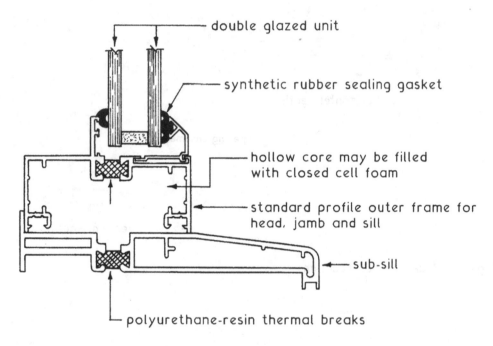

double glazed unit

synthetic rubber sealing gasket

hollow core may be filled with closed cell foam

standard profile outer frame for head, jamb and sill

sub-sill

polyurethane-resin thermal breaks

Figure 17.6 Section of aluminium window sill

Casement windows

uPVC casement windows

Manufactured from extruded uPVC profiles cut to size and thermally welded together, they include hollow aluminium sections within the frame jambs for strength and rigidity. They are predominantly white but are also available in a range of colours, including wood grain effect.

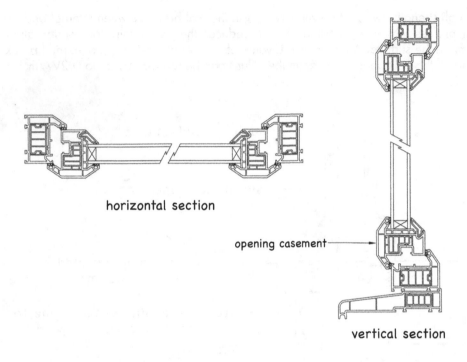

horizontal section

opening casement

vertical section

Figure 17.7 uPVC window sections

17.2 Sliding sash windows

This was the standard form of window up to the 1920s, available as either single vertical or double hung sash. This type of window can be specified in conservation areas by the local authority planning department for new build and replacement windows. Contemporary versions are available in aluminium and uPVC.

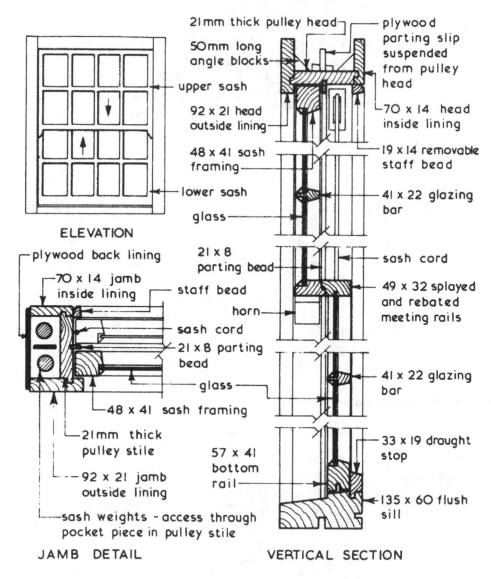

Figure 17.8 Typical double hung weight balanced window details

Sliding sash windows

Weight balance details

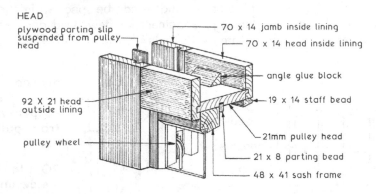

HEAD

plywood parting slip suspended from pulley head

70 x 14 jamb inside lining

70 x 14 head inside lining

angle glue block

92 X 21 head outside lining

19 x 14 staff bead

pulley wheel

21mm pulley head

21 x 8 parting bead

48 x 41 sash frame

Figure 17.9   Head

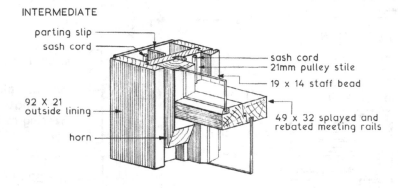

INTERMEDIATE

parting slip

sash cord

sash cord

21mm pulley stile

19 x 14 staff bead

92 X 21 outside lining

49 x 32 splayed and rebated meeting rails

horn

Figure 17.10   Intermediate

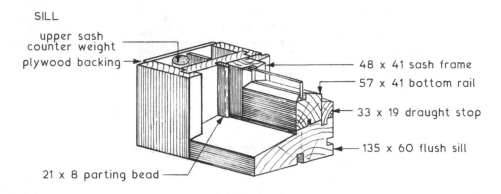

SILL

upper sash counter weight

plywood backing

48 x 41 sash frame

57 x 41 bottom rail

33 x 19 draught stop

135 x 60 flush sill

21 x 8 parting bead

Figure 17.11   Sill

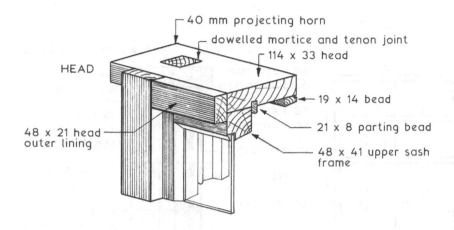

HEAD

40 mm projecting horn

dowelled mortice and tenon joint

114 x 33 head

19 x 14 bead

21 x 8 parting bead

48 x 41 upper sash frame

48 x 21 head outer lining

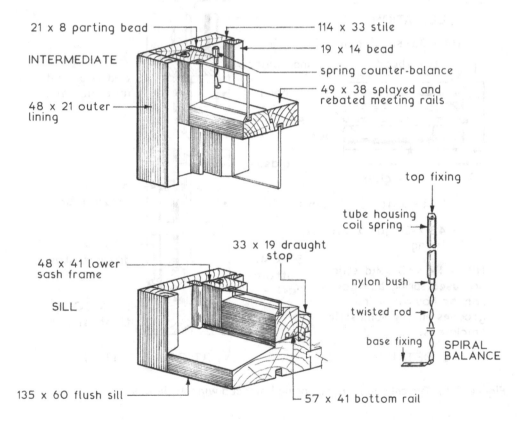

21 x 8 parting bead

INTERMEDIATE

114 x 33 stile

19 x 14 bead

spring counter-balance

49 x 38 splayed and rebated meeting rails

48 x 21 outer lining

top fixing

tube housing coil spring

nylon bush

twisted rod

base fixing

SPIRAL BALANCE

48 x 41 lower sash frame

33 x 19 draught stop

SILL

135 x 60 flush sill

57 x 41 bottom rail

Figure 17.12 Spring balanced sash details

Note: The spring balance can be housed within the stile. Stile thickness will need to be at least 60mm to accommodate the grooved recess.

Sliding sash windows

Spring balanced sash windows

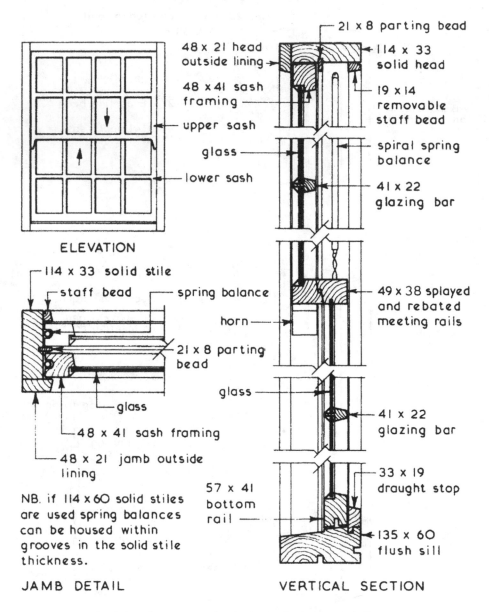

48 x 21 head outside lining

48 x 41 sash framing

upper sash

glass

lower sash

21 x 8 parting bead

114 x 33 solid head

19 x 14 removable staff bead

spiral spring balance

41 x 22 glazing bar

ELEVATION

114 x 33 solid stile

staff bead

spring balance

horn

21 x 8 parting bead

glass

48 x 41 sash framing

48 x 21 jamb outside lining

NB. if 114 x 60 solid stiles are used spring balances can be housed within grooves in the solid stile thickness.

JAMB DETAIL

49 x 38 splayed and rebated meeting rails

glass

41 x 22 glazing bar

33 x 19 draught stop

57 x 41 bottom rail

135 x 60 flush sill

VERTICAL SECTION

Figure 17.13 Typical double hung spring balanced window details

Horizontally sliding sash windows

These can be constructed in timber, metal, plastic or combinations of these materials with single or double glazing. A wide range of arrangements are available with two or more sliding sashes, which can have a vent light incorporated into the outer sliding sash.

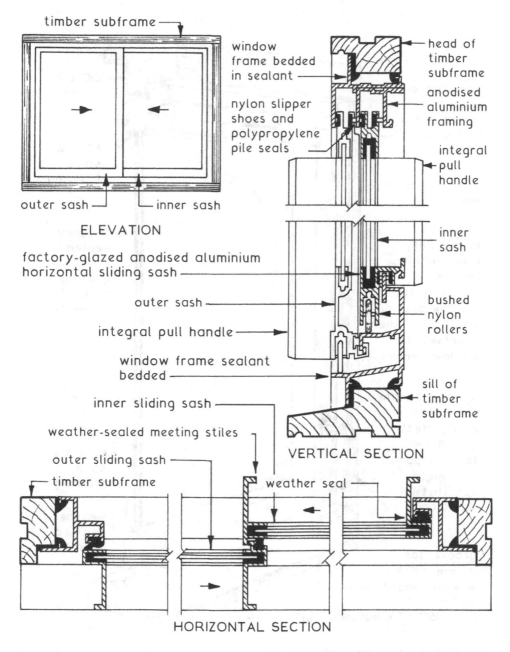

Figure 17.14 Typical horizontally sliding sash window details

17.3 Pivot windows

Like other windows these are available in timber, metal, plastic or in combinations of these materials. They can be constructed with centre jamb pivots enabling the sash to pivot or rotate in the horizontal plane or, alternatively, the pivots can be fixed in the head and sill of the frame so that the sash rotates in the vertical plane.

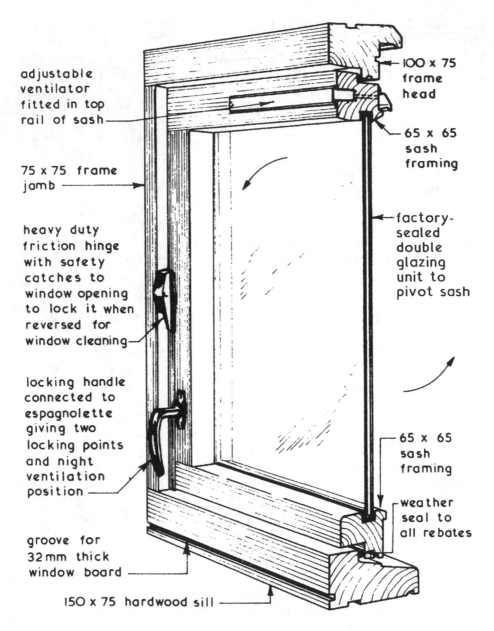

adjustable
ventilator
fitted in top
rail of sash

100 x 75
frame
head

65 x 65
sash
framing

75 x 75 frame
jamb

heavy duty
friction hinge
with safety
catches to
window opening
to lock it when
reversed for
window cleaning

factory-
sealed
double
glazing
unit to
pivot sash

locking handle
connected to
espagnolette
giving two
locking points
and night
ventilation
position

65 x 65
sash
framing

weather
seal to
all rebates

groove for
32 mm thick
window board

150 x 75 hardwood sill

Figure 17.15 Pivot window features

17.4 Bay windows

These can be defined as any window with sidelights that projects in front of the external wall and is supported by a sill height wall. Bay windows not supported by a sill height wall are called oriel windows. They can be of any window type, constructed from any of the usual window materials and are available in three plan formats, namely square, splay and circular or segmental. Timber corner posts can be boxed, solid or jointed, the latter being the common method.

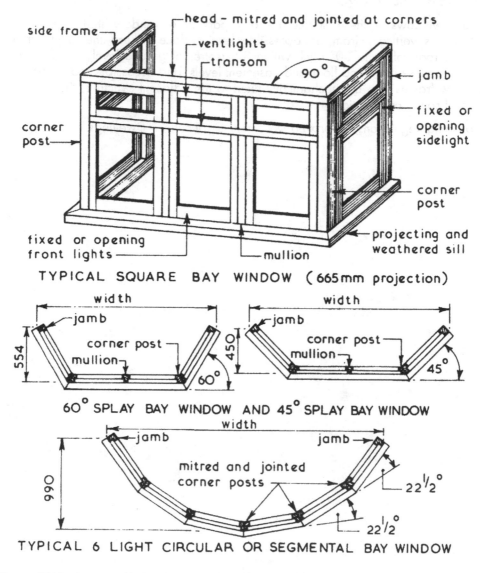

Figure 17.16 Common timber bay window layouts

17.5 Security

New dwellings; both houses and flats, are required to be fitted with windows that prevent unauthorised access from outside. For flats/apartments, this provision includes access from common areas, e.g. shared balconies and corridors with windows to individual living units.

A window conforming to an established standard of robust construction will incorporate features that are proven to reduce crime, e.g. key lockable casement fasteners and stays (unless designated an emergency exit route). An acceptable standard is to BS PAS 24.

Secure windows ~ these should be any part of a window that is less than 2.0 metres vertically from an accessible level surface (where the level surface includes roofs of up to 30° pitch within 3.5 metres of ground level). Affected windows are those at ground floor and basement levels, and accessible roof lights.
Window frames ~ to be mechanically fixed to the adjacent structure; that structure to be of sound construction.

Building Regulations, *Approved Document Q: Security – Dwellings.*

17.6 Window schedules

The main function of a schedule is to collect together all the necessary information for a particular group of components, such as windows, doors and drainage inspection chambers. There is no standard format for schedules but they should be easy to read, accurate and contain all the necessary information for their purpose. Schedules are usually presented in a tabulated format, which can be related to and read in conjunction with the working drawings.

Window manufacturers identify their products with a notation that combines figures with numbers. The objective is to simplify catalogue entries, specification clauses and schedules.

Notation will vary to some extent between the different joinery producers. The example of 313 CVC translates to:

3 = width divided into three units.
13 = first two dimensions of standard height, i.e. 1350mm.
C = casement.
V = ventlight.

Other common notations include:

N = narrow light.
P = plain (picture type window, i.e. no transom or mullion).
T = through transom.
S = sub-light, fixed.
VS = ventlight and sub-light.
F = fixed light.
B = bottom casement opening inwards.
RH/LH = right or lefthand as viewed from the outside.

WINDOW SCHEDULE – Sheet 1 of 1		Drawn By: RC		Date: 14/4/15		Rev.	
Contract Title & Number: Lane End Farm – H 341/80				Drg. Nos. C(31) 450–7			
Number	Type or catalogue ref.	Material	Overall size w x h	Glass	Hardware	Sill	
						External	Internal
2	213 CV	hardwood	1200 x 1350	sealed units as supplied with frames	supplied with casements	2 cos. plain tiles subsill	150×150×15 quarry tiles
4	309 CVC	ditto	1770 x 900	ditto	ditto	ditto	25mm thick softwood
4	313 CVC	ditto	1770 x 1350	ditto	ditto	sill of frame	ditto

Figure 17.17 Typical window schedule entries

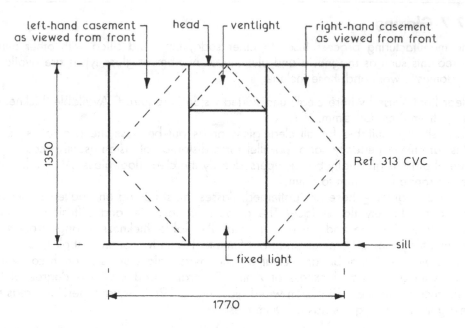

Figure 17.18 Manufacturers window notation

17.7 Glazing

The manufacturing process fuses together soda, lime and silica with other minor ingredients such as magnesia and alumina. A number of glass types are available for domestic work and these include:

Clear float ~ used where clear, undistorted vision is required. Available thicknesses range from 3mm to 25mm.

Clear sheet ~ suitable for all clear glass areas but because the two faces of the glass are never perfectly flat or parallel some distortion of vision usually occurs. This type of glass is gradually being superseded by the clear float glass. Available thicknesses range from 3mm to 6mm.

Translucent glass ~ these are patterned glasses, most having one patterned surface and one relatively flat surface. The amount of obscurity and diffusion obtained depends on the type and nature of pattern. Available thicknesses range from 4mm to 6mm for patterned glasses and from 5mm to 10mm for rough cast glasses.

Wired glass ~ obtainable as a clear polished wired glass or as a rough cast wired glass with a nominal thickness of 7mm. Generally used where a degree of fire resistance is required. Georgian wired glass has a 12mm square mesh whereas the hexagonally wired glass has a 20mm mesh.

Choice of glass

The main factors to be considered are:

1. Resistance to wind loadings.
2. Clear vision required.
3. Privacy.
4. Security.
5. Fire resistance.
6. Aesthetics.

Glazing terminology

Glazing ~ the act of fixing glass into a frame or surround in accordance with the recommendations contained in the BS 6262 series: Glazing for Buildings.

Timber surrounds ~ linseed oil putty is the traditional material for sealing and retaining glass in wooden frames. Putty should be protected with paint within two weeks of application. Alternatively, wood glazing beads mitred in the corners of the frame provide a weathering finish.

A general purpose putty is also available. This combines the properties of the two types.

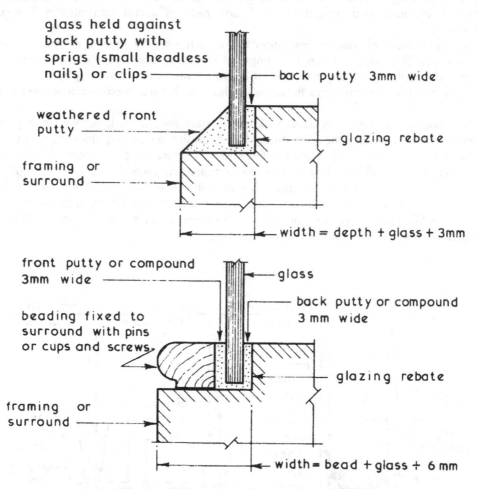

glass held against back putty with sprigs (small headless nails) or clips

back putty 3mm wide

weathered front putty

glazing rebate

framing or surround

width = depth + glass + 3mm

front putty or compound 3mm wide

glass

back putty or compound 3 mm wide

beading fixed to surround with pins or cups and screws

framing or surround

glazing rebate

width = bead + glass + 6 mm

Figure 17.19 Single glazed pane sealing and fixing methods

17.8 Double glazing

Double glazing is used instead of single glazing to reduce the rate of heat loss through windows and glazed doors. It also reduces sound transmission through windows.

In the context of thermal insulation this is achieved by having a small air or argon gas-filled space within the range of 6–20mm between the two layers of glass. The sealed double glazing unit will also prevent internal misting by condensation. If metal frames are used these should have a thermal break incorporated into their design.

All opening sashes in a double glazing system should be fitted with adequate weather seals to reduce the rate of heat loss through the opening clearance gap.

In the context of sound insulation three factors affect the performance of double glazing. First, good installation to ensure airtightness, second, the weight of glass used and third, the size of air space between the layers of glass. The heavier the glass used the better the sound insulation and the air space needs to be within the range of 50–300mm. Absorbent lining to the reveals within the air space will also improve the sound insulation properties of the system.

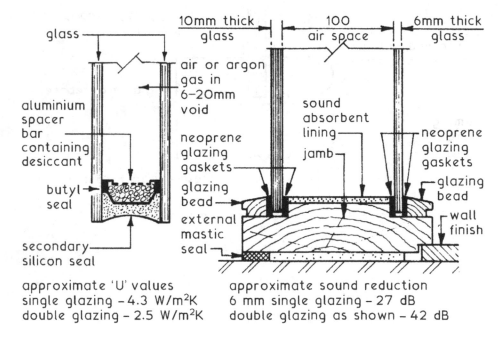

Figure 17.20 Comparison double glazed unit and double glazed frame

Low emissivity or 'low E' glass

Glass is specially manufactured with a surface coating to significantly improve its thermal performance. The surface coating has a dual function:

1. Allows solar short wave light radiation to penetrate a building.
2. Reflects long wave heat radiation losses back into a building.

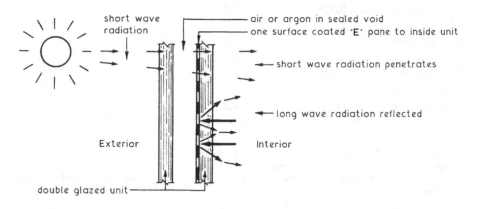

Figure 17.21 Low E glazing principles

Pyrolitic hard coat, applied on-line as the glass is made. Emissivity range, 0.15–0.20, e.g. Pilkington 'K'.

A sputtered soft coat applied after glass manufacture. Emissivity range, 0.05–0.10, e.g. Pilkington 'Kappafloat' and 'Suncool High Performance'.

Note: In relative terms, uncoated glass has a normal emissivity of about 0.90. Indicative U-values for multi-glazed windows of 4mm glass with a 16mm void width are shown in Table 17.1.

Inert gas fills

Argon is generally used as it is the least expensive and more readily available. Where krypton is used, the air gap need only be half that with argon to achieve a similar effect. Both gases have a lower thermal conductivity than air due to their greater density.

Densities (kg/m3):
*   Air = 1.20
*   Argon = 1.66
*   Krypton = 3.49

Double glazing

Table 17.1 Typical U values for multi-glazed windows

Glazing type	uPVC or wood frame	Metal frame
Double, air filled	2.7	3.3
Double, argon filled	2.6	3.2
Double, air filled Low E (0.20)	2.1	2.6
Double, argon filled Low E (0.20)	2.0	2.5
Double, air filled Low E (0.05)	2.0	2.3
Double, argon filled Low E (0.05)	1.7	2.1
Triple, air filled	2.0	2.5
Triple, argon filled	1.9	2.4
Triple, air filled Low E (0.20)	1.6	2.0
Triple, argon filled Low E (0.20)	1.5	1.9
Triple, air filled Low E (0.05)	1.4	1.8
Triple, argon filled Low E (0.05)	1.3	1.7

Notes:
- A larger void and thicker glass will reduce the U-value, and vice versa.
- Data for metal frames assumes a thermal break of 4mm.
- Hollow metal framing units can be filled with a closed cell insulant foam to considerably reduce U-values.

Spacers

Generally hollow aluminium with a desiccant or drying agent fill. The filling absorbs the initial moisture present in between the glass layers. Non-metallic spacers are preferred as aluminium is an effective heat conductor.

Approximate solar gains with ordinary float glass

'Low E' invisible coatings reduce the solar gain by up to one-third. Depending on the glass quality and cleanliness, about 10–15% of visible light reduction applies for each pane of glass.

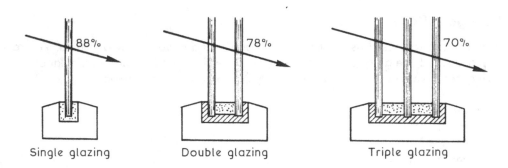

Single glazing   Double glazing   Triple glazing

Figure 17.22 Comparison of solar gains through glazing

17.9 Triple glazing

This has U-value potential of less than 1.0W/m²K.

- 'Low E' invisible metallic layer on one pane of double glazing gives a similar insulating value to standard triple glazing (see Table 17.1).
- Performance enhanced with blinds between wide gap panes.
- High quality ironmongery required due to weight of glazed frames.
- Improved sound insulation, particularly with heavier than air gap fill.

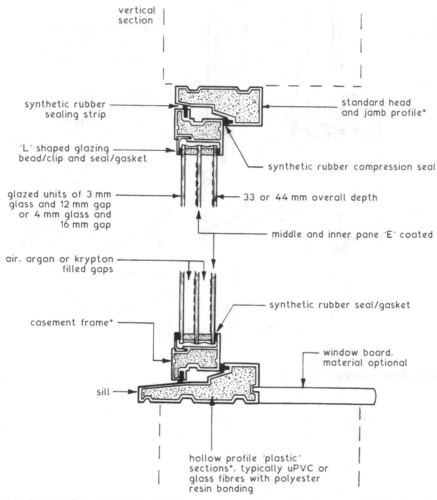

Figure 17.23 Triple glazed window details

Triple glazing

Triple glazed units with solar blinds

There are many manufacturers of triple glazed units, each with their own design profile. Some feature a wide gap between two of the glass panes to incorporate Venetian blinds. This type of curtaining is used to control solar radiation and heat gain. Glass specification and thickness may also vary depending on thermal and sound insulation requirements. Blinds can be actuated automatically in response to pre-set thermostatic control or overridden manually.

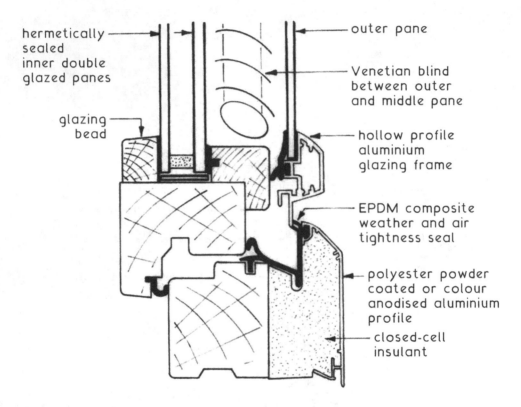

hermetically sealed inner double glazed panes

glazing bead

outer pane

Venetian blind between outer and middle pane

hollow profile aluminium glazing frame

EPDM composite weather and air tightness seal

polyester powder coated or colour anodised aluminium profile

closed-cell insulant

Figure 17.24 Solar blind within glazing

Table 17.2 Potential for sound insulation by glazing type

Glazing system	Max. sound reduction (dB)
Single glazed opening sash	20
Single glazed fixed light with weather strips	25
Double glazed, hermetically sealed with weather strips	35
Triple glazed, hermetically sealed, weather strips and blinds	45

Note: These units are considerably heavier than other window types, therefore they will need care in handling during site transportation. The manufacturer's purpose-made jamb fixings should be used.

17.10 Glazing health and safety

In critical locations, glazing must satisfy one of the following:

1. Breakage to leave only a small opening with small detachable particles without sharp edges.
2. Disintegrating glass must leave only small detached pieces.
3. Inherent robustness, e.g. polycarbonate composition. Annealed glass acceptable but with the limitations shown in Table 17.3.
4. Panes in small areas, <250mm wide and <0.5m^2 area, e.g. leaded lights (4mm annealed glass) and Georgian pattern (6mm annealed glass).
5. Protective screening as shown in Figure 17.26 with lower bar of screen <75mm above finished floor level.

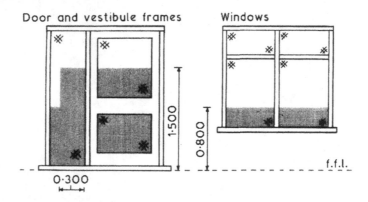

Figure 17.25 Critical locations

Table 17.3 Glazing limitations in critical areas

Thickness of annealed glass (mm)	Max. glazed area	
	Height (m)	Width (m)
8	1.100	1.100
10	2.250	2.250
12	3.000	4.500
15	No limit	No limit

Building Regulations, A.D. K4: *Protection Against Impact with Glazing.*

564

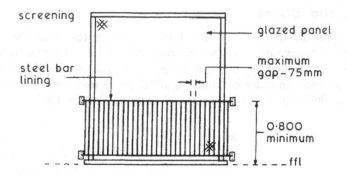

Figure 17.26 Protective guarding/screening

Manifestation

Commercial premises such as open plan offices, shops and showrooms often incorporate large walled areas of uninterrupted glass to promote visual depth, whilst dividing space or forming part of the exterior envelope. To prevent collision, glazed doors and walls must have prominent framing or intermediate transoms and mullions. An alternative is to position obvious markings at two levels. Glass doors could have large pull/push handles and/or IN and OUT signs in bold lettering. Other areas may be adorned with company logos, stripes, geometric shapes, etc.

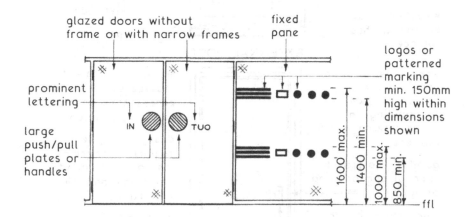

Figure 17.27 Manifestation critical locations

Building Regulations, A.D. K5.2: *Manifestation of Glazing.*
BS 6206: *Specification For Impact Performance Requirements for Flat Safety Glass and Safety Plastics for Use in Buildings.*
BS 6262 series: *Glazing for Buildings. Codes of Practice.*

17.11 External doors

Traditionally external doors were constructed from hardwoods for durability in a wide variety of types and styles set in hardwood rebated frames. The majority of external doors in new housing are composite doors and frames. The largest market for replacement doors in older buildings is uPVC doors and frames. For the commercial market aluminium doors with toughened glazing are standard.

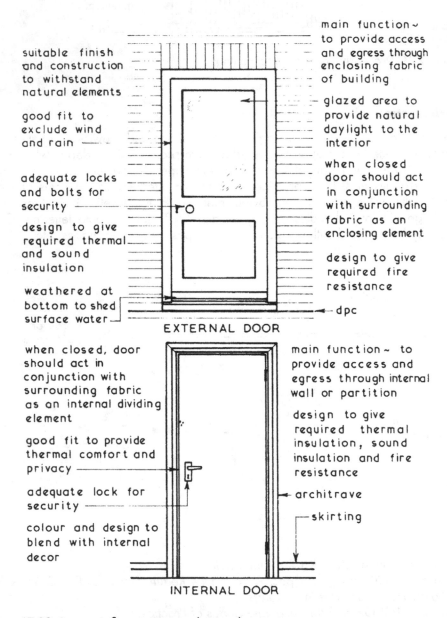

Figure 17.28 Door performance requirements

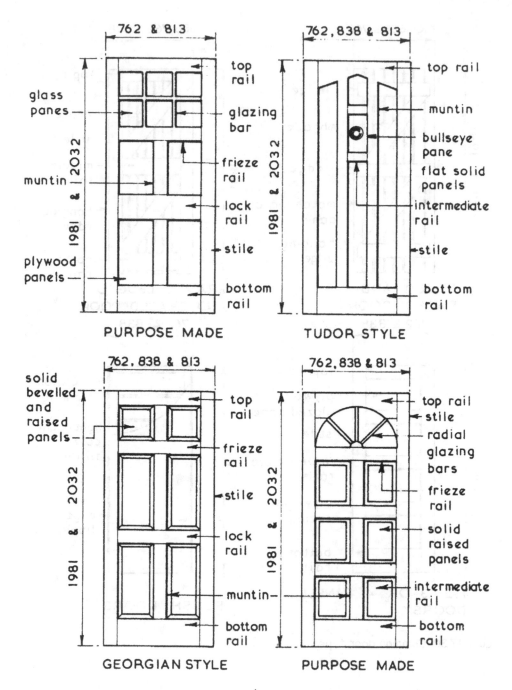

Figure 17.29 Common timber external panel door types

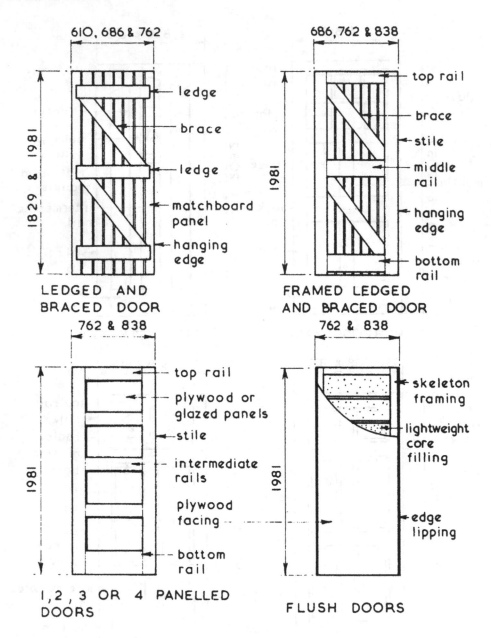

Figure 17.30 Common door types

Timber door frames

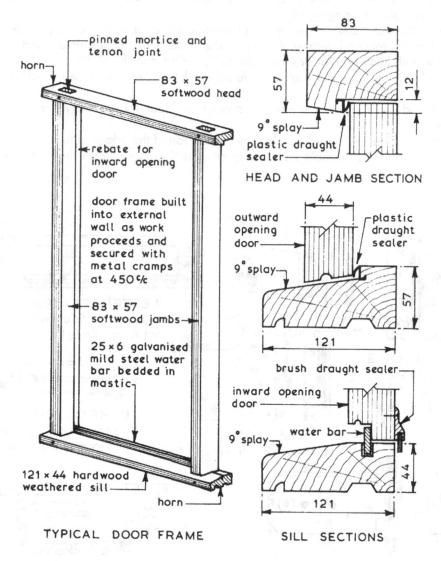

HEAD AND JAMB SECTION

SILL SECTIONS

TYPICAL DOOR FRAME

Figure 17.31 Timber external wall details

Secure doorset

A doorset (door, frame and hardware) should be sufficiently robust to incorporate construction features that are proven to reduce crime. Frames mechanically fixed to adjacent structure of sound construction. An acceptable standard is to BS PAS 24. Other satisfying standards are listed in Section 1.2 and Appendix B of AD Q.

External doors

Door hardware

Hinges or butts ~ these are used to fix the door to its frame or lining and to enable it to pivot about its hanging edge.

Locks, latches and bolts ~ the means of keeping the door in its closed position and providing the required degree of security. The handles and cover plates used in conjunction with locks and latches are collectively called door furniture.

Letter plates ~ fitted in external doors to enable letters, etc. to be deposited through the door.

Finger and kicking plates ~ used to protect the door fabric where there is high usage.

Draught excluders ~ to seal the clearance gap around the edges of the door.

Security chains ~ to enable the door to be partially opened and thus retain some security.

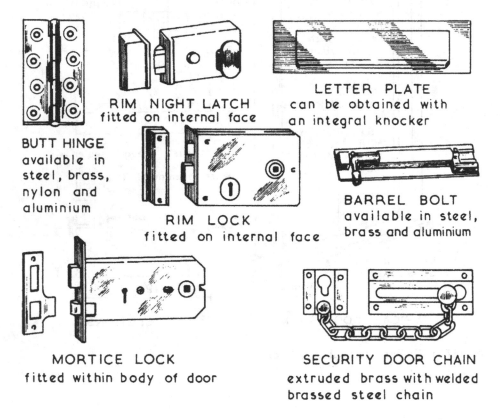

Figure 17.32 Common door locks for timber doors

17.12 Internal doors

These are similar in construction to the external doors but are usually thinner, so are lightweight and can be fixed to a lining. If heavy doors are specified these can be hung to frames in a similar manner to external doors. An alternative method is to use doorsets, which are usually storey height and supplied with pre-hung doors.

Flush doors of lightweight construction are suitable to access many interior situations. If there is an additional requirement for resistance to fire, the construction will include supplementary lining. Fire resisting and non-fire resisting doors can be produced with identical flush finishes, but the fire doors must be purposely labelled.

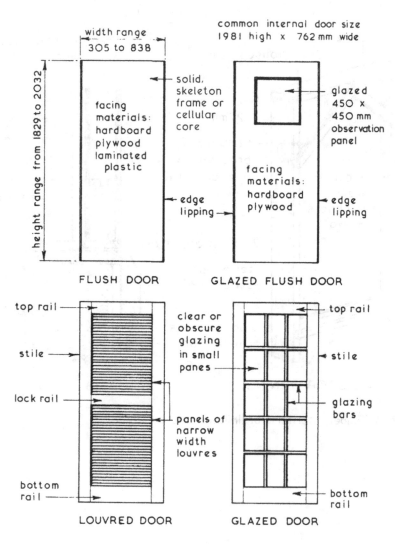

Figure 17.33 Common timber internal door types

Internal doors

Internal door construction

Typically, 35 or 40mm thickness.

Internal door frames/linings

The linings are sized to wall/partition thicknesses and surface finishes. Linings with planted stops are usually employed for lightweight domestic doors.
BS 4787: *Internal and External Wood Doorsets, Door Leaves and Frames. Specification for Dimensional Requirements.*

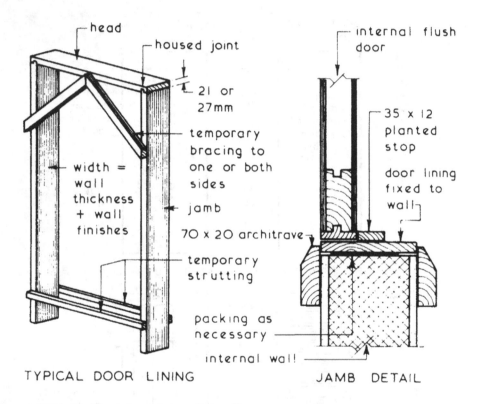

Figure 17.34 Timber internal door lining/frame

Doorsets

These are factory-produced, fully assembled, pre-hung doors that are supplied complete with frame, architraves and hardware. The doors may be hung to the frames using pin butts for easy door removal. Pre-hung door sets are available in standard and storey height versions and are suitable for all internal door applications with normal wall and partition thicknesses.

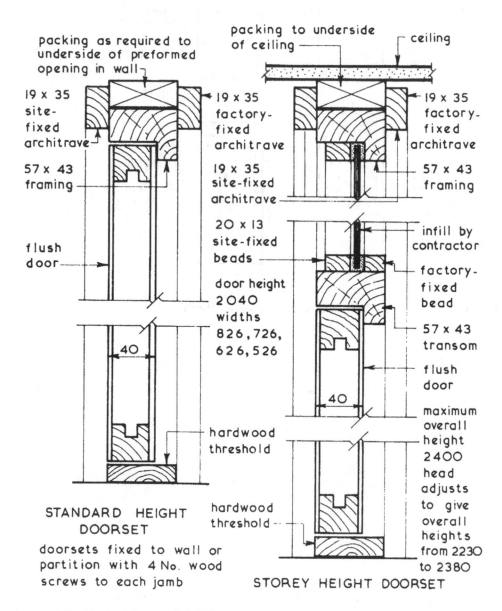

packing as required to underside of preformed opening in wall

19 x 35 site-fixed architrave

57 x 43 framing

flush door

40

STANDARD HEIGHT DOORSET

doorsets fixed to wall or partition with 4 No. wood screws to each jamb

packing to underside of ceiling — ceiling

19 x 35 factory-fixed architrave

19 x 35 site-fixed architrave

20 x 13 site-fixed beads

door height 2040 widths 826, 726, 626, 526

hardwood threshold

hardwood threshold

STOREY HEIGHT DOORSET

19 x 35 factory-fixed architrave

57 x 43 framing

infill by contractor

factory-fixed bead

57 x 43 transom

flush door

maximum overall height 2400 head adjusts to give overall heights from 2230 to 2380

40

Figure 17.35 Typical door set details

Internal doors

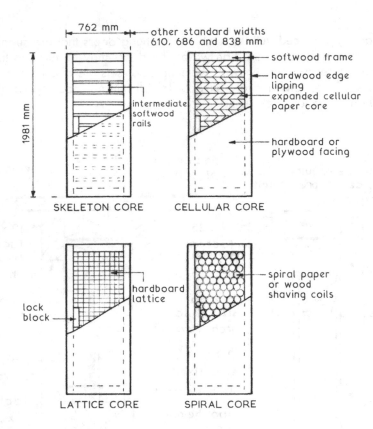

Figure 17.36 Internal flush door construction methods

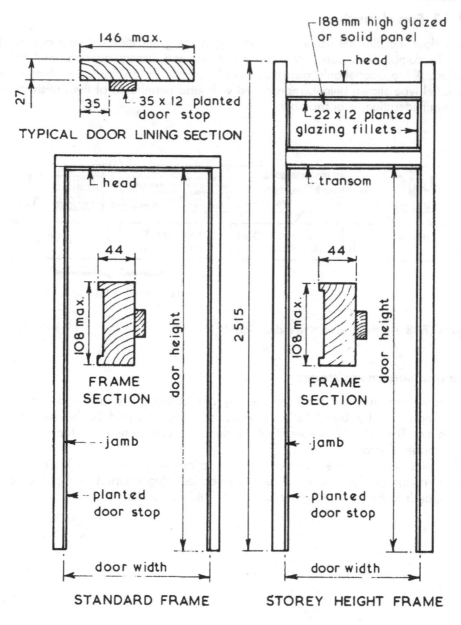

Figure 17.37 Internal door frames

17.13 Fire doors

A fire door includes the frame, ironmongery, glazing, intumescent core and smoke seal. To comply with European market requirements, ironmongery should be CE marked. A fire door should also be marked accordingly on the top or hinge side. The label type shown below, reproduced with kind permission of the British Woodworking Federation, is acceptable.

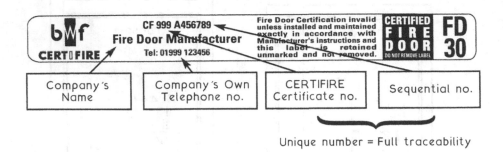

Figure 17.38 Manufacturers fire door label

Fire and smoke resistance

Fire doors are assessed for both integrity and smoke resistance and are coded accordingly, e.g. FD30 or FD30s. FD indicates a fire door and 30 the integrity time in minutes. The letter 's' denotes that the door or frame contains a facility to resist the passage of smoke.

Manufacturers produce doors of standard ratings – 30, 60 and 90 minutes, with higher ratings available to order. A colour-coded plug inserted in the door edge corresponds to the fire rating. See BS 8214, Table 1 for details.

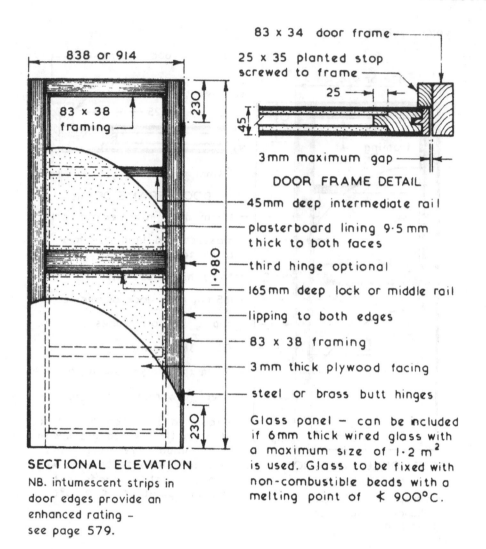

83 x 34 door frame

25 x 35 planted stop
screwed to frame

25

45

3mm maximum gap

DOOR FRAME DETAIL

838 or 914

230

83 x 38
framing

1·980

230

— 45mm deep intermediate rail

— plasterboard lining 9·5mm
thick to both faces

— third hinge optional

— 165mm deep lock or middle rail

— lipping to both edges

— 83 x 38 framing

— 3mm thick plywood facing

— steel or brass butt hinges

Glass panel – can be included
if 6mm thick wired glass with
a maximum size of 1·2 m²
is used. Glass to be fixed with
non-combustible beads with a
melting point of ≮ 900°C.

SECTIONAL ELEVATION
NB. intumescent strips in
door edges provide an
enhanced rating –
see page 579.

Figure 17.39 FD30 construction

Fire doors

Typical Details ~

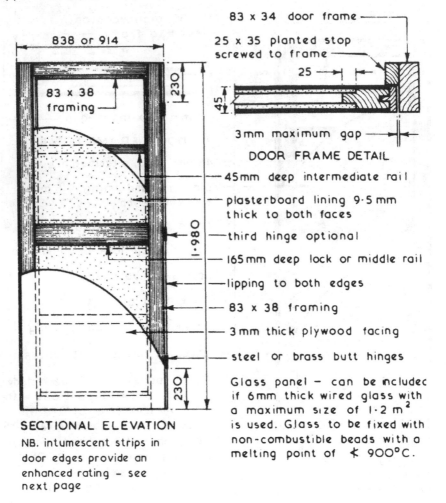

838 or 914

83 × 38 framing

230

1·980

230

SECTIONAL ELEVATION

NB. intumescent strips in door edges provide an enhanced rating – see next page

83 × 34 door frame

25 × 35 planted stop screwed to frame

25

45

3mm maximum gap

DOOR FRAME DETAIL

45mm deep intermediate rail

plasterboard lining 9·5 mm thick to both faces

third hinge optional

165mm deep lock or middle rail

lipping to both edges

83 × 38 framing

3mm thick plywood facing

steel or brass butt hinges

Glass panel – can be included if 6mm thick wired glass with a maximum size of 1·2 m² is used. Glass to be fixed with non-combustible beads with a melting point of ≮ 900°C.

Figure 17.40 FD60 construction
Note: BS 8214: *Code of Practice for Fire Door Assemblies.*

Intumescent strips

The intumescent core may be fitted to the door edge or the frame. In practice, most joinery manufacturers leave a recess in the frame where the seal is secured with rubber-based or PVA adhesive. At temperatures of about 150°C, the core expands to create a seal around the door edge. This remains throughout the fire-resistance period whilst the door can still be opened for escape and access purposes. The smoke seal will also function as an effective draught seal.

BS EN 1634-1: *Fire Resistance and Smoke Control Tests for Door, Shutter and Openable Window Assemblies and Elements of Building Hardware.*
BS EN 13501: *Fire Classification of Construction Products and Building Elements.*

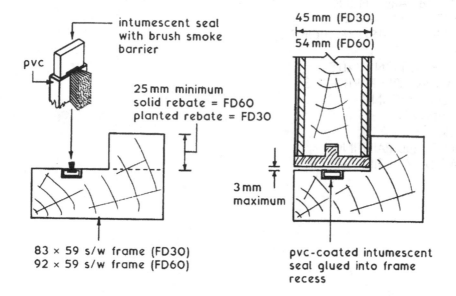

Figure 17.41 Intumescent strips in frame rebates

Fire door glazing

Apertures will reduce the potential fire resistance if not appropriately filled. Suitable material should have the same standard of fire performance as the door into which it is fitted. Fire-rated glass types:

- Embedded Georgian wired glass.
- Composite glass containing borosilicates and ceramics.
- Tempered and toughened glass.
- Glass laminated with reactive fire-resisting interlayers.

18 Internal elements
Ceilings

PLASTERBOARD CEILINGS
SUSPENDED CEILINGS
INTERNAL WALLS/PARTITIONS
PARTY WALLS
STUD PARTITION WALLS
DEMOUNTABLE PARTITIONS
SOLID PLASTERING
DRY LINING
STAIRS
PAINTS AND PAINTING
GLAZED WALL TILING

18.1 Plasterboard ceilings

For ceiling applications, the following types can be used.

Baseboard

1220 × 900 × 9.5mm thick for joist centres up to 400mm.
1220 × 600 × 12.5mm thick for joist centres up to 600mm.

Baseboard has square edges and can be plaster skim finished. Joints are reinforced with self-adhesive 50mm min. width glass fibre mesh scrim tape or the board manufacturer's recommended paper tape. These boards are also made with a metallised polyester foil backing for vapour check applications. The foil is to prevent any moisture produced in potentially damp situations, such as a bathroom or in warm roof construction, from affecting loft insulation and timber. Joints should be sealed with an adhesive metallised tape.

Wallboard

9.5, 12.5 and 15mm thicknesses, 900 and 1200mm widths and lengths of 1800 and 2400mm. Edges are either tapered for taped and filled joints for dry lining or square for skimmed plaster or textured finishes.
   Plasterboards should be fixed breaking joint to the underside of floor or ceiling joists with zinc plated (galvanised) nails or dry-wall screws at 150mm max. spacing. The junction at ceiling to wall is reinforced with glass fibre mesh scrim tape or a preformed plaster moulding.

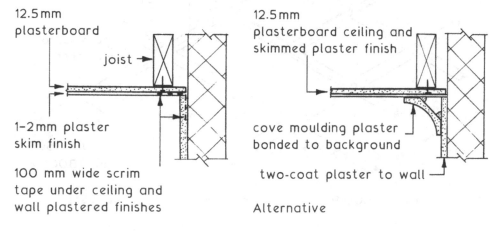

12.5 mm plasterboard
joist →
1–2 mm plaster skim finish
100 mm wide scrim tape under ceiling and wall plastered finishes

12.5 mm plasterboard ceiling and skimmed plaster finish
cove moulding plaster bonded to background
two-coat plaster to wall —
Alternative

Figure 18.1 Ceiling plasterboard and wall junction details

18.2 Suspended ceilings

These can be defined as ceilings that are fixed to a framework suspended from the main structure, thus forming a void between the two components. The basic functional requirements of suspended ceilings are as follows.

- Be easy to construct, repair, maintain and clean.
- So designed that an adequate means of access is provided to the void space for the maintenance of the suspension system, concealed services and/or light fittings.
- Provide any required sound and/or thermal insulation.
- Provide any required acoustic control in terms of absorption and reverberation.
- Provide, if required, structural fire protection to structural steel beams supporting a concrete floor and contain fire stop cavity barriers within the void at defined intervals.
- Conform with the minimum requirements set out in the Building Regulations governing the restriction of spread of flame over surfaces of ceilings and the exemptions permitting the use of certain plastic materials.
- Have flexural design strength in varying humidity and temperature.
- Have resistance to impact.
- Be designed on a planning module, preferably a 300mm dimensional coordinated system.

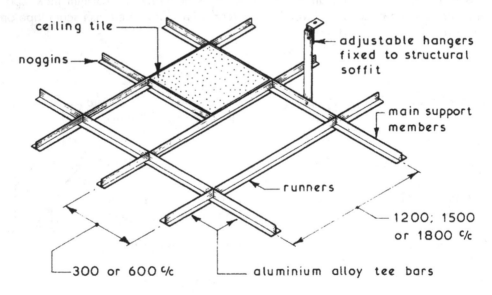

Figure 18.2 Typical suspended ceiling grid framework layout

Classification of suspended ceiling

There is no standard method of classification since some are classified by their function, such as illuminated and acoustic suspended ceilings, others are classified by the materials used and classification by method of construction is also a popular option. The latter method is simple, since most suspended ceiling types can be placed in one of three groups:

1. Jointless suspended ceilings.
2. Panelled suspended ceilings.
3. Decorative and open suspended ceilings.

Jointless suspended ceilings

These provide a continuous and jointless surface with the internal appearance of a conventional ceiling. They may be selected to fulfil fire resistance requirements or to provide a robust form of suspended ceiling. The two common ways of construction are a plasterboard or expanded metal lathing soffit with hand-applied plaster finish or a sprayed applied rendering with a cement base.

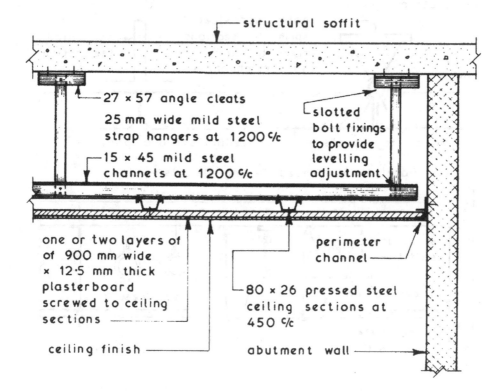

Figure 18.3 Jointless suspended ceiling

Suspended ceilings

Panelled suspended ceilings

These are the most popular form of suspended ceiling consisting of a suspended grid framework to which the ceiling covering is attached. The covering can be of a tile, tray, board or strip format in a wide variety of materials with an exposed or concealed supporting framework. Services such as luminaries can usually be incorporated within the system. Generally panelled systems are easy to assemble and install using a water level or laser beam for initial and final levelling. Provision for maintenance access can be easily incorporated into most systems and layouts.

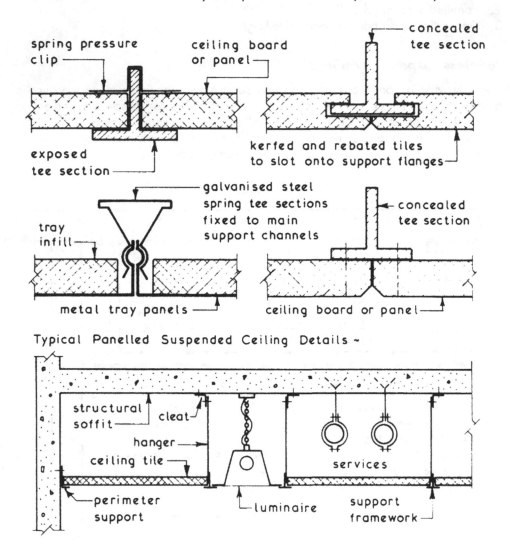

Typical Panelled Suspended Ceiling Details ~

Figure 18.4 Typical support details

Decorative and open suspended ceilings

These ceilings usually consist of an openwork grid or suspended shapes onto which the lights fixed at, above or below ceiling level can be trained, thus creating a decorative and illuminated effect. Many of these ceilings are purpose designed and built as opposed to the proprietary systems associated with jointless and panelled suspended ceilings.

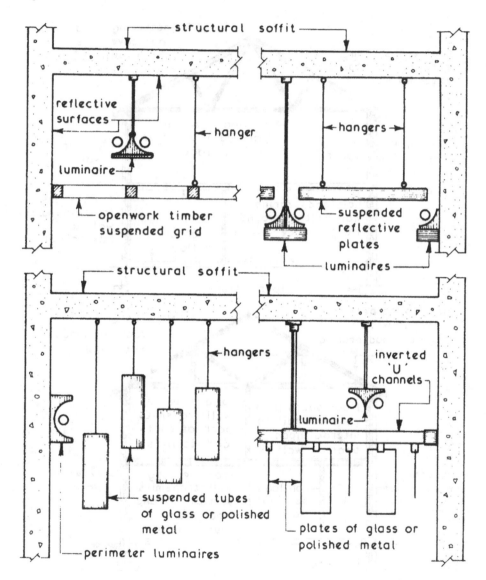

Figure 18.5 Typical open ceiling examples

18.3 Internal walls/partitions

There are two basic design concepts for internal walls: those that accept and transmit structural loads to the foundations are called load-bearing walls and those which support only their own weight and do not accept any structural loads are called non-load-bearing walls or partitions.

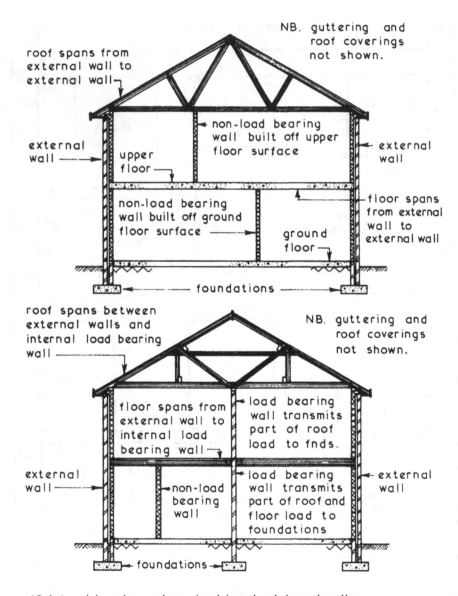

Figure 18.6 Load–bearing and non–load–bearing internal walls

Masonry walls

Internal brick walls

These are found in buildings up to the 1970s and can be load-bearing or non-load-bearing. For most two-storey buildings they are built in half-brick thickness in stretcher bond bonded into the inner leaf of the external wall.

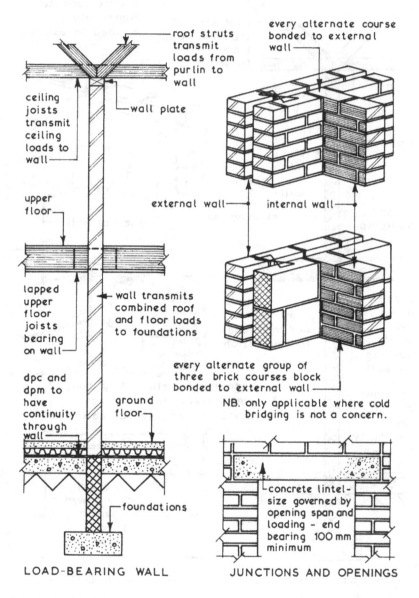

Figure 18.7 Typical brick internal wall details

Internal walls/partitions

Internal block walls

The most common form of internal solid walls, these can be load-bearing or non-load-bearing; the thickness and type of block to be used will depend upon the loadings it has to carry.

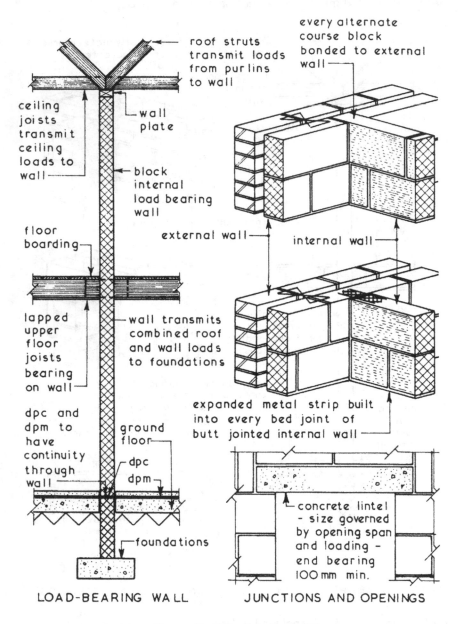

Figure 18.8 Mechanical bonding method for internal walls

An alternative to brick and block bonding is mechanical bonding by using wall profiles. These are quick and simple to install, provide adequate lateral stability, sufficient movement flexibility and will overcome the problem of thermal bridging where a brick partition would otherwise bond into a block inner leaf. They are also useful for attaching extension walls at right angles to existing masonry.

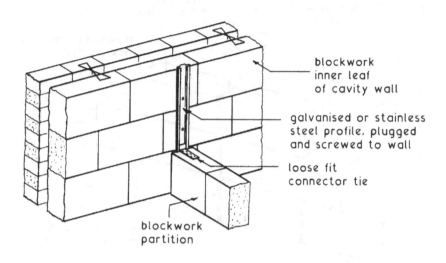

blockwork
inner leaf
of cavity wall

galvanised or stainless
steel profile, plugged
and screwed to wall

loose fit
connector tie

blockwork
partition

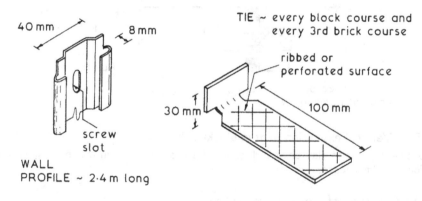

40 mm

8 mm

screw
slot

WALL
PROFILE ~ 2·4 m long

TIE ~ every block course and
every 3rd brick course

ribbed or
perforated surface

30 mm

100 mm

Figure 18.9 Mechanical bonding method for internal walls

Internal walls/partitions

Expansion joints

Used to prevent cracking due to thermal expansion/contraction and drying/shrinkage. Long internal block walls should have expansion joints at 6m spacings and within 3m of corners.

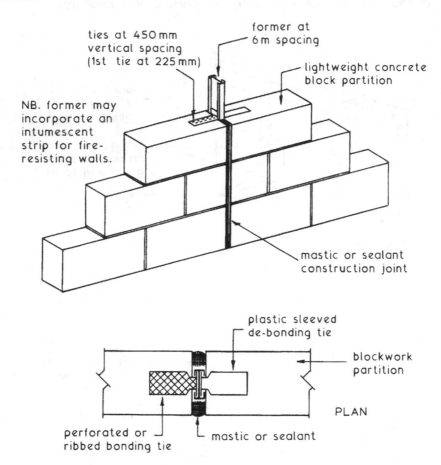

Figure 18.10 Vertical expansion joint in blockwork

Note: Movement joints in clay brickwork should be provided at 12m maximum spacing and 7.5–9m for calcium silicate.

BS EN 1996: *Design of Masonry Structures.*
PD 6697: *Recommendations for the Design of Masonry Structures.*

Reinforced bed joints

Specifically used in positions of high stress.

Reinforcement ~ expanded metal or wire mesh mortar cover with 13mm minimum thickness, 25mm to external faces.

Openings

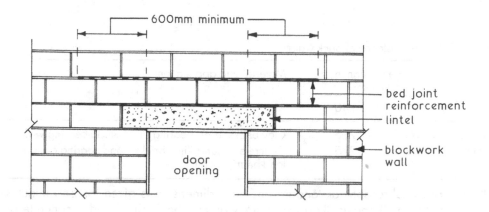

Figure 18.11 Typical application of bed joint reinforcement above a lintel

Concentrated load

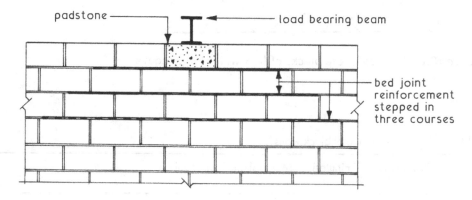

Figure 18.12 Typical application of bed joint reinforcement below a concentrated load

Differential movement ~ may occur where materials such as steel, brick, timber or dense concrete abut with or bear on lightweight concrete blocks. A smooth separating interface of two layers of non-compressible dpc or similar is suitable in this situation.

Internal walls/partitions

Fire resistant masonry walls

Figure 18.13 Solid brickwork

Table 18.1 Fire resistance of brick walls

	Fire resistance (minutes)		Material and application
	120	240	
T (mm)	102.5	215	Clay bricks. Load-bearing or non-load-bearing wall.
T (mm)	102.5	215	Concrete or sand/lime bricks. Load-bearing or non-load-bearing wall.

NB. For practical reasons a standard one-brick dimension is given for 240 minutes of fire resistance. Theoretically a clay brick wall can be 170mm and a concrete or sand/lime brick wall 200mm, finishes excluded.

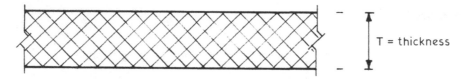

Figure 18.14 Solid concrete blocks of lightweight aggregate

Table 18.2 Fire resistance of block walls

	Fire resistance (minutes)			Material and application
	60	120	240	
T (mm)	100	130	200	Load-bearing, 2.8–3.5N/mm^2 compressive strength.
T (mm)	90	100	190	Load-bearing, 4.0–10N/mm^2 compressive strength.
T (mm)	75	100	140	Non-load-bearing, 2.8–3.5N/mm^2 compressive strength.
T (mm)	75	75	100	Non-load-bearing, 4.0–10N/mm^2 compressive strength.

NB. Finishes excluded.

18.4 Party walls

A party wall ~ separates different owners' buildings, i.e. a wall that stands astride the boundary line between property of different ownerships. It may also be solely on one owner's land but used to separate two buildings.

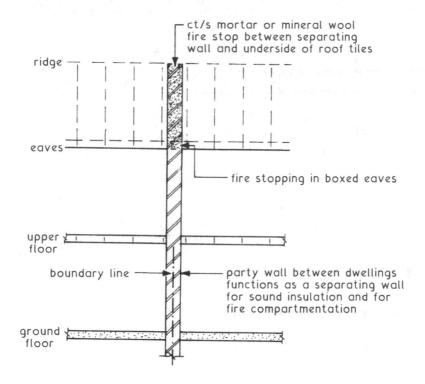

Figure 18.15 Typical party wall between two dwellings

Where an internal separating wall forms a junction with an external cavity wall, the cavity must be fire stopped by using a barrier of fire-resisting material. Depending on the application, the material specification is of at least 30 minutes' fire resistance. Between terraced and semi-detached dwellings the location is usually limited by the separating elements. For other buildings additional fire stopping will be required in constructional cavities such as suspended ceilings, rainscreen cladding and raised floors. The spacing of these cavity barriers is generally not more than 20m in any direction, subject to some variation as indicated in Volume 2 of Approved Document B.

Party Wall Act 1996.
Building Regulations, A.D. B: *Volumes 1 and 2: Fire safety.*
Building Regulations, A.D. E: *Resistance to the passage of sound.*

Party walls

Requirements for fire- and sound-resisting construction

Typical masonry construction

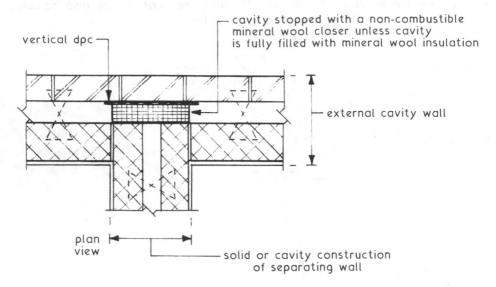

Figure 18.16 Typical masonry construction

Typical timber framed construction

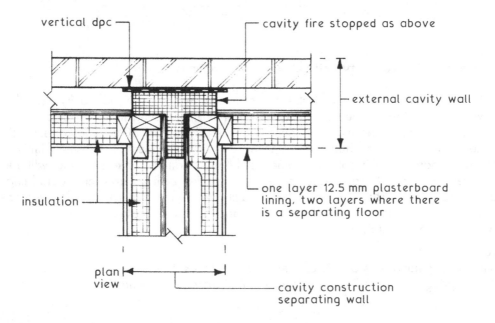

Figure 18.17 Typical timber framed construction

18.5 Stud partition walls

Timber

These are generally non-load-bearing, internal dividing walls that are easy to construct, lightweight, adaptable and can be clad and infilled with various materials to give different finishes and properties. The timber studs should be of prepared or planed material to ensure that the wall is of constant thickness with parallel faces. Stud spacings will be governed by the size and spanning ability of the facing or cladding material.

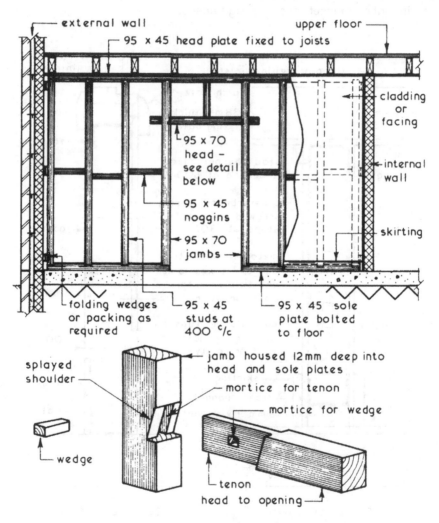

Figure 18.18 Typical timber stud partition wall details

Stud partition walls

NB. Although generally non-load-bearing, timber stud partitions may carry some of the load from the floor and roof structure. In these situations the vertical studs are considered struts.

Metal

These can be load-bearing and non-load-bearing partitions consisting of a framework of galvanised light gauge steel sections with vertical studs fixed between head and sole channels to which plasterboard linings are attached with self-tapping screws.

The void between the studs created by the two faces can be infilled with thermal or sound insulation to meet specific design needs.

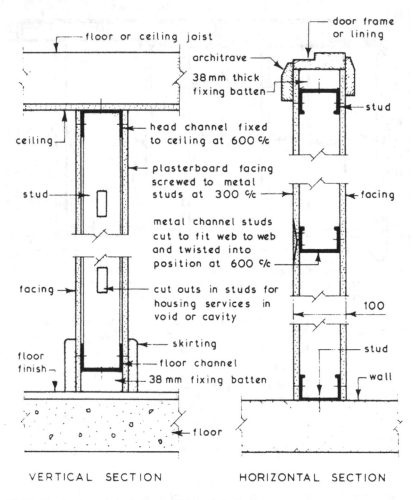

Figure 18.19 Typical metal stud partition details

Plasterboard linings

Plasterboard linings to stud-framed partition walls satisfies the Building Regulations, Approved Document B – Fire Safety, as a material of 'limited combustibility' with a Class 0 rating for surface spread of flame (Class 0 is better than Classes 1 to 4, as determined by BS 476-7). The plasterboard dry walling should completely protect any combustible timber components such as sole plates. The following guidance shows typical fire resistances. These apply to both metal and timber stud framing:

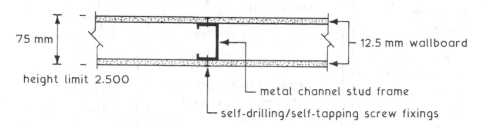

75 mm
height limit 2.500

12.5 mm wallboard
metal channel stud frame
self-drilling/self-tapping screw fixings

Figure 18.20 30-minute fire resistance

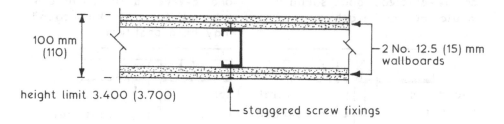

100 mm (110)
height limit 3.400 (3.700)

2 No. 12.5 (15) mm wallboards
staggered screw fixings

Figure 18.21 60(90)-minute fire resistance

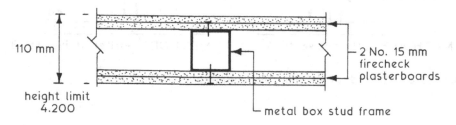

110 mm
height limit 4.200

2 No. 15 mm firecheck plasterboards
metal box stud frame

Figure 18.22 120-minute fire resistance

18.6 Demountable partitions

These are proprietary systems of pre-formed, interlocking panels including door and window sections that can be easily fixed in place (mounted) and removed (demounted), generally without damaging the fixed internal structure of the building.

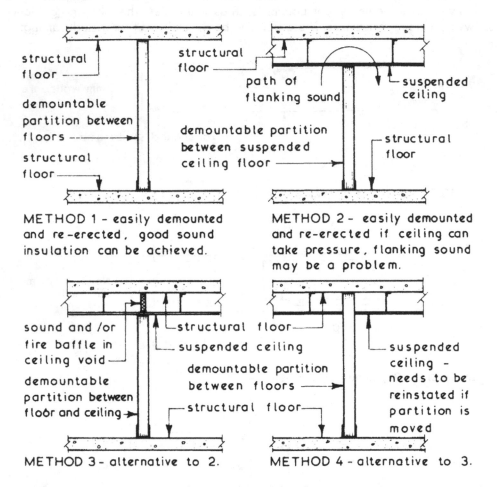

structural floor

demountable partition between floors

structural floor

METHOD 1 - easily demounted and re-erected, good sound insulation can be achieved.

structural floor

path of flanking sound

suspended ceiling

demountable partition between suspended ceiling floor

structural floor

METHOD 2 - easily demounted and re-erected if ceiling can take pressure, flanking sound may be a problem.

sound and /or fire baffle in ceiling void

demountable partition between floor and ceiling

structural floor

suspended ceiling

demountable partition between floors

structural floor

METHOD 3 - alternative to 2.

suspended ceiling - needs to be reinstated if partition is moved

METHOD 4 - alternative to 3.

Figure 18.23 Methods of installing demountable partitions

18.7 Solid plastering

This is the traditional wet system where dry plaster is mixed on site with water. It is applied in layers by hand with a trowel to a solid substrate to achieve a smooth and durable finish suitable for decorative treatments such as paint and wallpaper.

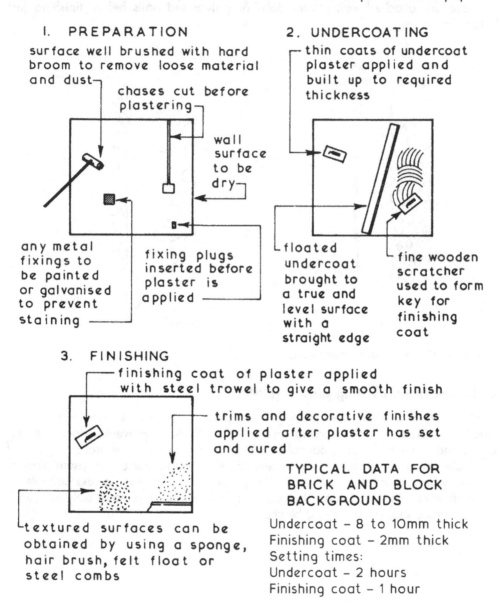

I. PREPARATION

surface well brushed with hard broom to remove loose material and dust

chases cut before plastering

wall surface to be dry

any metal fixings to be painted or galvanised to prevent staining

fixing plugs inserted before plaster is applied

2. UNDERCOATING

thin coats of undercoat plaster applied and built up to required thickness

floated undercoat brought to a true and level surface with a straight edge

fine wooden scratcher used to form key for finishing coat

3. FINISHING

finishing coat of plaster applied with steel trowel to give a smooth finish

trims and decorative finishes applied after plaster has set and cured

textured surfaces can be obtained by using a sponge, hair brush, felt float or steel combs

TYPICAL DATA FOR BRICK AND BLOCK BACKGROUNDS

Undercoat – 8 to 10mm thick
Finishing coat – 2mm thick
Setting times:
Undercoat – 2 hours
Finishing coat – 1 hour

Figure 18.24 Typical method of application

Solid plastering

Non-porous background such as steel or glazed surfaces require application of a bonding agent to improve plaster adhesion. A wire mesh or expanded metal surface attachment may also be required with metal lathing plaster as the undercoat.

Expanded metal angle beads are used at all external wall angles to provide a straight edge in both directions.

These are attached with plaster dabs or galvanised nails before finishing just below the nosing.

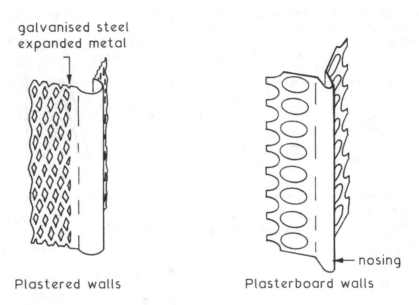

Figure 18.25 External corner beads

Lightweight renovating plaster (LRP)

This consists of cement, lime and expanded perlite, generally proportioned 1:1:6. Additives include a waterproofing agent, salt inhibitor to prevent surface efflorescence and synthetic fibres to control shrinkage and to improve flexural strength. LRP is an alternative undercoat plaster to cement and sand render and gypsum plasters for use where renovating previously damp situations. Gypsum is less suitable in these situations as it is naturally hygroscopic. LRP can be applied directly to sound, dry substrates such as bricks and blocks.

18.8 Dry lining

Dry lining can be applied to solid walls or stud partition walls, instead of using wet undercoat plaster to form a straight even surface to receive the finishing coat of plaster. For solid walls 2.4m high × 1.2m wide × 9.5mm thick plasterboards are bonded to the wall with an adhesive; in a process called 'dot and dab' the boards are lined and levelled to create a straight flat surface. There are two types of board that can be used: square edge and tapered edge. If the boards are to be finished with wet finishing plaster then square edge boards are used with a glass-fibre scrim tape applied over the joint before being plastered. Alternatively, tapered edge boards can be used to provide a flush, seamless surface without a coat of finishing plaster, this is obtained by filling the tapered joint with a special filling plaster, applying a joint tape over the filling and finishing with a thin layer of joint filling plaster, the edge of which is feathered out using a jointing tool. When the joint is dry it is sanded flat.

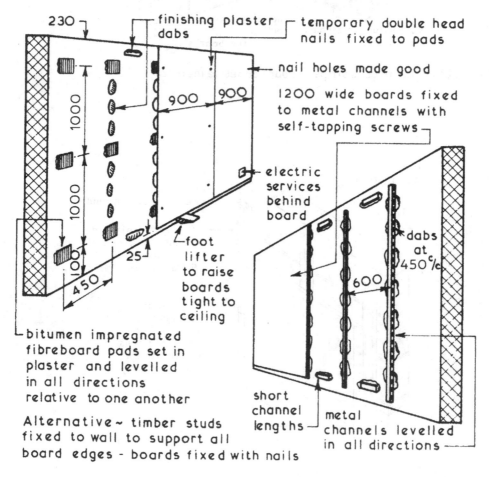

Figure 18.26 Dot and dab methods of fixing wallboards

Dry lining

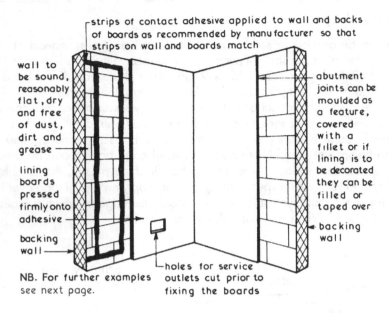

strips of contact adhesive applied to wall and backs of boards as recommended by manufacturer so that strips on wall and boards match

wall to be sound, reasonably flat, dry and free of dust, dirt and grease

lining boards pressed firmly onto adhesive

backing wall

abutment joints can be moulded as a feature, covered with a fillet or if lining is to be decorated they can be filled or taped over

backing wall

holes for service outlets cut prior to fixing the boards

NB. For further examples see next page.

Figure 18.27 Continuous strips of board adhesive method

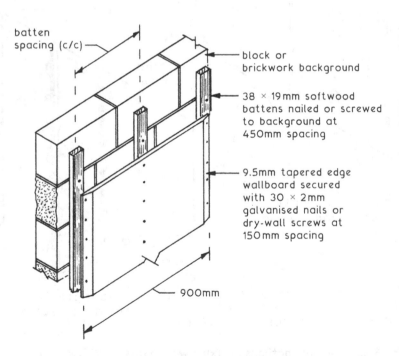

batten spacing (c/c)

block or brickwork background

38 × 19mm softwood battens nailed or screwed to background at 450mm spacing

9.5mm tapered edge wallboard secured with 30 × 2mm galvanised nails or dry-wall screws at 150mm spacing

900mm

Figure 18.28 Fixing taper edge boards to timber battens

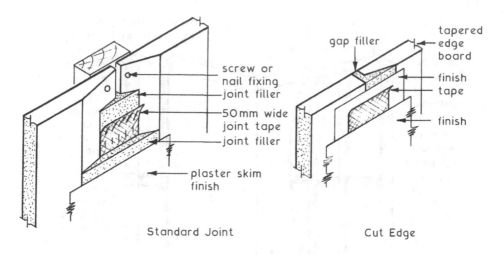

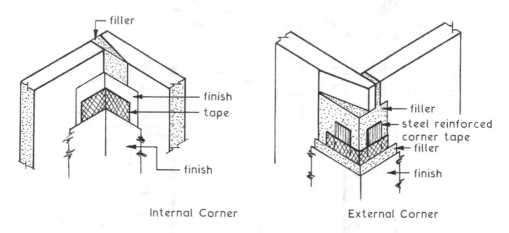

Figure 18.29 Taping and filling taper edge wallboards

The main advantage of dry lining walls is that the drying out period required with wet finishes is eliminated. By careful selection and fixing of some dry lining materials it is possible to improve the thermal insulation properties of a wall. Dry linings can be fixed direct to the backing by means of a recommended adhesive or they can be fixed to a suitable arrangement of wall battens.

Plasterboard types

BS EN 520: *Gypsum Plasterboards. Definitions, Requirements and Test Methods.*

Dry lining

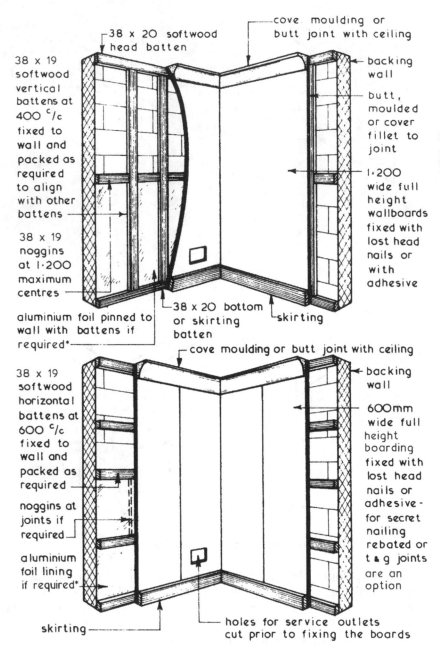

38 x 20 softwood head batten

38 x 19 softwood vertical battens at 400 $^c/c$ fixed to wall and packed as required to align with other battens

38 x 19 noggins at 1·200 maximum centres

aluminium foil pinned to wall with battens if required*

38 x 20 bottom or skirting batten

cove moulding or butt joint with ceiling

backing wall

butt, moulded or cover fillet to joint

1·200 wide full height wallboards fixed with lost head nails or with adhesive

skirting

cove moulding or butt joint with ceiling

38 x 19 softwood horizontal battens at 600 $^c/c$ fixed to wall and packed as required

noggins at joints if required

aluminium foil lining if required*

skirting

backing wall

600mm wide full height boarding fixed with lost head nails or adhesive – for secret nailing rebated or t & g joints are an option

holes for service outlets cut prior to fixing the boards

*alternatively use polythene sheet as a vapour check.

Figure 18.30 Typical example using plywood or similar boarding

604

18.9 Stairs

All stairs, landings, balustrades and handrails must comply with the appropriate requirements set out in Part K of the Building Regulations.

Timber stairs

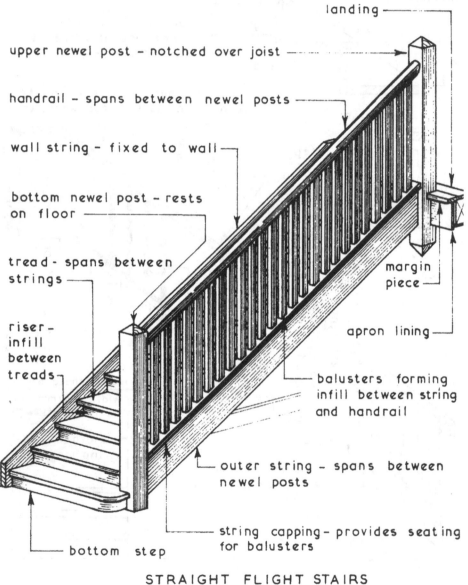

landing

upper newel post – notched over joist

handrail – spans between newel posts

wall string – fixed to wall

bottom newel post – rests on floor

tread – spans between strings

riser – infill between treads

margin piece

apron lining

balusters forming infill between string and handrail

outer string – spans between newel posts

string capping – provides seating for balusters

bottom step

STRAIGHT FLIGHT STAIRS

Figure 18.31 Stair constituent parts

Stairs

All dimensions quoted are the minimum required for domestic stairs exclusive to one dwelling, as given in Approved Document K, unless stated otherwise. (A.D. K does not give a minimum dimension for stair width.)

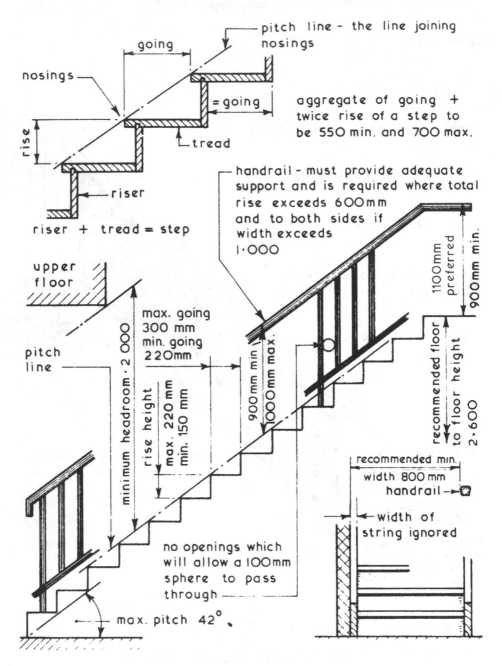

pitch line - the line joining nosings

going

nosings

= going

tread

riser

riser + tread = step

rise

aggregate of going + twice rise of a step to be 550 min. and 700 max.

handrail - must provide adequate support and is required where total rise exceeds 600mm and to both sides if width exceeds 1·000

upper floor

pitch line

minimum headroom · 2 000

max. going 300 mm min. going 220mm

rise height max. 220 mm min. 150 mm

900 mm min 1000mm max.

1100mm preferred

900mm min.

recommended floor to floor height 2·600

recommended min. width 800 mm

handrail

width of string ignored

no openings which will allow a 100mm sphere to pass through

max. pitch 42°

Figure 18.32 Stair terminology

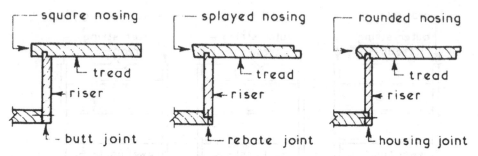

square nosing — tread — riser — butt joint

splayed nosing — tread — riser — rebate joint

rounded nosing — tread — riser — housing joint

NB. nosing types and joints are interchangeable between step formats.

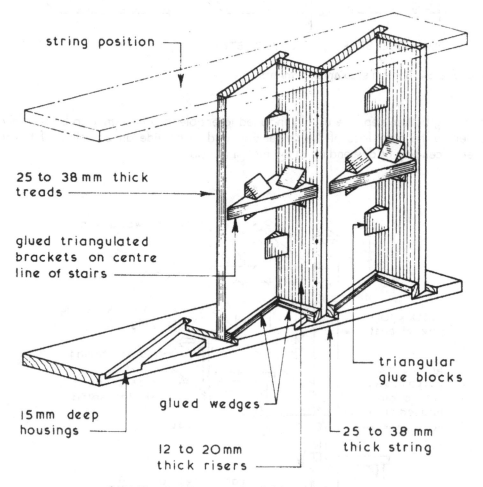

string position

25 to 38 mm thick treads

glued triangulated brackets on centre line of stairs

triangular glue blocks

15 mm deep housings

glued wedges

25 to 38 mm thick string

12 to 20 mm thick risers

STAIR FLIGHT CONSTRUCTION

Figure 18.33 Step formats

Stairs

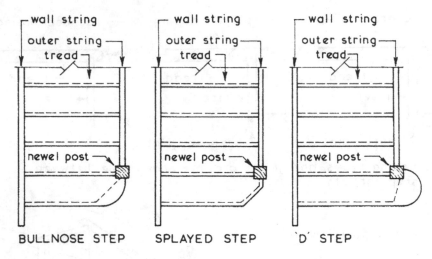

Figure 18.34 Bottom step arrangements

Projecting bottom steps are usually included to enable the outer string to be securely jointed to the back face of the newel post and to provide an easy line of travel when ascending or descending at the foot of the stairs.

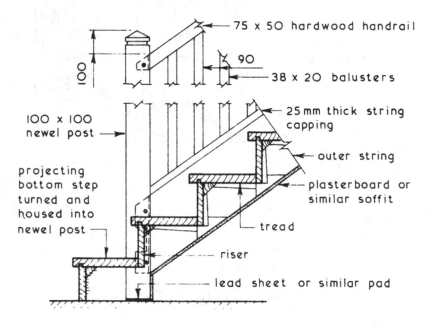

Figure 18.35 Typical detail at bottom newel post

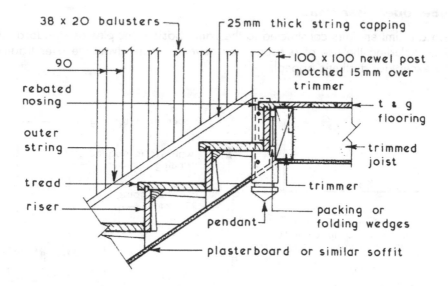

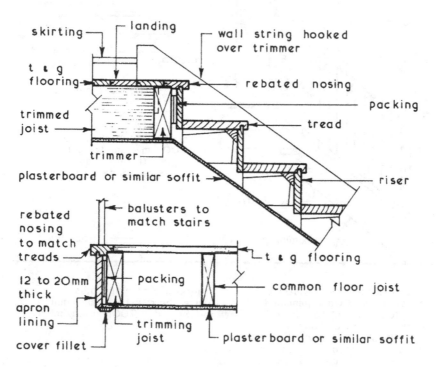

Figure 18.36 Typical detail at top newel post and landing and stairwell

Stairs

Timber open riser stairs

These are timber stairs constructed to the same basic principles as standard timber stairs, excluding the use of a riser. They have no real advantage over traditional stairs except for aesthetic appeal.

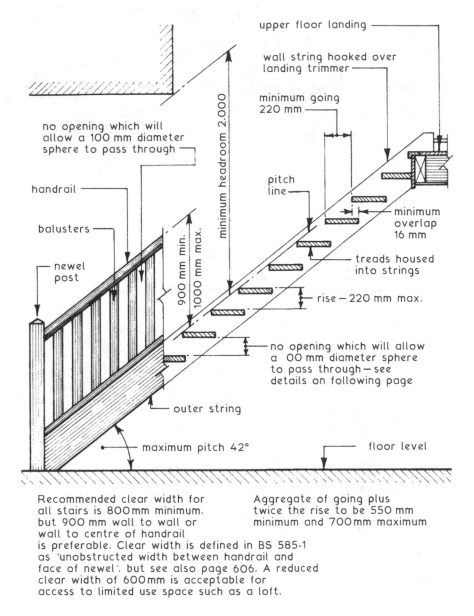

Recommended clear width for all stairs is 800 mm minimum, but 900 mm wall to wall or wall to centre of handrail is preferable. Clear width is defined in BS 585-1 as 'unobstructed width between handrail and face of newel', but see also page 606. A reduced clear width of 600 mm is acceptable for access to limited use space such as a loft.

Aggregate of going plus twice the rise to be 550 mm minimum and 700 mm maximum

Figure 18.37 Typical requirements for stairs in a small residential building

Because of the legal requirement of not having a gap between any two consecutive treads through which a 100mm diameter sphere can pass, and the limitation relating to the going and rise, it is generally not practicable to have a completely riserless stair for residential buildings. Using minimum dimensions a very low pitch of approximately 27.5° would result and, by choosing an acceptable pitch, a very thick tread would have to be used to restrict the gap to 100mm.

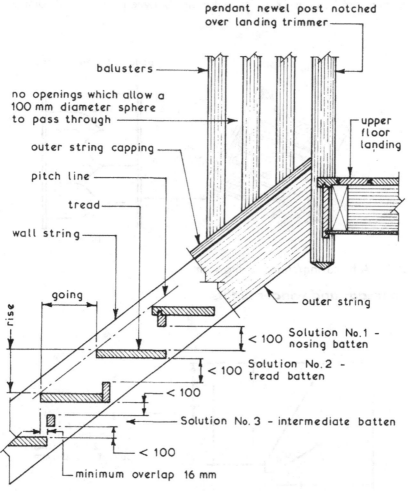

treads and battens housed and fixed into strings

Figure 18.38 Possible solutions

Alternating tread stairs

Also known as 'space saver stairs' these are suitable only for a straight flight to a domestic loft conversion. The economic use of space is achieved by a very steep pitch of about 60° and opposing overlapping treads. Pitch and tread profile differ

Stairs

considerably from other stairs, but they are acceptable to Building Regulations by virtue of 'familiarity and regular use' by the building occupants.

Additional requirements are:

- Non-slip tread surface.
- Handrails to both sides.
- Minimum going 220mm.
- Maximum rise 220mm (2 + rise) + (going) between 550 and 700mm.

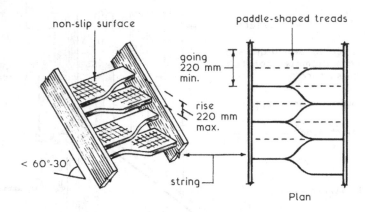

Figure 18.39 Alternating/space saving stairs

Tapered treads and winder flights

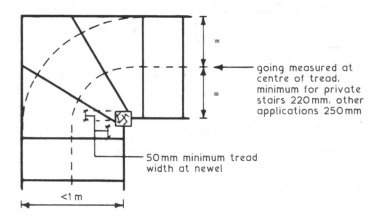

Figure 18.40 Quarter turn stair flight

For stair widths greater than 1m, the going is measured at 270mm from each side of the stair.

Additional requirements:

- Going of tapered treads not less than the going of parallel treads in the same stair.
- Curved landing lengths measured on the stair centre line.
- Twice the rise plus the going, 550 to 700 mm.
- Uniform going for consecutive tapered treads.

Stair landings

Straight flight stairs are simple and easy to construct and install but, by the introduction of intermediate landings, stairs can be designed to change direction of travel and be more compact in plan than the straight flight stairs.

Landings can be detailed for a 90° change of direction (quarter space landing) or a 180° change of direction (half space landing) and can be introduced at any position between the two floors being served by the stairs.

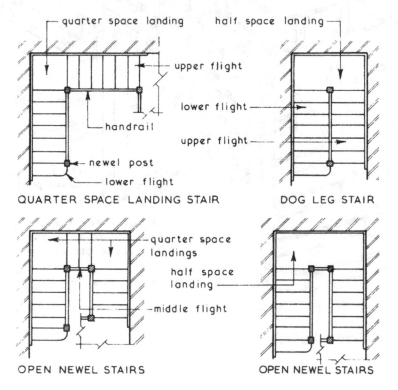

Figure 18.41 Typical landing layouts

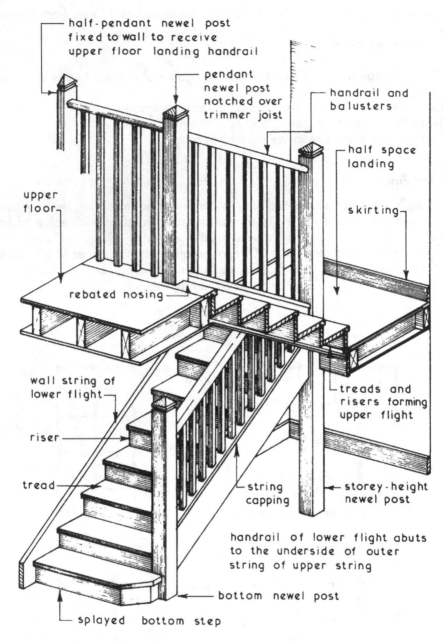

half-pendant newel post
fixed to wall to receive
upper floor landing handrail

pendant
newel post
notched over
trimmer joist

handrail and
balusters

half space
landing

upper
floor

skirting

rebated nosing

wall string of
lower flight

treads and
risers forming
upper flight

riser

tread

string
capping

storey-height
newel post

handrail of lower flight abuts
to the underside of outer
string of upper string

bottom newel post

splayed bottom step

Figure 18.42 Typical dog leg or string over string stairs

Precast concrete stairs

Used predominantly for non-domestic buildings, they consist of precast concrete units of straight stair flights and landings. Due to concrete's inherent fire resistance they are used in fire escape stairways as standard.

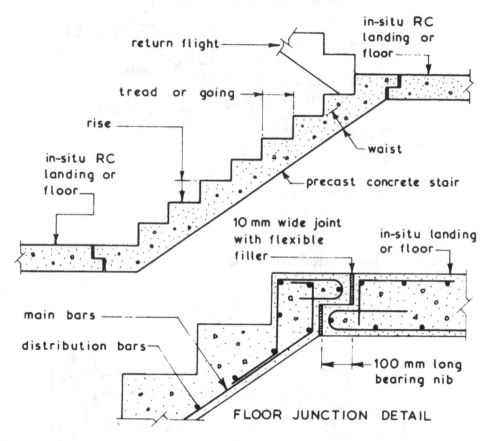

Figure 18.43 Typical example – straight flight stairs

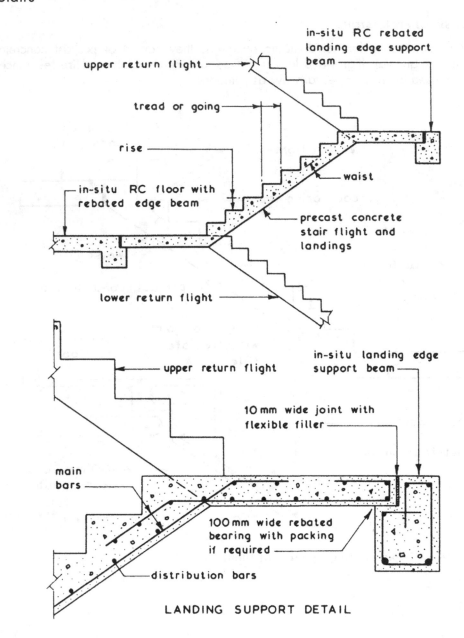

upper return flight

in-situ RC rebated landing edge support beam

tread or going

rise

in-situ RC floor with rebated edge beam

waist

precast concrete stair flight and landings

lower return flight

upper return flight

in-situ landing edge support beam

10 mm wide joint with flexible filler

main bars

100 mm wide rebated bearing with packing if required

distribution bars

LANDING SUPPORT DETAIL

Figure 18.44 Typical example – cranked slab stairs

18.10 Paints and painting

The main functions of paint are to provide:

* Surface protection to building materials and components.
* Surface decoration to building materials and components.

Paint composition

Paints can be categorised as water based or oil based.

Paint components:

* Binder: this is the liquid vehicle or medium that dries to form the surface film and can be composed of linseed oil, drying oils, synthetic resins or water.
* Pigment: this provides the body, colour, durability, opacity and corrosion protection properties of the paint.
* Solvents and thinners: these are materials that can be added to a paint to alter its viscosity. This increases workability and penetration. Water is used for emulsion paint and white spirit or turpentine for oil paint.
* Drier: accelerates drying by absorbing oxygen from the air and converting by oxidation to a solid.

Oil-based paints

These are available in priming, undercoat and finishing grades. The latter can be obtained in a wide range of colours and finishes such as matt, semi-matt, eggshell, satin, gloss and enamel. Polyurethane paints have a good hardness and resistance to water and cleaning. Oil-based paints are suitable for most applications if used in conjunction with the correct primer and undercoat.

Water-based paints

Most of these are called emulsion paints, the various finishes available being obtained by adding to the water medium additives such as alkyd resin and polyvinyl acetate (PVA). Finishes include matt, eggshell, semi-gloss and gloss. Emulsion paints are easily applied, quick drying, can be obtained with a washable finish and are suitable for most applications.

Surface preparation

Application can be to almost any surface providing the surface preparation and sequence of paint coats are suitable. The manufacturer's specification and/or the recommendations of BS 6150 (painting of buildings) should be followed.

Paints and painting

Timber

To ensure a good adhesion of the paint film all timber should have a moisture content of less than 18%. The timber surface should be prepared using an abrasive paper to produce a smooth surface brushed and wiped free of dust, with any grease having been removed with a suitable spirit. Careful treatment of knots is essential either by sealing with two coats of knotting or, in extreme cases, cutting out the knot and replacing with sound timber. The stopping and filling of cracks and fixing holes with putty or an appropriate filler should be carried out after the application of the priming coat. Each coat of paint must be allowed to dry hard and be rubbed down with a fine abrasive paper before applying the next coat. On previously painted surfaces, if the paint is in a reasonable condition the surface will only require cleaning and rubbing down before repainting; when the paint is in a poor condition it will be necessary to remove completely the layers of paint and then prepare the surface as described above for new timber.

In new work the basic buildup of paint coats consists of:

1. Priming coats: these are used on unpainted surfaces to obtain the necessary adhesion and to inhibit corrosion of ferrous metals. New timber should have the knots treated with a solution of shellac or other alcohol-based resin called knotting prior to the application of the primer.
2. Undercoats: these are used on top of the primer after any defects have been made good with a suitable stopper or filler. The primary function of an undercoat is to give the opacity and buildup necessary for the application of the finishing coat(s).
3. Finish: applied directly over the undercoating in one or more coats to impart the required colour and finish.

Plaster

The essential requirement of the preparation is to ensure that the plaster surface is perfectly dry, smooth and free of defects before applying any coats of paint. Plasters that contain lime can be alkaline and such surfaces should be treated with an alkali-resistant primer when the surface is dry, before applying the final coats of paint. Because of the high degree of suction on newly plastered walls the first coat of emulsion should be watered down by 50% (this is called a mist coat) to stabilise the wall suction for following coats.

Paint defects

These may be due to poor or incorrect preparation of the surface, poor application of the paint and/or chemical reactions. The general remedy is to remove all the affected paint and carry out the correct preparation of the surface before applying new coats of paint in the correct manner. Most paint defects are visual and therefore an accurate diagnosis of the cause must be established before any remedial treatment is undertaken.

Typical paint defects:

- Bleeding, staining and disruption of the paint surface by chemical action, usually caused by applying an incorrect paint over another. Remedy is to remove affected paint surface and repaint with correct type of overcoat paint.
- Blistering, usually caused by poor presentation allowing resin or moisture to be entrapped, the subsequent expansion causing the defect. Remedy is to remove all the coats of paint and ensure that the surface is dry before repainting.
- Blooming, mistiness usually on high gloss or varnished surfaces due to the presence of moisture during application. It can be avoided by not painting under these conditions. Remedy is to remove affected paint and repaint.
- Chalking, powdering of the paint surface due to natural ageing or the use of poor-quality paint. Remedy is to remove paint if necessary, prepare surface and repaint.
- Cracking and crazing, usually due to unequal elasticity of successive coats of paint. Remedy is to remove affected paint and repaint with compatible coats of paint.
- Flaking and peeling, can be due to poor adhesion, presence of moisture, painting over unclean areas or poor preparation. Remedy is to remove defective paint, prepare surface and repaint.
- Grinning, due to poor opacity of paint film, allowing paint coat below or background to show through. This could be the result of poor application, incorrect thinning or the use of the wrong colour. Remedy is to apply further coats of paint to obtain a satisfactory surface.

Painting

Paint can applied by:

1. Brush: the correct type, size and quality of brush needs to be selected and used. To achieve a first-class finish by means of brush application requires a high degree of skill.
2. Spray: as with brush application a high degree of skill is required to achieve a good finish. Generally compressed air sprays or airless sprays are used for building works.

3. Roller: simple and inexpensive method of quickly and cleanly applying a wide range of paints to flat and textured surfaces. Roller heads vary in size from 50 to 450mm wide with various covers such as sheepskin, synthetic pile fibres, mohair and foamed polystyrene. All paint applicators must be thoroughly cleaned after use.

The main objectives of applying coats of paint to a surface are preservation, protection and decoration to give a finish which is easy to clean and maintain. To achieve these objectives the surface preparation and paint application must be adequate. The preparation of new and previously painted surfaces should ensure that prior to painting the surface is smooth, clean, dry and stable.

18.11 Glazed wall tiling

Internal glazed wall tiles are usually made to the various specifications under BS EN 14411: *Ceramic Tiles. Definitions, Classification, Characteristics, Evaluation of Conformity and Marking.*

The body of the tile can be made from ball-clay, china clay, china stone, flint and limestone. The material is usually mixed with water to the desired consistency, shaped and then fired in a tunnel oven at a high temperature (1150°C) for several days to form the unglazed biscuit tile. The glaze pattern and colour can now be imparted onto the biscuit tile before the final firing process at a temperature slightly lower than that of the first firing (1050°C) for about two days.

Typical internal glazed wall tiles and fittings

Modular sizes: 100 × 100 × 5mm thick and
200 × 100 × 6.5mm thick.
Non-modular sizes: 152 × 152 × 5–8mm thick and
108 × 108 × 4 and 6.5mm thick.
Other sizes: 200 × 300, 250 × 330, 250 × 400, 300 × 450, 300 × 600 and
   330 × 600mm.
Fittings: wide range available, particularly in the non-modular format.

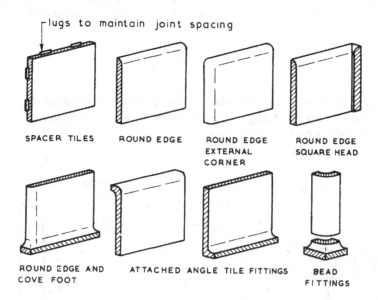

Figure 18.45 Non–standard tile formats

Glazed wall tiling

Bedding of internal wall tiles

Generally, glazed internal wall tiles are considered to be inert in the context of moisture and thermal movement. Therefore, if movement of the applied wall tile finish is to be avoided, attention must be given to the background and the method of fixing the tiles.

Backgrounds

These are usually of a cement rendered or plastered surface and should be flat, dry, stable, firmly attached to the substrate and sufficiently established for any initial shrinkage to have taken place. The flatness of the background should be not more than 3mm in 2m for the thin bedding of tiles and not more than 6mm in 2m for thick bedded tiles.

Fixing wall tiles

There are two methods in general use:

Thin bedding ~ lightweight, internal glazed wall tiles fixed dry using a recommended adhesive, which is applied to the wall in small areas 1m^2 at a time with a notched trowel, the tile being pressed into the adhesive.

Thick bedding ~ cement mortar within the mix range of 1:3 to 1:4 can be used or a proprietary adhesive, either by buttering the backs of the tiles, which are then pressed into position, or by rendering the wall surface to a thickness of approximately 10mm and then applying thin bedded tiles to the rendered wall surface within two hours.

Grouting

When the wall tiles have set, the joints can be grouted by rubbing into the joints a grout paste either using a sponge or brush. Most grouting materials are based on cement with inert fillers and are used neat.

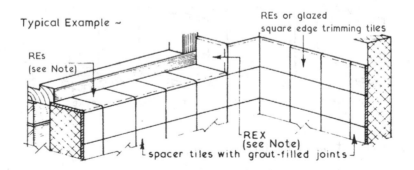

Figure 18.46 Use of radiused edge tiles
Note: The alternative treatment at edges is application of a radiused profile plastic trimming to standard spacer tiles.

19 DRAINAGE

Opening remarks

Drainage is the above and below ground services to remove effluent from a building. There are two main forms of effluent:

Surface water ~ effluent collected from surfaces such as roofs and paved areas. It is considered to be clean and can be discharged direct into an approved water course or soakaway.

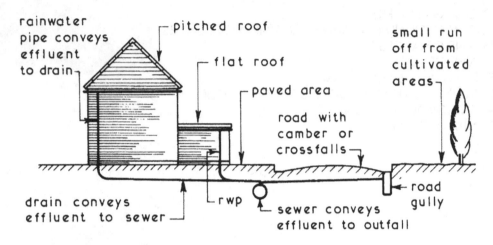

Figure 19.1 Surface water drain principles

Foul or soil water ~ effluent contaminated by domestic or trade waste. It will require treatment to render it clean before it can be discharged into an approved water course.

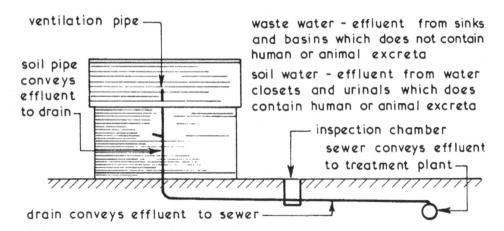

Figure 19.2 Foul water drain principles

19.1 Below ground drainage – drains and sewers

Drains

These can be defined as below ground pipes conveying surface water or foul water within the curtilage of that building. They are the responsibility of the building owner.

Sewers

These can be defined as below ground pipes that collect the discharge from a number of drains and convey it to the final outfall. Shared and public sewers are maintained by the local water and sewerage authority.

Design principles

*   To provide a drainage system which is simple, efficient and economical by laying the drains to a gradient that will render them self-cleansing by gravity and will convey the effluent to a sewer without danger to health or giving nuisance.
*   To provide a drainage system that will comply with the minimum requirements given in AD H of the Building Regulations.

There must be an access point at a junction unless each run can be cleared from another access point.

Drainage systems

There are three types of drainage arrangement:

Separate system

In this system surface water discharge from the building is conveyed in separate drains and sewers to that of foul water discharges. The surface water sewer terminates at a river or water course whilst the foul discharge terminates at a sewage treatment plant.

Combined system

This is the simplest and least expensive system to design and install but since all forms of discharge are conveyed in the same sewer all the rainwater and foul effluent must be treated unless a sea outfall is used to discharge the untreated effluent.

BS EN 752: *Drain and Sewer Systems Outside Buildings*.

Below ground drainage – drains and sewers

Partially separate system

In this compromise system there are two drains, one to convey only surface water and a foul drain that also carries a proportion of the surface water where it is a more practical and efficient design.

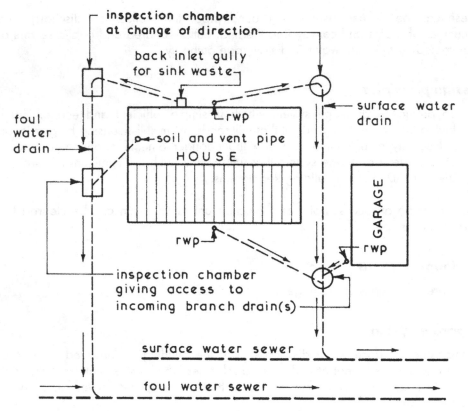

All junctions should be oblique and in direction of flow

Figure 19.3 Typical basic requirements

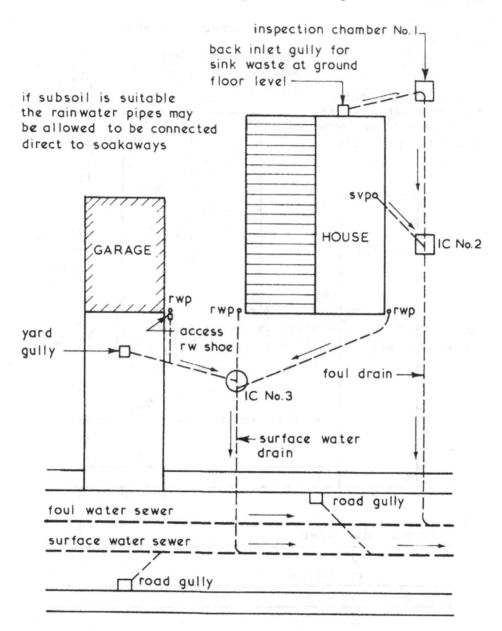

Figure 19.4 Separate drainage system

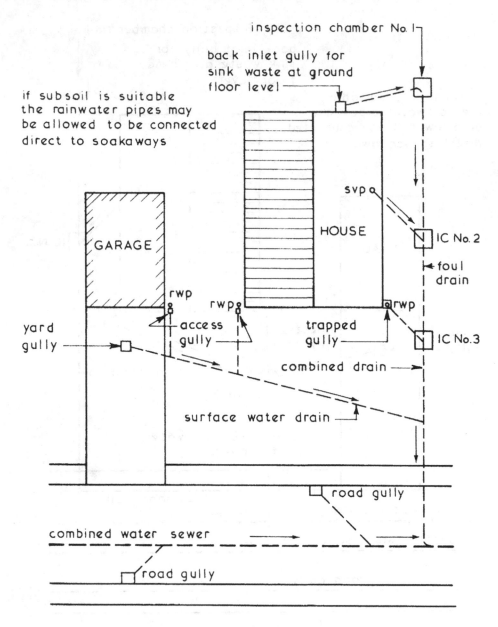

inspection chamber No. 1

back inlet gully for
sink waste at ground
floor level

if subsoil is suitable
the rainwater pipes may
be allowed to be connected
direct to soakaways

svp

HOUSE

GARAGE

IC No. 2

foul
drain

rwp

rwp

rwp

yard
gully

access
gully

trapped
gully

IC No. 3

combined drain

surface water drain

road gully

combined water sewer

road gully

Figure 19.5 Combined drainage system

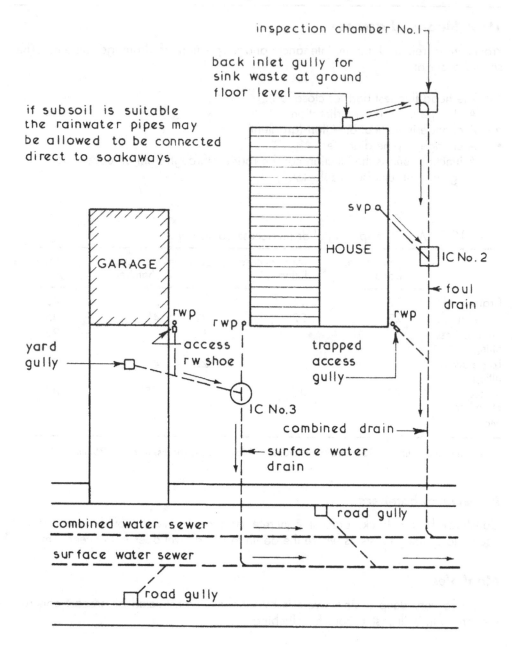

Figure 19.6 Partially separate drainage system

19.2 Means of access

Provision is required for maintenance and inspection of drainage systems. This should occur at:

- The head (highest part) or close to it.
- A change in horizontal direction.
- A change in vertical direction (gradient).
- A change in pipe diameter.
- A junction, unless the junction can be rodded through from an access point.
- Long straight runs (see Table 19.1).

Table 19.1 Maximum spacing of drain access points (m)

From	Small access fitting	Large access fitting	Junction	Inspection chamber	Manhole
Drain head	12	12		22	45
Rodding eye	22	22	22	45	45
Small access fitting			12	22	22
Large access fitting			22	45	45
Inspection chamber	22	45	22	45	45
Manhole				45	45

Note: Small access fitting is 150mm dia. or 150mm × 100mm. Large access fitting is 225mm × 100mm.

Inspection chambers

Constructed from brick, precast concrete or plastic, these provide a means of access to drainage systems where the depth to invert level does not exceed 1m.

Manholes

Access to very deep drain/sewers is by manholes constructed of precast concrete sections with built-in step irons for climbing.

Rodding eyes

These may be used at the head of a drain run.
   Building Regulations, *Approved Document H1: Foul Water Drainage*.

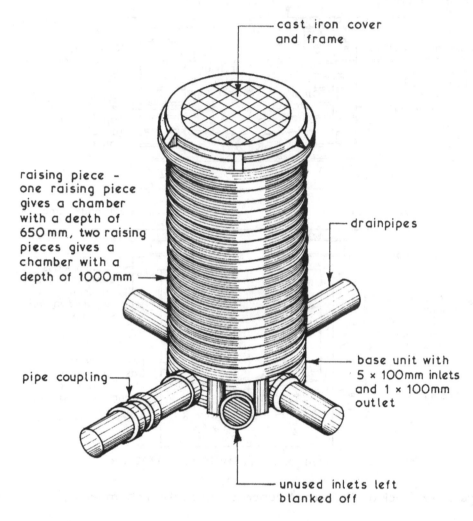

cast iron cover
and frame

raising piece -
one raising piece
gives a chamber
with a depth of
650 mm, two raising
pieces gives a
chamber with a
depth of 1000 mm

drainpipes

base unit with
5 × 100 mm inlets
and 1 × 100 mm
outlet

pipe coupling

unused inlets left
blanked off

Figure 19.7 uPVC inspection chamber

Means of access

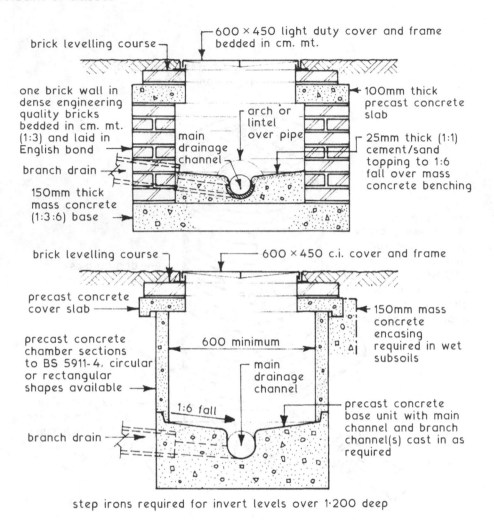

brick levelling course

600 × 450 light duty cover and frame bedded in cm. mt.

one brick wall in dense engineering quality bricks bedded in cm. mt. (1:3) and laid in English bond

100mm thick precast concrete slab

arch or lintel over pipe

main drainage channel

25mm thick (1:1) cement/sand topping to 1:6 fall over mass concrete benching

branch drain

150mm thick mass concrete (1:3:6) base

brick levelling course

600 × 450 c.i. cover and frame

precast concrete cover slab

150mm mass concrete encasing required in wet subsoils

precast concrete chamber sections to BS 5911- 4, circular or rectangular shapes available

600 minimum

main drainage channel

1:6 fall

precast concrete base unit with main channel and branch channel(s) cast in as required

branch drain

step irons required for invert levels over 1·200 deep

Figure 19.8 Brick and pre-cast concrete inspection chambers

632

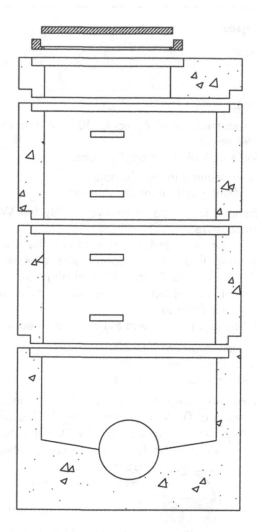

Figure 19.9 Deep manhole in PCC sections with step irons

19.3 Drainage pipes

Legacy materials include:

- Clayware.
- Cast iron.
- Asbestos fibre pipes.

Modern drainage is predominantly uPVC up to 300mm diameter and precast concrete for larger diameter sewers.

Sizes for normal domestic foul water applications:

<20 dwellings = 100mm nominal inside diameter
20–150 dwellings = 150mm nominal inside diameter

Exceptions: 75mm diameter for waste or rainwater only (no WCs); 150mm diameter minimum for a public sewer.

Other situations can be assessed by summating the Discharge Units from appliances and converting these to an appropriate diameter stack and drain, see BS EN 12056–2 (stack) and BS EN 752 (drain). Gradient will also affect pipe capacity and, when combined with discharge calculations, provides the basis for complex hydraulic theories.

The simplest correlation of pipe size and fall is represented in Maguire's rule:

4" (100mm) pipe, minimum gradient 1 in 40.
6" (150mm) pipe, minimum gradient 1 in 60.
9" (225mm) pipe, minimum gradient 1 in 90.

The Building Regulations, Approved Document H1 provides more scope and relates to foul water drains running at 0.75 proportional depth. See Diagram 9 and Table 6 in Section 2 of the Approved Document. Fig. 19.10 illustrates a typical half bore design application. The void provides for free flow with limited spare capacity for later connections.

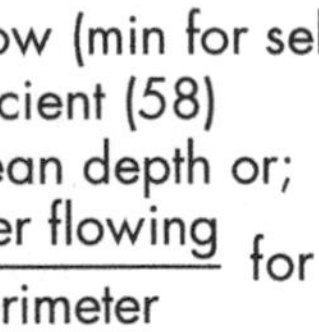

150 mm (0·15m)
area of water flowing
0·5 proportional depth
wetted perimeter

Applying the Chezy formula for gradient calculations:

$$v = c\sqrt{m \times i}$$

where: v = velocity of flow (min for self-cleansing = 0.8m/s)
c = Chezy coefficient (58)
m = hydraulic mean depth or;

$$\frac{\text{area of water flowing}}{\text{wetted perimeter}} \text{ for 0.5 p.d.} = \text{diam}/4$$

i = inclination or gradient as a fraction $1/x$

Selecting a velocity of 1m/s as a margin of safety over the minimum:

$$1 = 58\sqrt{0.15 / 4 \times i}$$

i = 0.0079 where i = 1/x
So, x = 1/0.0079 = 126, i.e. a minimum gradient of 1 in 126

Figure 19.10 Proportional depth of flow in a pipe

Soundness testing

There are two methods to test the soundness of a drain: air test and water test.

Air test

Equipment: Manometer and two drain stoppers, one with tube attachment.
Test ~ 100mm water gauge to fall no more than 25mm in five mins. Or, 50mm water gauge to fall no more than 12mm in five mins.

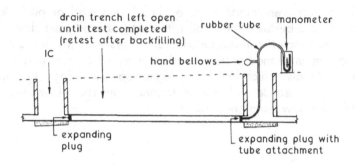

Figure 19.11 Air testing of drain run

Water test

Equipment: drain stopper, test bend, extension pipe.
Test ~ 1.5m head of water to stand for two hours and then topped up. Leakage over the next 30 minutes should be minimal, i.e. 100mm pipe, 005 litres per metre, which equates to a drop of 6.4mm/m in the extension pipe, and 150mm pipe, 008 litres per metre, which equates to a drop of 4.5mm/m in the extension pipe.

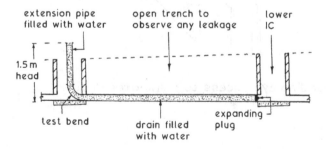

Figure 19.12 Water testing of drain run

19.4 Surface water drainage

Roof gutters

A roof must be designed with a suitable gradient/fall towards the surface water collection channel or gutter, which, in turn, is connected to vertical rainwater pipes that convey the collected discharge to the drainage system.

Materials

Traditionally gutters were made of wood lined with lead or pitch, later cast iron and then painted asbestos cement, all with cast iron rainwater down pipes (RWP). The most common material in use today is uPVC systems, which are simple to install and have low maintenance costs. A wide variety of colours and styles are available. Other materials, which feature on higher spec buildings, are powder coated aluminium alloy or seamless aluminium formed from sheet aluminium on a roll on site by machine to any length required.

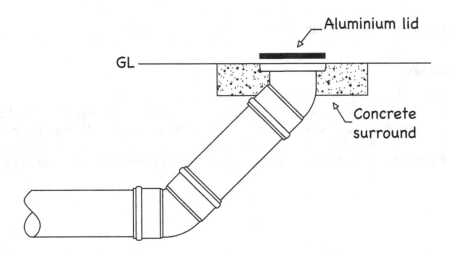

Figure 19.13 Rodding eye access to rainwater drain

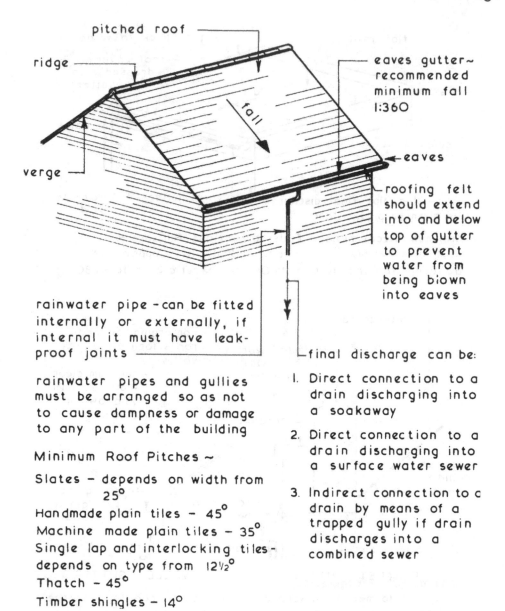

pitched roof

ridge

eaves gutter~
recommended
minimum fall
1:360

fall

verge

eaves

roofing felt
should extend
into and below
top of gutter
to prevent
water from
being blown
into eaves

rainwater pipe -can be fitted
internally or externally, if
internal it must have leak-
proof joints

final discharge can be:

rainwater pipes and gullies
must be arranged so as not
to cause dampness or damage
to any part of the building

Minimum Roof Pitches ~

Slates – depends on width from
25°
Handmade plain tiles – 45°
Machine made plain tiles – 35°
Single lap and interlocking tiles –
depends on type from 12½°
Thatch – 45°
Timber shingles – 14°

1. Direct connection to a
drain discharging into
a soakaway

2. Direct connection to a
drain discharging into
a surface water sewer

3. Indirect connection to a
drain by means of a
trapped gully if drain
discharges into a
combined sewer

Figure 19.14 Rainwater/surface water drainage from a pitched roof

Surface water drainage

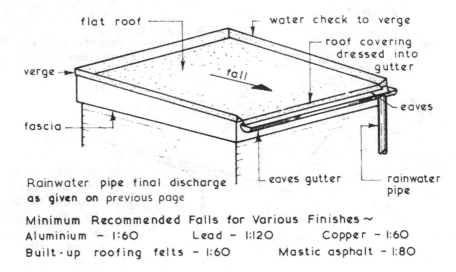

Rainwater pipe final discharge as given on previous page

Minimum Recommended Falls for Various Finishes ~
Aluminium – 1:60   Lead – 1:120   Copper – 1:60
Built-up roofing felts – 1:60   Mastic asphalt – 1:80

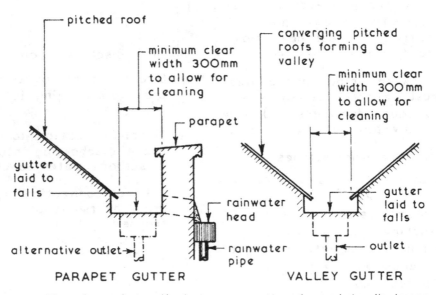

PARAPET GUTTER

gutter formed to discharge into internal rainwater pipes or to external rainwater pipes via outlets through the parapet

VALLEY GUTTER

gutter formed to discharge into internal rainwater pipes or to external rainwater pipes sited at the gable ends

Figure 19.15 Typical drain RWP connection

Typical Eaves Details ⌣

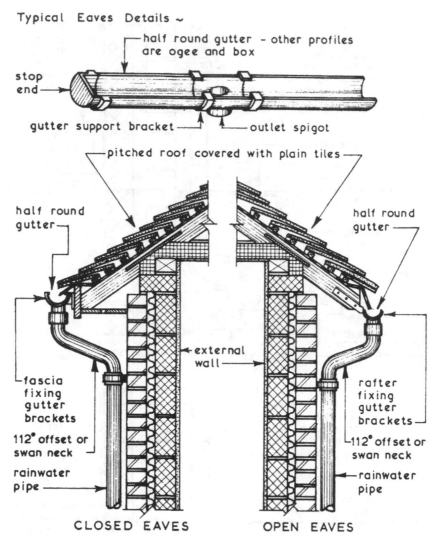

half round gutter - other profiles are ogee and box

stop end →

gutter support bracket —

outlet spigot

pitched roof covered with plain tiles

half round gutter

half round gutter

fascia fixing gutter brackets

rafter fixing gutter brackets

← external wall →

112° offset or swan neck

112° offset or swan neck

rainwater pipe

rainwater pipe

CLOSED EAVES

OPEN EAVES

For details of rainwater pipe connection to drainage see next page.

Note: Details show established construction with a block cavity closer course. New construction to have insulation continuity to prevent cold bridging.

Figure 19.16 Typical eaves details

Surface water drainage

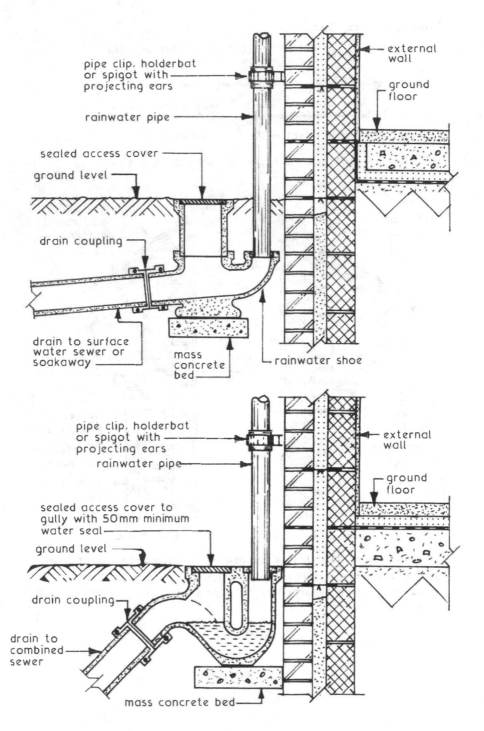

Figure 19.17 Rainwater pipe and gully details

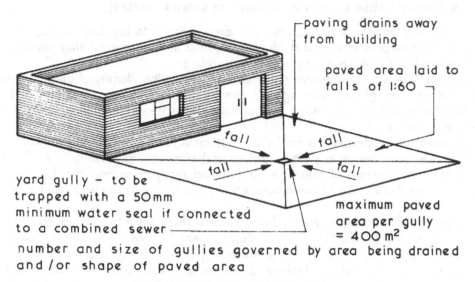

paving drains away
from building

paved area laid to
falls of 1:60

fall   fall

fall   fall

yard gully – to be
trapped with a 50mm
minimum water seal if connected
to a combined sewer

maximum paved
area per gully
= 400 m²

number and size of gullies governed by area being drained
and/or shape of paved area

YARD GULLY COLLECTION

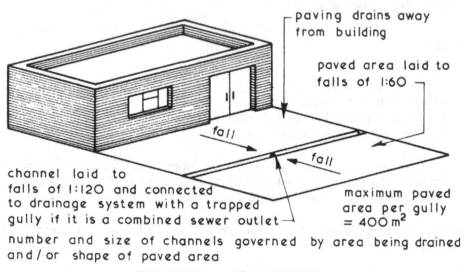

paving drains away
from building

paved area laid to
falls of 1:60

fall

fall

channel laid to
falls of 1:120 and connected
to drainage system with a trapped
gully if it is a combined sewer outlet

maximum paved
area per gully
= 400 m²

number and size of channels governed by area being drained
and/or shape of paved area

CHANNEL COLLECTION

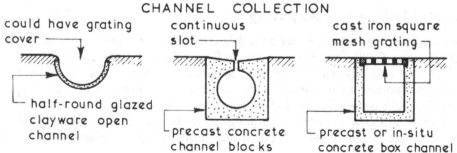

could have grating
cover

half-round glazed
clayware open
channel

continuous
slot

precast concrete
channel blocks

cast iron square
mesh grating

precast or in-situ
concrete box channel

Figure 19.18 Surface water paved areas

19.5 Sustainable urban drainage systems (SUDS)

These cover a variety of applications that are designed to regulate surface water run-off. Their purpose is to control the level of water in the ground during periods of intense rainfall, thereby reducing the risk of flooding.

Problem: Growth in urbanisation, increase in population, density of development, extreme weather, global warming and climate change – these are some of the factors that contribute to a rise in ground water table levels and higher volumes of surface water run-off.

Solution: Before developing land and undertaking major refurbishment projects, it is necessary to design a surface and ground water control system that relieves potential concentrations of rainwater. This should replicate or improve the natural site drainage that existed before groundworks (foundations, basements, etc.) disturb the subgrade.

Applications: Installation systems that manage surface water by attenuation and filtration as required by the Flood Water Management Act. Measures may include site treatment, containment and rainwater harvesting (see *Building Services Handbook*) before surface water can be discharged into surface water sewers. SUDS controls, such as swales and retention ponds, are another option, but space for these is limited or non-existent in urban situations.

Objectives:

- Improved amenity.
- Improved water resource quality.
- Reduced exposure to flooding.
- Minimised dispersal through foul water drainage systems.
- Regulated natural flow conditions.

Flood and Water Management Act 2010 established a SUDS approval body (SAB) and amended the automatic right to connect surface water drains to a public surface water sewer, connection subject to conditional standards.

See also: *Building Services Handbook*, Part 8 - F. Hall and R. Greeno (2017), Routledge.

Concrete block permeable paving

Unlike traditional impermeable surfaces that direct surface water to drainage channels and pipes, CBPP filters and removes pollutants before dispersal. Hydrocarbons are degraded and digested by naturally occurring microbes in the sub-base, therefore oil interceptors are not required. Silt traps are also not needed as the system retains silt. Block pavers filter and clean surface water before it is accommodated in the sub-base and slowly released into the ground. In less naturally draining subsoils the retained water percolates into perforated drainage pipes before flowing on to a drainage discharge system.

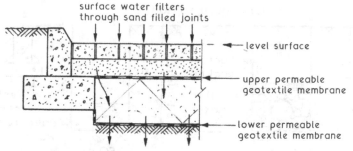

surface water filters
through sand filled joints

level surface

upper permeable
geotextile membrane

lower permeable
geotextile membrane

Subgrades of good drainage medium

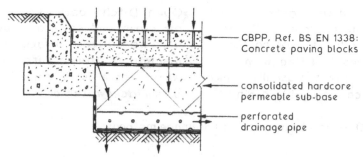

CBPP, Ref. BS EN 1338:
Concrete paving blocks

consolidated hardcore
permeable sub-base

perforated
drainage pipe

Subgrades composed of material with limited drainage

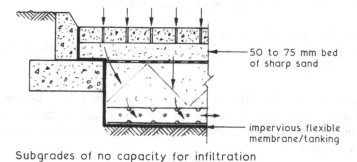

50 to 75 mm bed
of sharp sand

impervious flexible
membrane/tanking

Subgrades of no capacity for infiltration

Figure 19.19 Permeable block paving installations

Sustainable urban drainage systems (SUDS)

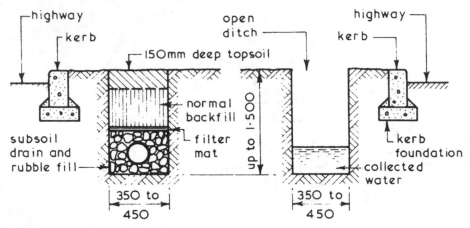

Subsoil Drain – acts as a cut off drain and can be formed using perforated or porous drainpipes. If filled with rubble only it is usually called a French or rubble drain.

Open Ditch – acts as a cut off drain and could also be used to collect surface water discharged from a rural road where there is no raised kerb or surface water drains.

Figure 19.20 Typical highway subsoil drainage methods

Highway drainage

The stability of a highway or road relies on two factors:

1. Strength and durability of upper surface.
2. Strength and durability of subgrade, which is the subsoil on which the highway construction is laid.

The above can be adversely affected by water; therefore it may be necessary to install two drainage systems. One system (subsoil drainage) to reduce the flow of subsoil water through the subgrade under the highway construction and one system of surface water drainage.

Road drainage

This consists of laying the paved area or road to a suitable cross fall or gradient to direct the run-off of surface water towards the drainage channel or gutter. This is usually bounded by a kerb which helps to convey the water to the road gullies connected to a surface water sewer.

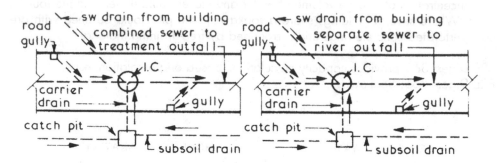

Figure 19.21 Surface water drainage systems

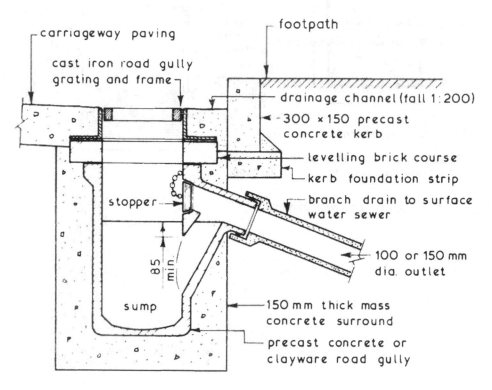

Figure 19.22 Typical road gully detail

Soakaways

These provide a means for collecting and controlling the seepage of rainwater into suitable granular subsoils. They are not suitable in clay subsoils. Soakaways should preferably be lower than adjacent buildings and no closer than 5m to any building,

Sustainable urban drainage systems (SUDS)

as concentration of a large volume of water any closer could undermine the foundations. Where several buildings share a soakaway, the pit should be lined with precast perforated concrete rings and surrounded by free-draining material.

BRE Digest 365 provides capacity calculations based on percolation tests.
BRE Digest 365: Soakaways.

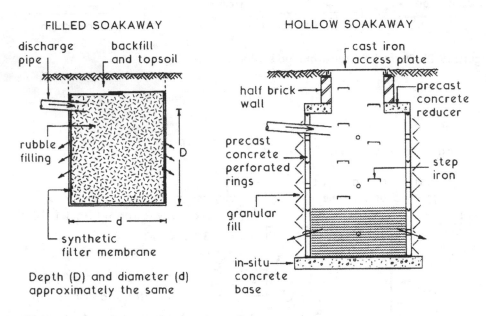

Figure 19.23 Typical soakaway examples

19.6 Above ground drainage

This is a system composed of drainage fittings and pipework to remove waste water from a building by means of gravity to an underground drain and sewer system. Three types of pipework arrangements are available to ensure efficient removal of waste water to avoid blockages or loss of trap seal.

Single stack system

This system is only possible when the sanitary appliances are closely grouped around the discharge stack. The slope and distance of the branch connections must be kept within the design limitations given in Figure 19.24 to prevent loss of trap seal on the sanitary appliances.

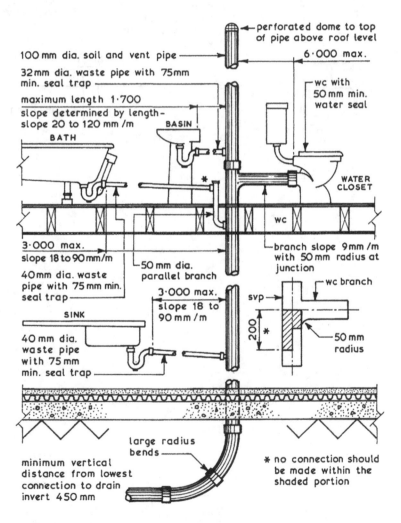

Figure 19.24 Typical single stack system details

Above ground drainage

Air admittance valve (AAV)

A pressure-sensitive device fitted at the top of an internal stub discharge stack that senses when negative pressure occurs in the discharge stack and automatically opens to allow air into the stack to maintain a neutral pressure. It then closes afterwards.

Internal AAV must be fitted above the spill level of the uppermost appliance connected to the stack, usually the wash basin.

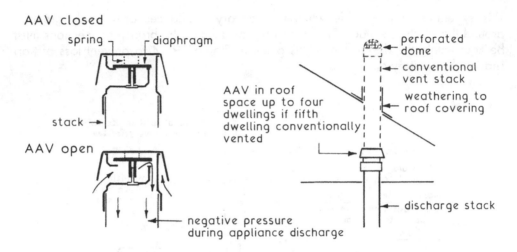

Figure 19.25 AAV principles

Ventilated stack systems

This is the traditional method of preventing loss of trap seal by providing a vent pipe to prevent self-siphonage of the appliance trap seal. The advantage of this system is that it is a 100% reliable system because there are no moving parts to fail. The disadvantage is that there is considerably more pipework to install than other options.

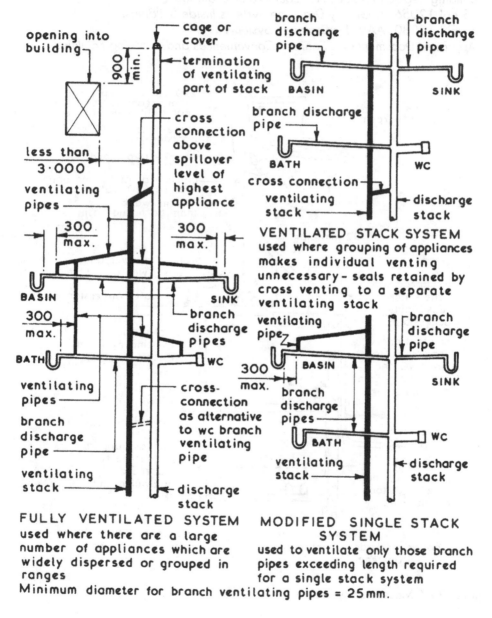

VENTILATED STACK SYSTEM used where grouping of appliances makes individual venting unnecessary – seals retained by cross venting to a separate ventilating stack

FULLY VENTILATED SYSTEM used where there are a large number of appliances which are widely dispersed or grouped in ranges

MODIFIED SINGLE STACK SYSTEM used to ventilate only those branch pipes exceeding length required for a single stack system

Minimum diameter for branch ventilating pipes = 25 mm.

Figure 19.26 Ventilated stack systems

Above ground drainage

Macerator pumped effluent

This is a common method of connecting remote sanitary fittings such as bathrooms in loft or basement conversions to the above ground drainage stack. The macerator unit liquefies solid waste that can be pumped vertically and horizontally. Multiple sanitary fittings can be connected to the macerator unit.

Building Regulations A.D. H1: *Foul Water Drainage*.
BS EN 12056–2: *Gravity Drainage Systems Inside Buildings*.
BS EN 12380: *AAVs for Drainage Systems*.
Approved Document G4: Sanitary Conveniences and Washing Facilities.

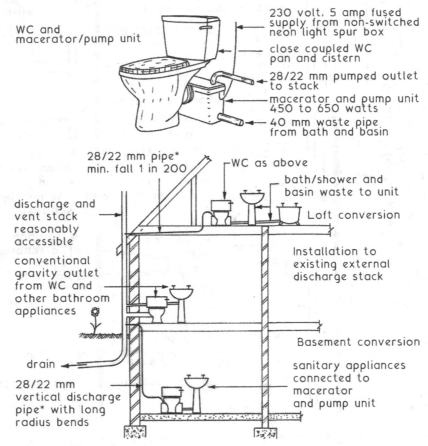

WC and macerator/pump unit

230 volt, 5 amp fused supply from non-switched neon light spur box
close coupled WC pan and cistern
28/22 mm pumped outlet to stack
macerator and pump unit 450 to 650 watts
40 mm waste pipe from bath and basin

28/22 mm pipe* min. fall 1 in 200
WC as above
bath/shower and basin waste to unit
Loft conversion

discharge and vent stack reasonably accessible

Installation to existing external discharge stack

conventional gravity outlet from WC and other bathroom appliances

Basement conversion

drain

sanitary appliances connected to macerator and pump unit

28/22 mm vertical discharge pipe* with long radius bends

* Note: Pipe diameter to suit number of appliances, 28 or 22 mm copper outside diameter or equivalent in stainless steel or polypropylene.

Figure 19.27 Macerator system

19.7 Sanitary fittings

These are fixed fittings used for sanitation purposes.

Food preparation

Sinks are used for the preparation of food with a potable water supply, most usually in stainless steel but also common in fireclay and GRP.

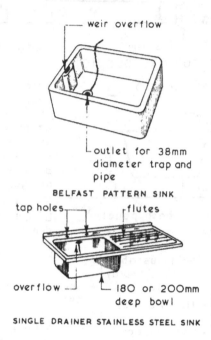

Figure 19.28 Sinks

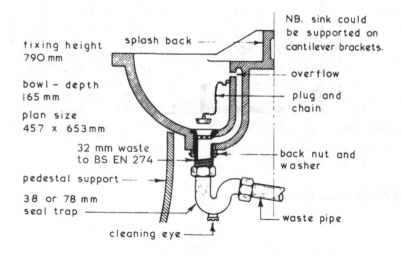

Figure 19.29 Ceramic wash basins (BS 1188)

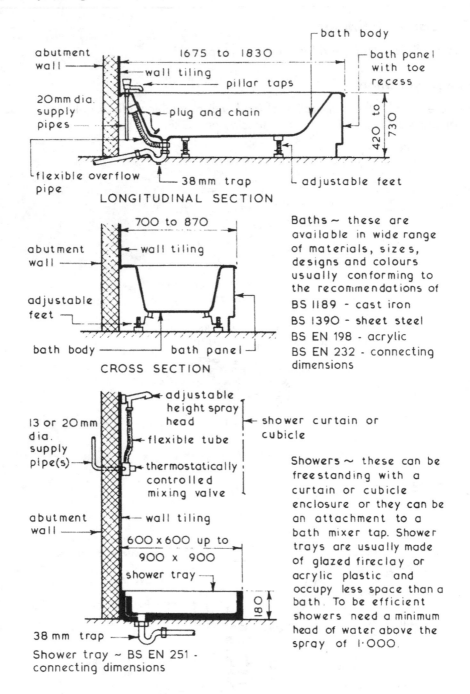

LONGITUDINAL SECTION

bath body

1675 to 1830

abutment wall

wall tiling

pillar taps

bath panel with toe recess

20 mm dia. supply pipes

plug and chain

420 to 730

flexible overflow pipe

38 mm trap

adjustable feet

CROSS SECTION

700 to 870

abutment wall

wall tiling

adjustable feet

bath body

bath panel

Baths ~ these are available in wide range of materials, sizes, designs and colours usually conforming to the recommendations of

BS 1189 - cast iron
BS 1390 - sheet steel
BS EN 198 - acrylic
BS EN 232 - connecting dimensions

13 or 20 mm dia. supply pipe(s)

adjustable height spray head

flexible tube

thermostatically controlled mixing valve

shower curtain or cubicle

abutment wall

wall tiling

600 x 600 up to 900 x 900

shower tray

180

38 mm trap

Shower tray ~ BS EN 251 - connecting dimensions

Showers ~ these can be freestanding with a curtain or cubicle enclosure or they can be an attachment to a bath mixer tap. Shower trays are usually made of glazed fireclay or acrylic plastic and occupy less space than a bath. To be efficient showers need a minimum head of water above the spray of 1·000.

Figure 19.30 Bath and shower installations

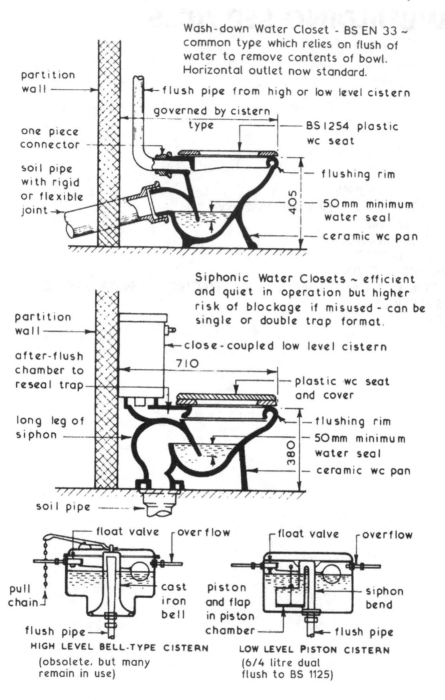

Figure 19.31 Typical WC installations

20 BUILDING SERVICES

COLD WATER SUPPLY
DOMESTIC HOT WATER SYSTEMS (DHW)
SPACE HEATING
ELECTRICAL SERVICES
POWER CIRCUITS
LIGHTING CIRCUITS
TELECOMMUNICATION SERVICES AND ELECTRONIC INSTALLATIONS
GAS SERVICES
FLUES

20.1 Cold water supply

An adequate supply of drinking-quality cold water should be provided to every residential building and a drinking water tap installed within the building. The installation should be designed to prevent waste, undue consumption, misuse, contamination of general supply, be protected against corrosion and frost damage and be accessible for maintenance activities. Mains cold water to the site boundary is the responsibility of the water supply authority. Thereafter, it is the responsibility of the building owner.

Wholesome drinking water from the mains is also known as potable water.

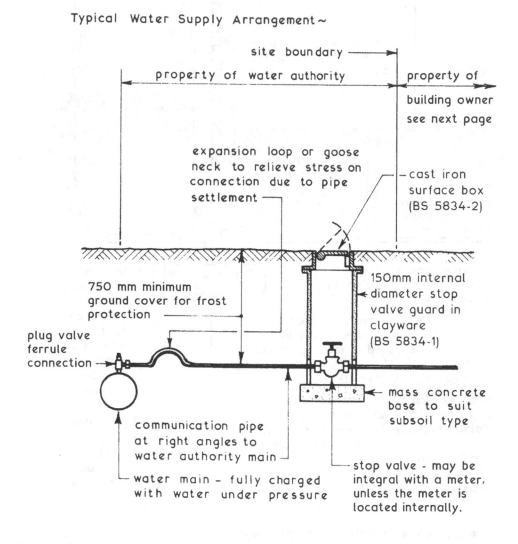

Typical Water Supply Arrangement ~

site boundary

property of water authority

property of building owner see next page

expansion loop or goose neck to relieve stress on connection due to pipe settlement

cast iron surface box (BS 5834-2)

750 mm minimum ground cover for frost protection

150mm internal diameter stop valve guard in clayware (BS 5834-1)

plug valve ferrule connection

mass concrete base to suit subsoil type

communication pipe at right angles to water authority main

water main – fully charged with water under pressure

stop valve - may be integral with a meter, unless the meter is located internally.

Figure 20.1 Typical water supply arrangement

Cold water supply

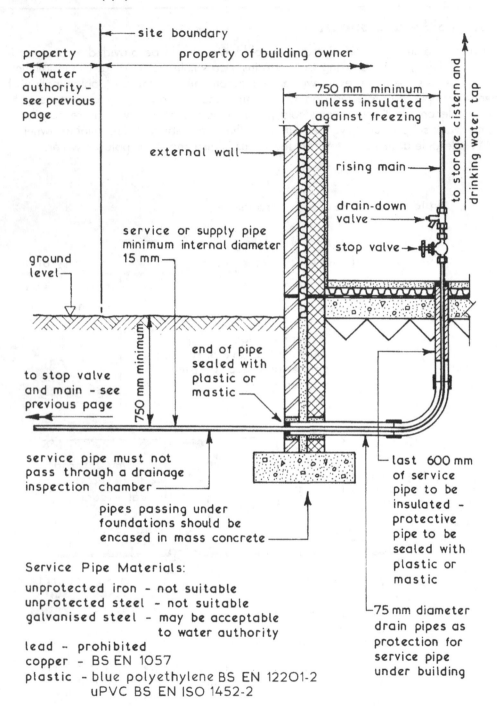

Figure 20.2 Water service pipe entry into building

site boundary

property of water authority - see previous page

property of building owner

750 mm minimum unless insulated against freezing

to storage cistern and drinking water tap

external wall

rising main

drain-down valve

stop valve

service or supply pipe minimum internal diameter 15 mm

ground level

750 mm minimum

to stop valve and main - see previous page

end of pipe sealed with plastic or mastic

service pipe must not pass through a drainage inspection chamber

pipes passing under foundations should be encased in mass concrete

last 600 mm of service pipe to be insulated - protective pipe to be sealed with plastic or mastic

75 mm diameter drain pipes as protection for service pipe under building

Service Pipe Materials:

unprotected iron - not suitable
unprotected steel - not suitable
galvanised steel - may be acceptable
                   to water authority
lead - prohibited
copper - BS EN 1057
plastic - blue polyethylene BS EN 12201-2
          uPVC BS EN ISO 1452-2

External water meters at the stop tap

External water meters with an integrated isolating stop valve are generally installed for all new water services close to the property boundary.

Cold water systems

When planning or designing any water installation the basic physical laws must be considered:

1. Water is subject to the force of gravity and will find its own level.
2. To overcome friction within the conveying pipes water, which is stored prior to distribution, will need to be under pressure and this is normally achieved by storing the water at a level above the level of the outlets. The vertical distance between these levels is usually called the head.
3. Water becomes less dense as its temperature is raised, therefore warm water will always displace colder water whether in a closed or open circuit.

There are two primary cold water systems; direct and indirect.

Direct cold water system

Potable cold water is supplied to the outlets direct from the mains at mains pressure; the only storage requirement is a small capacity cistern to feed the hot water storage tank. These systems are suitable for districts which have high level reservoirs with a good supply and pressure.

The main advantage is that drinking water is available from all cold water outlets. Disadvantages include lack of reserve in case of supply cut off, risk of back siphonage due to negative mains pressure and a risk of reduced pressure during peak demand periods.

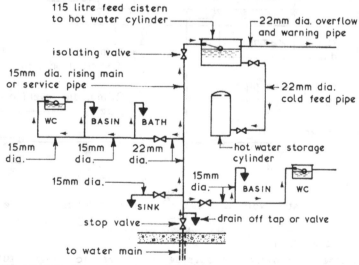

NB. All pipe sizes given are outside diameters for copper tube.

Figure 20.3 Typical direct cold water system

Cold water supply

Indirect cold water system

Only the kitchen sink is supplied from the mains with potable cold water, all other cold water outlets are fed indirectly from a large cold water storage cistern usually in the loft space.

This system requires more pipework than the direct system, but it reduces the risk of back siphonage and provides a reserve of water should the mains supply fail or be cut off. The local water authority will stipulate the system to be used in their area.

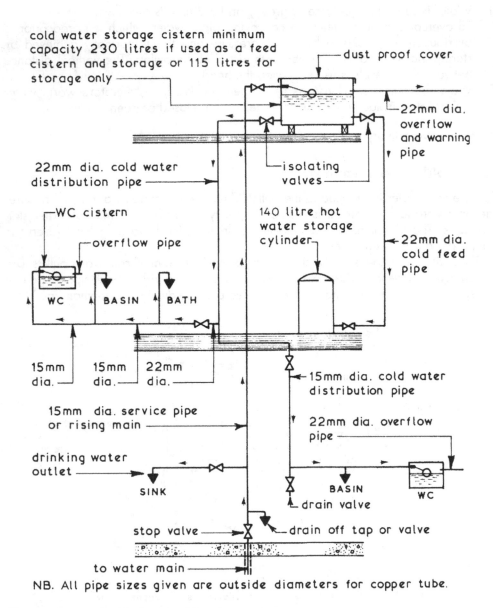

NB. All pipe sizes given are outside diameters for copper tube.

Figure 20.4 Typical indirect cold water system

20.2 Domestic hot water systems (DHW)

Indirect system

In these systems the central heating circuit is used to indirectly heat hot water in a hot water cylinder. Hot water is allowed to circulate through a coil of copper pipe heat exchanger in the cylinder when the cylinder thermostat calls for heat. A diverter valve is then operated to allow or shut off circulation within the cylinder. When the tap is turned off the flow ceases, which is sensed by the flow switch which then shuts the boiler off.

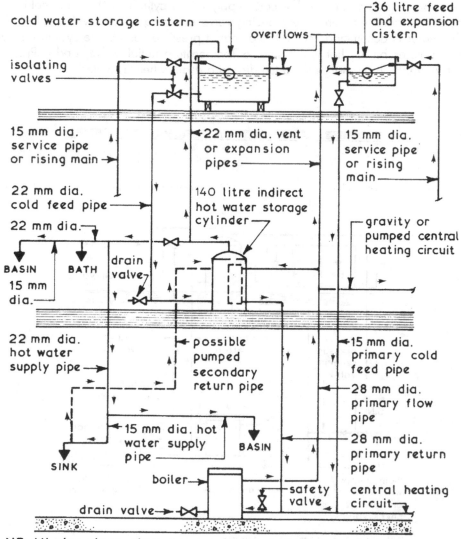

NB. All pipe sizes given are outside diameters for copper tube.

Figure 20.5 Typical indirect hot water system

Domestic hot water systems (DHW)

Mains pressure systems

Combination boilers are the most common mains fed DHW system. The boiler features two separate heat exchangers in the boiler casing, one for DHW and the other for central heating. When a hot water tap is turned on a flow switch in the boiler is activated that fires the boiler to the water flows around the DHW heat exchanger, which instantaneously heats the water. This type of system is not suitable where there is a high demand for hot water due to restricted flow rates.

Unvented hot water cylinders

These are suitable where the demand for hot water is high as a ready supply of stored hot water is available. The cold supply to the cylinder is from the cold main through a strainer and pressure reducing valve and single check valve. Expansion of the water is accommodated within a sealed pressure vessel. The system has two safety features on the hot water side: a Temperature Relief Valve, and a Pressure Relief Valve. Both discharge via a tun dish to a drain when activated.

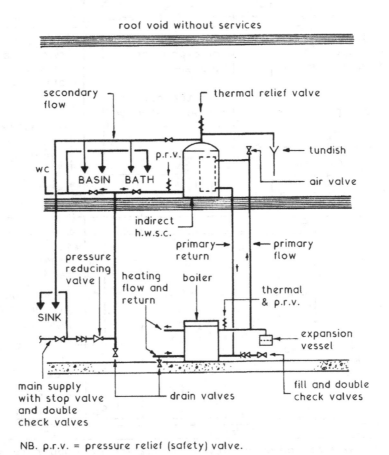

NB. p.r.v. = pressure relief (safety) valve.

Figure 20.6 Unvented hot water system

Hot water system temperatures

Hot water has to be at 60°C to 65°C to prevent the growth of water-borne pathogens such as legionella pneumophila (legionnaires' disease) but at this temperature could cause scalding.

Distributed hot water is not less than 55°C.
Supply to outlets is approximately 50°C.

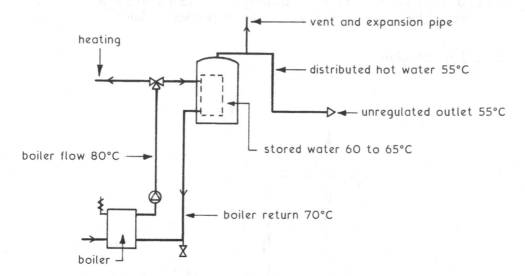

Figure 20.7 Temperature profile within an indirect system

Thermostatic mixing valves

These are used to temper the hot water outlet temperature to prevent scalding. Scalding is unlikely for most people, but there is a risk where appliances are used by young children, the elderly and those with loss of sensory perception.

Domestic hot water systems (DHW)

Safe hot water outlet temperatures:

- Bath 43°C
- Shower 40°C
- Wash basin 40°C
- Bidet 37°C
- Sink 48°C

Installation should be as close as possible to the hot water outlet to control water temperature. Thermostatic mixing valves can be manually set to blend the hot and cold water supplies at a fixed discharge temperature. These valves contain an automated fast shutdown facility in the event of a supply failure.

Typical installation

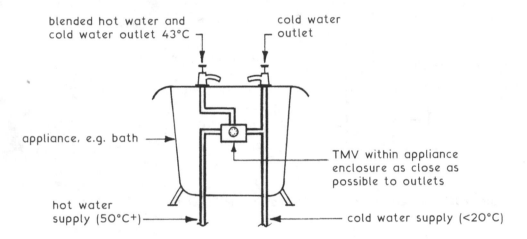

Figure 20.8   Thermostatic mixing valve to a bath

Building Regulations, *Approved Document G3: Hot water supply and systems.*
  BS EN 1111: *Sanitary tapware, thermostatic mixing valves (PN 10).* General technical specification.
  BS EN 1287: *Sanitary tapware, low pressure thermostaticmixing valves.* General technical specification.

Flow controls

There are essentially two types of valves to control water flow in a plumbing installation:

- Gate valves for low pressure supplies.
- Stop valves for high pressure supplies.

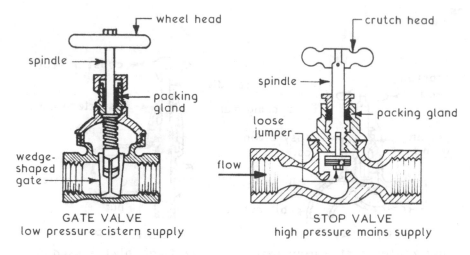

Figure 20.9 **Valve types**

Float operated valves are used for cold water storage cisterns (CWSC).

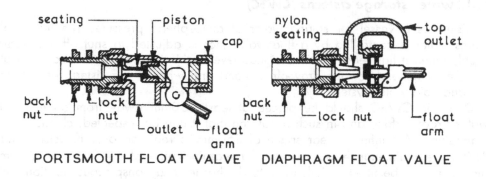

Figure 20.10 **Float valve construction**

Domestic hot water systems (DHW)

Taps

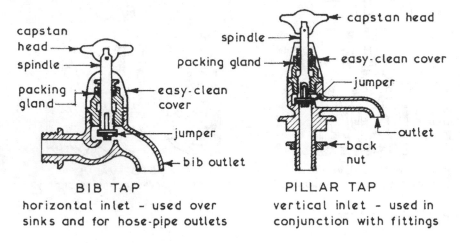

BIB TAP
horizontal inlet – used over
sinks and for hose-pipe outlets

PILLAR TAP
vertical inlet – used in
conjunction with fittings

Figure 20.11 Tap types

Cold water storage cisterns (CWSC)

A cistern is a vessel for storing water at atmospheric pressure. The inflow of mains water is controlled by a float valve that is adjusted to shut off the water supply when it has reached the designed level within the cistern. The capacity of the cistern depends on the draw-off demand and whether the cistern feeds both hot and cold water systems.

Domestic CWSC should be placed at least 750mm away from an external wall or roof surface and in such a position that it can be inspected, cleaned and maintained. A minimum clear space of 350mm is required over the cistern for float valve maintenance. An overflow or warning pipe of not less than 22mm diameter must be fitted to fall away to discharge in a conspicuous position. All draw-off pipes must be fitted with a gate valve positioned as near to the cistern as possible.

Cisterns for domestic applications are made from injection moulded plastic, whereas large cisterns for commercial buildings may be sectional GRP panels.

If the cistern and its associated pipework are to be housed in a cold area such as a roof they should be insulated against freezing.

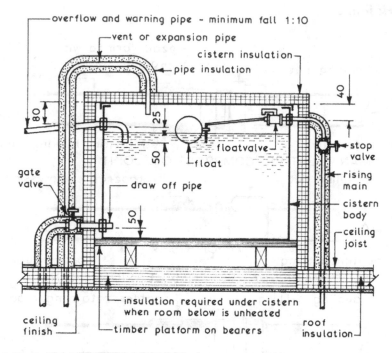

Figure 20.12 Typical CWSC details

Indirect hot water cylinders

These cylinders are a form of heat exchanger where the primary circuit of hot water from the boiler flows through a coil or annulus within the storage vessel and transfers the heat to the water stored within.

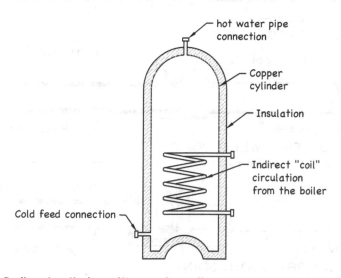

Figure 20.13 Indirect cylinder with annular coil

665

Domestic hot water systems (DHW)

Pipework joints

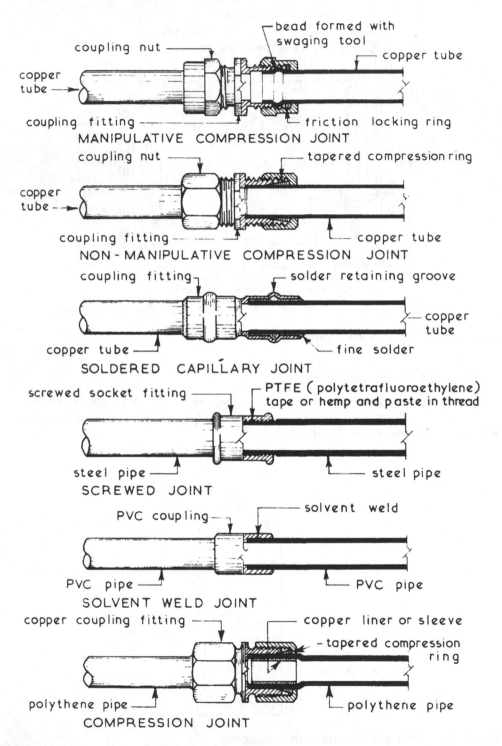

Figure 20.14 Various pipe connection methods

20.3 Space heating

The majority of domestic space heating systems are hydronic or hot water systems which feature a centralised heat source, a heat distribution network and heat emitters.

Two pipe system

This system features flow and return pipework in a closed circuit (two pipes). Heat is circulated around the distribution pipework (22–15mm) from the boiler by a pump through Flow pipework to each emitter (radiator) through the emitter and back to the boiler via the Return pipework. There is an approximate drop in temperature of 10–12° between the Flow and Return.

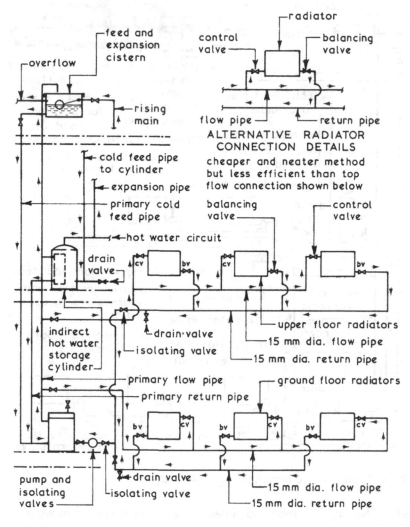

Figure 20.15 Typical 2 pipe heating system layout

Space heating

Manifold systems

Micro bore system

This is still a two-pipe system but uses small bore tubes (6 to 12mm dia.) with an individual flow and return pipe to each heat emitter or radiator from a 22mm Flow and Return manifold. The small bore and flexible pipework make this system easy to install but it requires a more powerful pump than that used in the traditional small bore systems. The heat emitter or radiator valves can be as used for the one- or two-pipe small bore systems; alternatively, a double entry valve can be used.

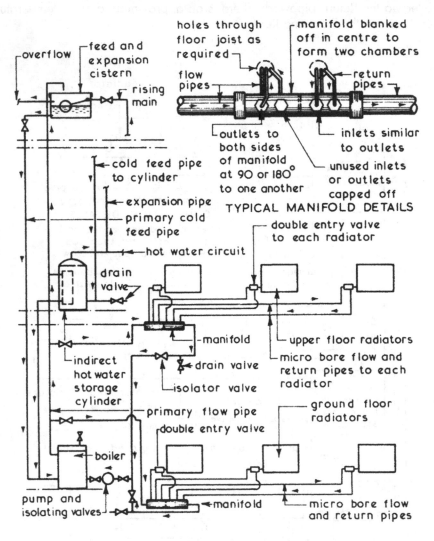

Figure 20.16 Typical micro bore heating system layout

Underfloor heating is a manifold system using lower temperature water than conventional radiator systems.

System controls

A heating system must have effective controls and be able to be commissioned and adjusted to use no more power than is 'reasonable' (Approved Document L1A).

Boiler ~ must be fitted with a thermostat to control the temperature of the hot water leaving the boiler.

Heat Emitters or Radiators ~ fitted with thermostatically controlled radiator valves to control flow of hot water to the radiators to keep room at desired temperature.

Programmer and Room Thermostat ~ The Programmer is a time switch to control at which times of day, and on which days, the heating is to be turned on. The Room Thermostat signals the boiler to fire when the room temperature is below that set on the Thermostat. These two controls can be combined in a single unit.

Cylinder thermostat ~ Controls the temperature of stored water.

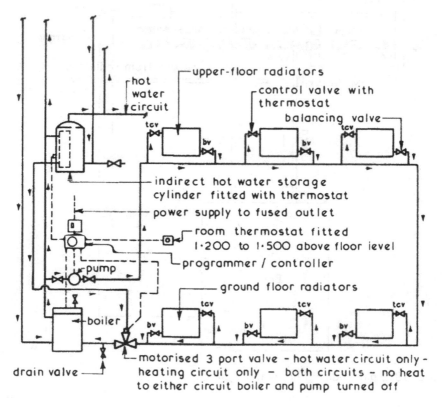

Figure 20.17 Heating system controls

Space heating

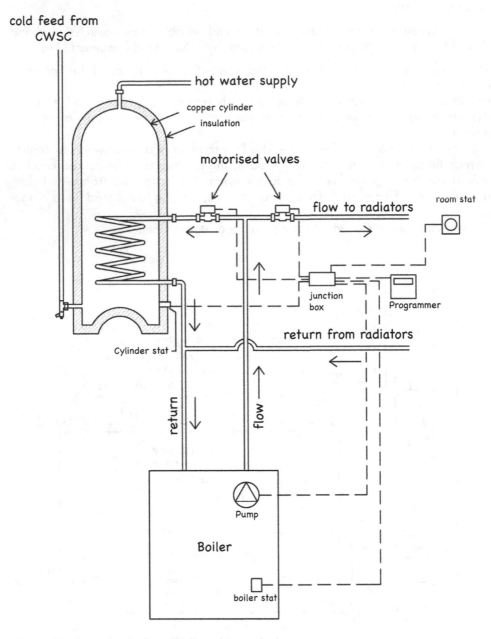

cold feed from
CWSC

hot water supply

copper cylinder

insulation

motorised valves

flow to radiators

room stat

junction box

Programmer

return from radiators

Cylinder stat

return

flow

Pump

Boiler

boiler stat

Figure 20.18 S plan indirect hot water system

20.4 Electrical services

Electrical supply equipment

In the UK electricity is generated mainly from gas, coal and nuclear, with renewable sources such as wind, biomass and hydro. The electrical supply to a domestic installation is 230 volt single phase and is designed with the following safety objectives:

1. Proper circuit protection to earth to avoid shocks to occupant.
2. Prevention of current leakage.
3. Prevention of outbreak of fire.

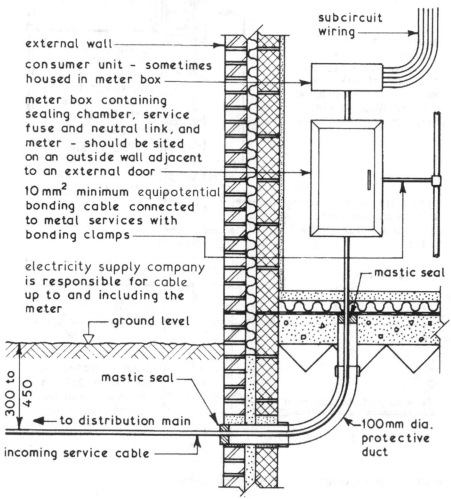

external wall

consumer unit – sometimes housed in meter box

meter box containing sealing chamber, service fuse and neutral link, and meter – should be sited on an outside wall adjacent to an external door

10 mm² minimum equipotential bonding cable connected to metal services with bonding clamps

electricity supply company is responsible for cable up to and including the meter

ground level

mastic seal

to distribution main

incoming service cable

subcircuit wiring

mastic seal

100 mm dia. protective duct

300 to 450

NB. For alternative arrangement of supply intake see following page.

Figure 20.19 Typical electrical supply intake details

Electrical services

Electrical supply intake

Although the electrical supply intake can be terminated in a meter box situated within a dwelling, most supply companies prefer to use the external meter box to enable the meter to be read without the need to enter the premises.

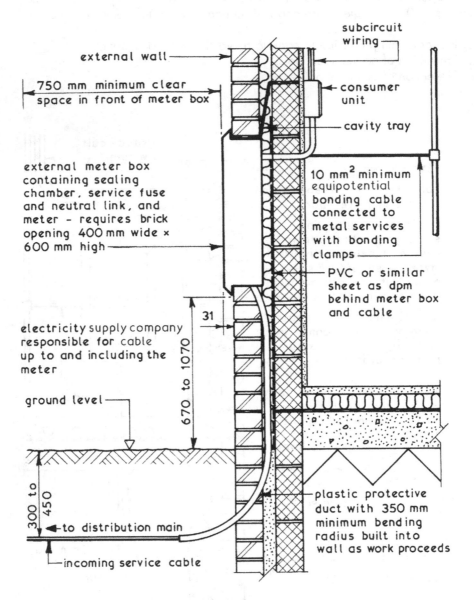

Figure 20.20 Typical electrical supply intake details

The supply cable, main fuse and meter are the responsibility of the supplier, all equipment thereafter is the responsibility of the building owner.

Consumer's power supply control unit

This control unit, conveniently abbreviated to 'consumer unit', contains the main double pole isolating switch controlling the Line (Live) and Neutral conductors, called bus bars. These connect to the circuit protection devices for individual circuits.

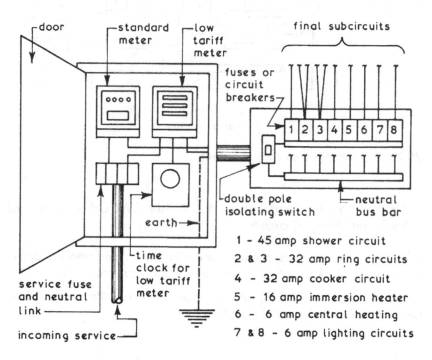

Figure 20.21 Typical meter box and consumer unit layout

It also contains the earth bar, plus a range of individual circuit overload safety protection devices. By historical reference this unit is sometimes referred to as a fuse box, but modern variants are far more sophisticated. Overload protection is provided by miniature circuit breakers attached to the live or phase bar. Additional protection is provided by a split load residual current device (RCD) dedicated specifically to any circuits that could be used as a supply to equipment outdoors, e.g. power sockets on a ground floor ring final circuit.

RCD ~ a type of electromagnetic switch or solenoid which disconnects the electricity supply when a surge of current or earth fault occurs. See Part 11 of the *Building Services Handbook* for more details.

Electrical services

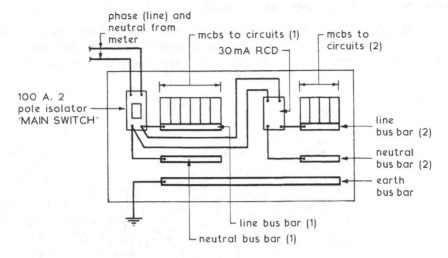

Figure 20.22 Typical split load consumer unit

Note that with an overhead supply, the MAIN SWITCH is combined with a 100 mA RCD protecting all circuits.

Circuits (1) to fixtures, i.e. lights, cooker, immersion heater and smoke alarms.
Circuits (2) to socket outlets that could supply portable equipment outdoors.

Electric cables ~ these are made up of copper or aluminium wires called conductors surrounded by an insulating material such as PVC.

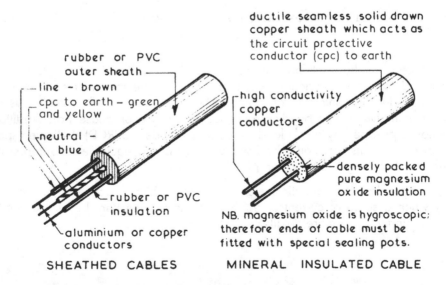

Figure 20.23 Two types of industrial cables

Wiring systems

Rewireable systems housed in horizontal conduits can be cast into the structural floor slab or sited within the depth of the floor screed. To ensure that such a system is rewireable, draw-in boxes must be incorporated at regular intervals with not more than two right angle boxes to be included between draw-in points. Vertical conduits can be surface mounted or housed in a chase cut into a wall, provided the depth of the chase is not more than one-third of the wall thickness. A horizontal, non-rewireable system can be housed within the depth of the timber joists to a suspended floor whereas vertical cables can be surface mounted or housed in a length of conduit as described for rewireable systems.

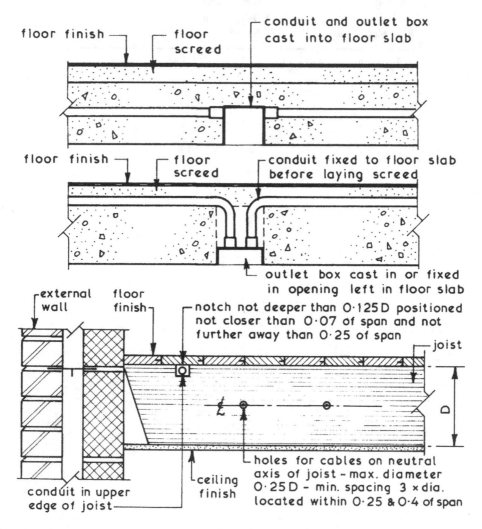

Figure 20.24 Typical cable distribution methods

Electrical services

Cable sizing

The size of a conductor wire can be calculated taking into account the maximum current the conductor will have to carry (which is limited by the heating effect caused by the resistance to the flow of electricity through the conductor) and the voltage drop that will occur when the current is carried. For domestic electrical installations the Twin and Earth minimum cable specifications shown in Table 20.1 are usually suitable.

Table 20.1   Typical twin and earth cable/circuit sizing

Circuit	L & N Conductor cross sectional area
Lighting	1.0 or 1.5mm^2
Radial*	4.0–10.0mm^2
Ring Main (Power)	2.5mm^2

*   Typically immersion heater/shower/cooker circuits

20.5 Power circuits

There are two types of circuit for power: ring and radial. For ring circuits the number of socket outlets is unlimited but a separate circuit must be provided for every 100m^2 of floor area. To conserve wiring, spur outlets can be used as long as the total number of spur outlets does not exceed the total number of outlets connected to the ring and that there are not more than two outlets per spur.

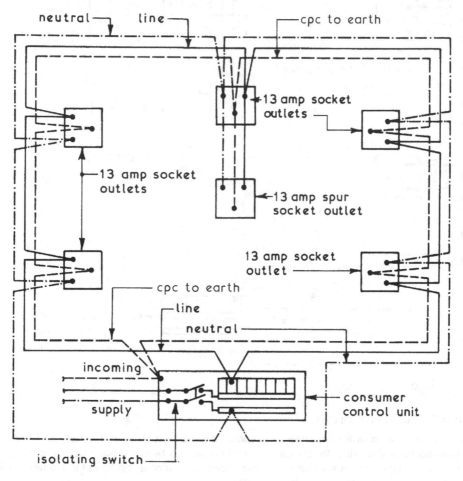

Figure 20.25 Typical ring final circuit wiring diagram

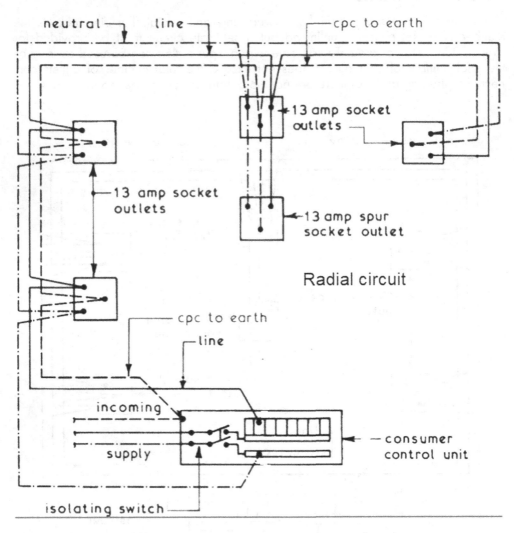

Figure 20.26 Radial circuit

A radial circuit starts at the consumer unit, connects to each socket outlet in turn, and ends at the final socket outlet. All of the socket outlets will have two twin and earth cables, one coming in from the previous socket or consumer unit, and another going out to the next socket except the final one where the cable terminates.

20.6 Lighting circuits

These are usually wired by the loop-in method using a line, neutral and circuit protective conductor to earth cable with a 6amp MCB. In calculating the rating of a lighting circuit an allowance of 100 watts per outlet should be used. More than one lighting circuit should be used for each installation so that in the event of a circuit failure some lighting will be in working order.

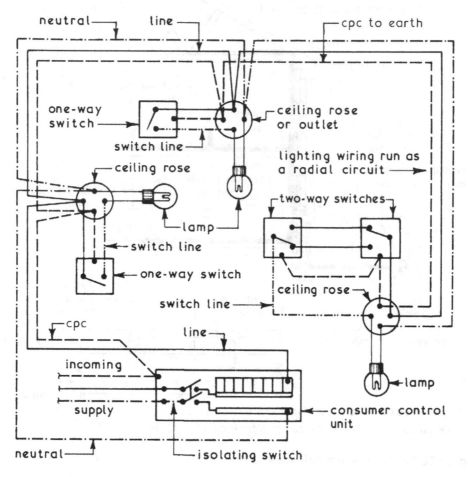

Figure 20.27 Typical lighting circuit wiring diagram

20.7 Telecommunication services and electronic installations

A typical installation will provide connection from a common external terminal chamber via underground ducting to a terminal distribution box within the building. Internal distribution is through service voids within the structure or attached trunking.

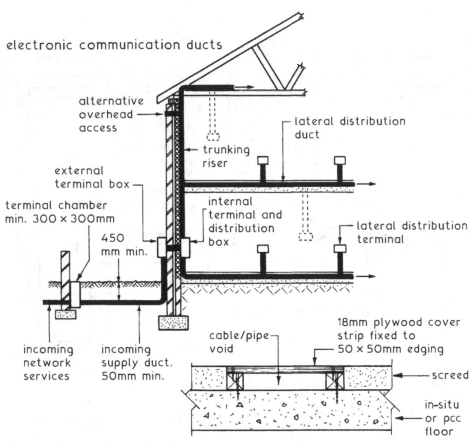

electronic communication ducts

alternative overhead access

lateral distribution duct

trunking riser

external terminal box

terminal chamber min. 300 × 300mm

450 mm min.

internal terminal and distribution box

lateral distribution terminal

incoming network services

incoming supply duct. 50mm min.

cable/pipe void

18mm plywood cover strip fixed to 50 × 50mm edging

screed

in-situ or pcc floor

Typical lateral distribution duct
(see also Part 15 – *Building Services Handbook*).

Figure 20.28 Telecommunication installation

20.8 Gas services

The supply, appliances and installation must comply with the safety requirements made under the Gas Safety (Installation and Use) Regulations, 1998, and Part J of the Building Regulations.

Typical Gas Supply Arrangement ~

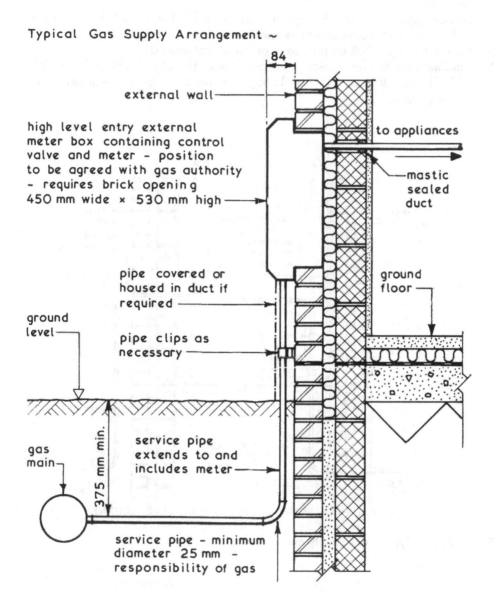

Figure 20.29 Typical gas supply arrangement

Gas services

Gas service pipes

1. Whenever possible the service pipe should enter the building on the side nearest to the main.
2. A service pipe must not pass under the foundations of a building.
3. No service pipe must be run within a cavity but it may pass through a cavity by the shortest route.
4. Service pipes passing through a wall or solid floor must be enclosed by a sleeve or duct that is end sealed with mastic.
5. No service pipe shall be housed in an unventilated void.
6. Suitable materials for service pipes are copper (BS EN 1057) and steel (BS EN 10255). Polyethylene (BS EN 1555) is normally used underground but not above ground.

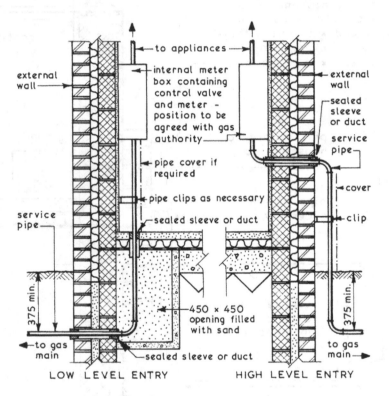

Figure 20.30 Typical gas supply arrangement

Gas fires

For domestic use these generally have a low energy rating of less than 7kW net input and must be installed in accordance with minimum requirements set out in Part J of the Building Regulations. Most gas fires connected to a flue are designed to provide radiant and convection heating whereas the room sealed balanced flue appliances are primarily convector heaters.

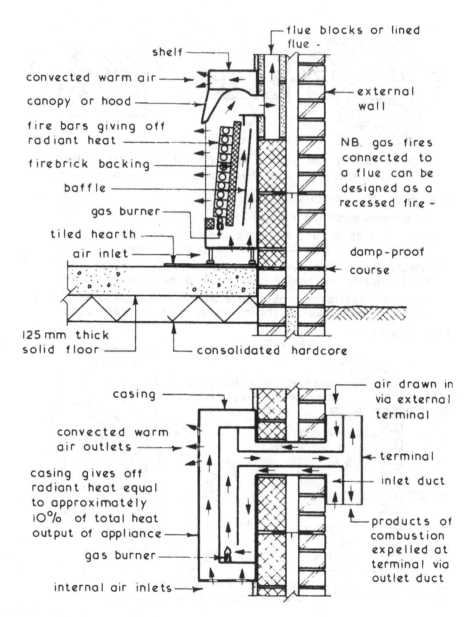

Figure 20.31 Typical established examples

20.9 Flues

A flue ~ a passage for the discharge of the products of combustion to the outside air and can be formed by means of a chimney, special flue blocks or by using a flue pipe.

In all cases the type and size of the flue, as recommended in Approved Document J, BS EN 1806 and BS 5440, will meet the requirements of the Building Regulations.

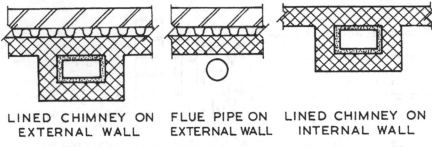

LINED CHIMNEY ON   FLUE PIPE ON   LINED CHIMNEY ON
EXTERNAL WALL   EXTERNAL WALL   INTERNAL WALL

Flue Size Requirements:

1. No dimension should be less than 63 mm.

2. Flue for a decorative appliance should have a minimum dimension measured across the axis of 175 mm.

3. Flues for gas fires – min. area = $12000\,mm^2$ if round, $16500\,mm^2$ if rectangular and having a minimum dimension of 90 mm.

4. Any other appliance should have a flue with a cross sectional area at least equal to the outlet size of the appliance.

Flue Blocks ~

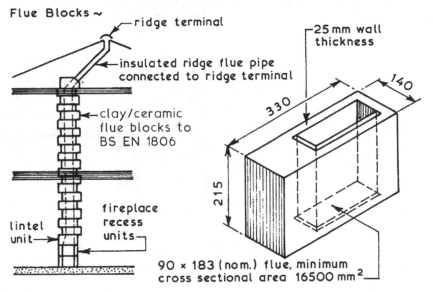

Figure 20.32 Typical single gas fire flues

In traditional open fireplaces for coal fires with flue to chimney termination, the main functions of a chimney and flue are to:

1 Induce an adequate supply of air for the combustion of the fuel being used.
2 Remove the products of combustion.

Approved Document J recommends that all flues should be lined with approved materials so that the minimum size of the flue so formed will be 200mm diameter or a square section of equivalent area. Flues should also be terminated above the roof level as shown in Figure 20.32, with a significant increase where combustible roof coverings such as thatch or wood shingles are present.

21 INSULATION AND U-VALUES

Opening remarks

The Building Regulations, Approved Document L, Conservation of fuel and power, is in four parts, each dealing with acceptable energy efficiency standards for domestic and non-domestic buildings to new buildings and alterations to existing buildings with the objective of limiting the emission of carbon dioxide and other burnt gases into the atmosphere.

AD L 1A New Dwellings
AD L 1B Existing Dwellings
AD L 2A New Buildings Other than Dwellings
AD L 2B Existing Buildings Other than Dwellings

21.1 U-values

The thermal insulation of external elements of construction is measured in terms of thermal transmittance rate, otherwise known as the U-value. It is the amount of heat energy in Watts transmitted through one square metre of construction for every degree Kelvin between external and internal air temperature, i.e. W/m^2K.

U-values are unlikely to be entirely accurate, due to:

- Variable atmospheric conditions: solar radiation, humidity and prevailing winds.
- Construction quality including thermal bridging.

Nevertheless, calculation of the U-value for a particular element of construction will provide guidance as to whether the structure is thermally acceptable.

The U-value is calculated by taking the reciprocal of the summed thermal resistances (R) of the component parts of an element of construction:

$$U = \frac{1}{\Sigma R} = W/m^2K$$

Thermal resistance (R) is expressed in m^2K/W. The higher the resistance value, the better a component's insulation. Conversely, the lower the U-value the better the insulative properties of the structure.

Thermal resistances are a combination of the different structural, surface and air space components that make up an element of construction.

Typically:

$$U = \frac{1}{Rso + R1 + R2 + Ra + R3 + R4etc.. + Rsi}$$

Where: Rso = Outside or external surface resistance.
R1, R2, etc. = Thermal resistance of structural components.
Ra = Air space resistance, e.g. wall cavity.
Rsi = Internal surface resistance.

The thermal resistance of a structural component (R1, R2, etc.) is calculated by dividing its thickness (L) by its thermal conductivity (λ)

$$R = \frac{L(m)}{\lambda\left(\frac{W}{mK}\right)} = m^2K/W$$

Example 1

A 102mm brick with a conductivity of 0.84W/mK has a thermal resistance (R) of: 0.102 + 0.840 = 0.121m²K/W.

Example 2

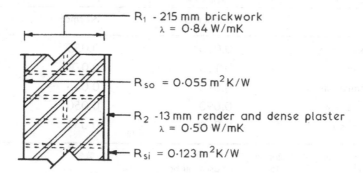

R_1 - 215 mm brickwork
λ = 0.84 W/mK

R_{so} = 0.055 m^{2}K/W

R_2 -13 mm render and dense plaster
λ = 0.50 W/mK

R_{si} = 0.123 m^{2}K/W

Figure 21.1 215mm solid brick wall

Note: The effect of mortar joints in the brickwork can be ignored, as both components have similar density and insulative properties.

Rso = 0.055
Rsi = 0.123
R1 = 0.215 ÷ 0.84 = 0.256
R2 = 0.013 ÷ 0.50 = 0.026

$$U = \frac{1}{Rso + R1 + R2 + Rsi}$$

$$U = \frac{1}{0.055 + 0.256 + 0.026 + 0.123} = 2.17 W/m^2 K$$

Typical values for surface resistances

Internal surface resistances (Rsi):

- Walls – 0.123
- Floors or ceilings for upward heat flow – 0.104
- Floors or ceilings for downward heat flow – 0.148
- Roofs (flat or pitched) – 0.104

U-values

Table 21.1 External surface resistances (Rso)

Surface	Exposure		
	Sheltered	Normal	Severe
Wall – high emissivity	0.080	0.055	0.030
Wall – low emissivity	0.110	0.070	0.030
Roof – high emissivity	0.070	0.045	0.020
Roof – low emissivity	0.090	0.050	0.020
Floor – high emissivity	0.070	0.040	0.020

Sheltered – town buildings to 3 storeys.
Normal – town buildings 4 to 8 storeys and most suburban premises.
Severe – > 9 storeys in towns.
> 5 storeys elsewhere and any buildings on exposed coasts and hills.

Air space resistances (Ra):

- Pitched or flat roof space – 0.180
- Behind vertical tile hanging – 0.120
- Cavity wall void – 0.180
- Between high and low emissivity surfaces – 0.300
- Unventilated/sealed – 0.180

Emissivity relates to the heat transfer across and from surfaces by radiant heat emission and absorption effects. The amount will depend on the surface texture, the quantity and temperature of air movement across it, the surface position or orientation and the temperature of adjacent bodies or materials. High surface emissivity is appropriate for most building materials. An example of low emissivity would be bright aluminium foil on one or both sides of an air space.

21.2 Thermal conductivity (λ) of typical building materials

Table 21.2 Thermal conductivity of typical building materials

Material	Density (kg/m^3)	Conductivity (λ) (W/mK)
WALLS:		
Boarding (hardwood)	700	0.18
(softwood)	500	0.13
Brick outer leaf	1700	0.84
... inner leaf	1700	0.62
Calcium silicate board	875	0.17
Ceramic tiles	2300	1.30
Concrete	2400	1.93
	2200	1.59
	2000	1.33
	1800	1.13
(lightweight)	1200	0.38
(reinforced)	2400	2.50
Concrete block (lightweight)	600	0.18
(medium weight)	1400	0.53
Cement mortar (protected)	1750	0.88
(exposed)	1750	0.94
Fibreboard	350	0.08
Gypsum plaster (dense)	1300	0.57
Gypsum plaster (lightweight)	600	0.16
Plasterboard	950	0.16
Tile hanging	1900	0.84
Rendering	1300	0.57
Sandstone	2600	2.30
Wall ties (st/st)	7900	17.00
ROOFS:		
Aerated concrete slab	500	0.16
Asphalt	1900	0.60
Bituminous felt in 3 layers	1700	0.50
Sarking felt	1700	0.50
Stone chippings	1800	0.96
Tiles (clay)	2000	1.00
.... (concrete)	2100	1.50

(Continued)

Thermal conductivity (λ) of typical building materials

Table 21.2 (Cont.)

Material	Density (kg/m³)	Conductivity (λ) (W/mK)
Woodwool slab	500	0.10
FLOORS:		
Cast concrete	2000	1.33
Hardwood block/strip	700	0.18
Plywood/particle board	650	0.14
Screed	1200	0.41
Softwood board	500	0.13
Steel tray	7800	50.00
INSULATION:		
Expanded polystyrene board	20	0.035
Mineral wool batt/slab	25	0.038
Mineral wool quilt	12	0.042
Phenolic foam board	30	0.025
Polyurethane board	30	0.025
Urea formaldehyde foam	10	0.040
GROUND:		
Clay/silt	1250	1.50
Sand/gravel	1500	2.00
Homogeneous rock	3000	3.50

21.3 Floor insulation

Manufacturers usually provide tables where specific U-values can be achieved by adopting their materials; these are useful where standard forms of construction are adopted. The values contain appropriate allowances for variable heat transfer due to different components in the construction.

U-value calculation of a solid ground floor

The example below shows the tabulated data for a solid ground floor with embedded insulation of λ = 0.03W/mK.

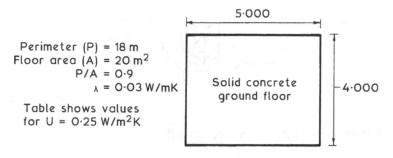

Perimeter (P) = 18 m
Floor area (A) = 20 m^2
P/A = 0·9
λ = 0·03 W/mK

Table shows values
for U = 0·25 W/m^{2}K

Figure 21.2 Floor area and perimeter ratio example

Typical table for floor insulation is given in Table 21.3.

Table 21.3 Insulation floor and perimeter ratios

P/A	0–020	0–025	0–030*	0–035	0–040	0–045	W/mK
1.0	61	76	91	107	122	137	mm
0.9*	60	75	**90**	105	120	135	..
0.8	58	73	88	102	117	132	..
0.7	57	71	85	99	113	128	..
0.6	54	68	82	95	109	122	..
0.5	51	64	77	90	103	115	..

90 mm of insulation required.
BS EN ISO 6946: *Building components and building elements. Thermal resistance and thermal transmittance. Calculation method.*
BS EN ISO 13,370: *Thermal performance of buildings. Heat transfer via the ground. Calculation methods.*

Floor insulation

Various applications to different ground floor situations are considered in BS EN ISO 13,370. The following is an example of a solid concrete slab in direct contact with the ground.

Floor section
Perimeter = 18m (exposed)
Floor area = 20m^2
λ for 90mm insulation = 0.03W/mK
Characteristic floor dimension = B1
B1 = Floor area ÷ (1/2 exp.perimeter)
B1 = 20 ÷ 9 = 2.222m

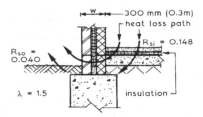

Figure 21.3 Possible heat loss path through a floor

Formula to calculate total equivalent floor thickness for uninsulated and insulated all over floor:

$$dt = w + \lambda(Rsi + Rf + Rso)$$

where: dt = total equivalent floor thickness (m)
w = wall thickness (m)
λ = thermal conductivity of soil (W/mK)
Rsi = internal surface resistance (m^{2}K/W)
Rf = insulation resistance (0.09 ÷ 0.03 = 3m^{2}K/W)
Rso = external surface resistance (m^{2}K/W)

Uninsulated: dt = 0.3 + 1.5(0.148 + 0 + 0.04) = 0.582m
Insulated: dt = 0.3 + 1.5(0.148 + 3 + 0.04) = 5.082m

Formulae to calculate U-values

Uninsulated or poorly insulated floor, dt < B1

$$U = (2\lambda) \div [(\pi B1) + dt] \times \ln[(\pi B1 \div dt) + 1]$$

Well insulated floor, dt ≥ B1

$$U = \lambda \div [(0.457 \times B1) + dt]$$

where:

U = thermal transmittance coefficient (W/m2/K)
λ = thermal conductivity of soil (W/mK)
B1 = characteristic floor dimension (m)
dt = total equivalent floor thickness (m)
ln = natural logarithm

Uninsulated floor

$$U = (2 \times 1.5) \div [(3.142 \times 2.222) + 0.582] \times \ln[(3.142 \times 2.222) \div 0.582 + 1]$$

$$U = 0.397 \times \ln 12.996 = 1.02 W/m^2 K$$

Insulated floor

$$U = 1.5 \div [(0.457 \times 2.222) + 5.082] = 1.5 \div 6.097 = 0.246 W/m^2 K$$

Note: Compares with the tabulated figure of 0.250W/m²K on page 693.

Wall U-value calculation: proportional area method

Proportional Area Method (Wall)

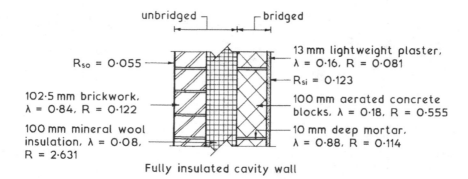

Figure 21.4 Proportional area method (wall)

Floor insulation

A standard block with mortar is $450 \times 225mm = 101{,}250mm^2$
A standard block format of $440 \times 215mm = 94{,}600mm^2$
The area of mortar per block $= 6650mm^2$
Proportional area of mortar $= 6.57\%$ (0066)
Therefore the proportional area of blocks $= 93.43\%$ (0.934)

Table 21.4 Tabulated wall U-value calculation

Thermal resistances (R)					
Outer leaf + insulation (unbridged)		Inner leaf (unbridged)		Inner leaf (bridged)	
Rso	0.055	blocks	0.555	Mortar	0.114
brickwork	0.122	plaster	0.081	plaster	0.081
insulation	2.631	Rsi	0.123	Rsi	0.123
total	2.808	total	0.759	Total	0.318
$\times 100\% = 2.808$		$\times 93.43\% = 0.709$		$\times 6.57\% = 0.021$	

$$U = \frac{1}{\Sigma R} = \frac{1}{2.808 + 0.709 + 0.021} = 0.283 W/m^2 K$$

This proportional method can be applied to calculate the U-value of any composite construction element.

21.4 Thermal bridging

This is where there is a zone of lower thermal resistance within an external element. The rate of heat loss in this zone will be greater than the rest of the element, i.e. thermal bridge. Surface temperatures will be lower in this zone and condensation could occur.

Typical thermal bridges in older buildings

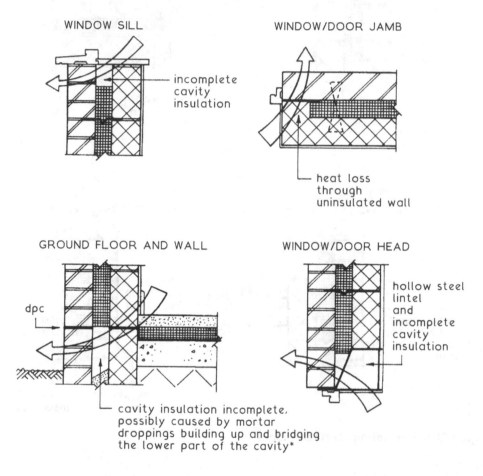

WINDOW SILL

incomplete cavity insulation

WINDOW/DOOR JAMB

heat loss through uninsulated wall

GROUND FLOOR AND WALL

dpc

cavity insulation incomplete, possibly caused by mortar droppings building up and bridging the lower part of the cavity*

WINDOW/DOOR HEAD

hollow steel lintel and incomplete cavity insulation

*Cavity should extend down at least 225mm below the level of the lowest dpc (A.D. C: Section 5).

Figure 21.5 Heat loss paths/thermal bridges

Thermal bridging

Modern preventative details

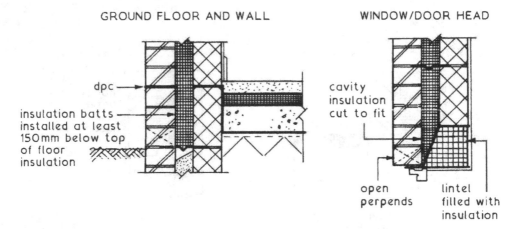

WINDOW SILL

cavity insulation
to underside of
window board

WINDOW/DOOR JAMB

proprietary cavity
closer and
insulated dpc

lightweight
insulation
blocks

full or
part full
cavity
insulation

GROUND FLOOR AND WALL

dpc

insulation batts
installed at least
150mm below top
of floor
insulation

WINDOW/DOOR HEAD

cavity
insulation
cut to fit

open
perpends

lintel
filled with
insulation

Figure 21.6 Eliminating thermal bridges

Air infiltration

Heating costs will increase if cold air is allowed to penetrate peripheral gaps and
breaks in the continuity of construction. Furthermore, heat energy will escape
through structural breaks and the following are prime situations for treatment:

1. Loft hatch.
2. Services penetrating the structure.
3. Opening components in windows, doors and rooflights.
4. Gaps between dry lining and masonry walls.

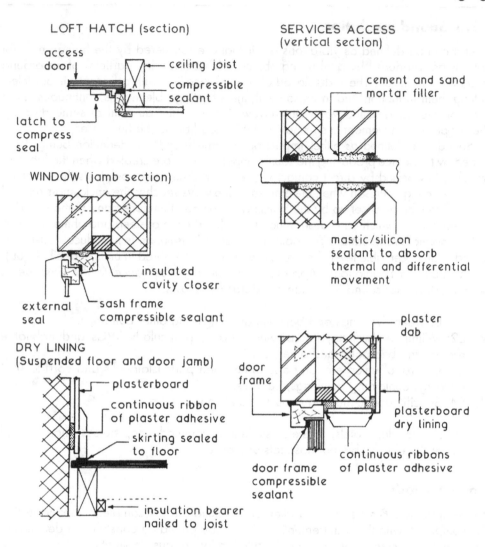

LOFT HATCH (section)

access door

ceiling joist

compressible sealant

latch to compress seal

WINDOW (jamb section)

insulated cavity closer

external seal

sash frame compressible sealant

DRY LINING
(Suspended floor and door jamb)

plasterboard

continuous ribbon of plaster adhesive

skirting sealed to floor

insulation bearer nailed to joist

SERVICES ACCESS
(vertical section)

cement and sand mortar filler

mastic/silicon sealant to absorb thermal and differential movement

plaster dab

door frame

plasterboard dry lining

continuous ribbons of plaster adhesive

door frame compressible sealant

Figure 21.7 Eliminating air infiltration

21.5 Sound insulation

Sound can be defined as vibrations of air that are registered by the human ear. All sounds are produced by a vibrating object which causes tiny particles of air around it to move in unison. These displaced air particles collide with adjacent air particles, setting them in motion and in unison with the vibrating object. This continuous chain reaction creates a sound-wave which travels through the air until at some distance the air particle movement is so small that it is inaudible to the human ear.

Sounds are defined as either impact or airborne sound, the definition being determined by the source producing the sound. Impact sounds are created when the fabric of structure is vibrated by direct contact whereas airborne sound only sets the structural fabric vibrating in unison when the emitted sound-wave reaches the enclosing structural fabric. The vibrations set up by the structural fabric can therefore transmit the sound to adjacent rooms, which can cause annoyance, disturbance of sleep and of the ability to hold a normal conversation. The objective of sound insulation is to reduce transmitted sound to an acceptable level, the intensity of which is measured in units of decibels (dB).

The Building Regulations, *Approved Document E: Resistance to the passage of sound*, establishes sound insulation standards as follows:

1. E1: Between dwellings and between dwellings and other buildings.
2. E2: Within a dwelling, i.e. between rooms, particularly WCs and habitable rooms, and bedrooms and other rooms.
3. E3: Control of reverberation noise in common parts (stairwells and corridors) of buildings containing dwellings, i.e. flats.
4. E4: Specific applications to acoustic conditions in schools.

Note: E1 includes hotels, hostels, student accommodation, nurses' homes and homes for the elderly, but not hospitals and prisons.

Robust details

The robustdetails® scheme is an alternative solution to pre-completion sound testing for complying with the requirements of A.D. E by providing construction details that provide the necessary level of sound insulation for various applications.

The Approved Document to Building Regulation E2 provides for internal walls and floors located between a bedroom or a room containing a WC and other rooms to have a reasonable resistance to airborne sound. Impact sound can be improved by provision of a carpet.

Wall sound insulation

Separating wall types:

*   Solid masonry.
*   Cavity masonry.
*   Masonry between isolating panels.
*   Timber frame.

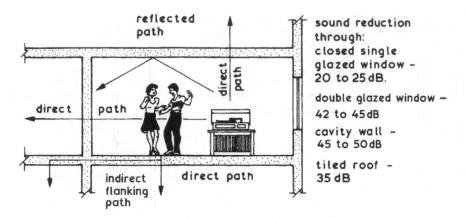

Figure 21.8 Typical sources and transmission of sound

Absorbent material, quilting of unfaced mineral fibre batts with a minimum density of $10kg/m^3$, is located in the cavity or frames (Table 21.8).

Separating floors

Types:

- Concrete with soft covering.
- Concrete with floating layer.
- Timber with floating layer.

Resilient layers:

1. 25mm paper faced mineral fibre, density $36kg/m^3$.
   - Timber floor – paper faced underside.
   - Screeded floor – paper faced upper side to prevent screed from entering layer.

2. Screeded floor only:
   - 13mm pre-compressed expanded polystyrene (EPS) board, or 5mm extruded polyethylene foam of density $30–45kg/m^3$, laid over a levelling screed for protection.

BS EN 29,052–1: *Acoustics. Method for the Determination of Dynamic Stiffness. Materials Used under Floating Floors in Dwellings.*

Sound insulation

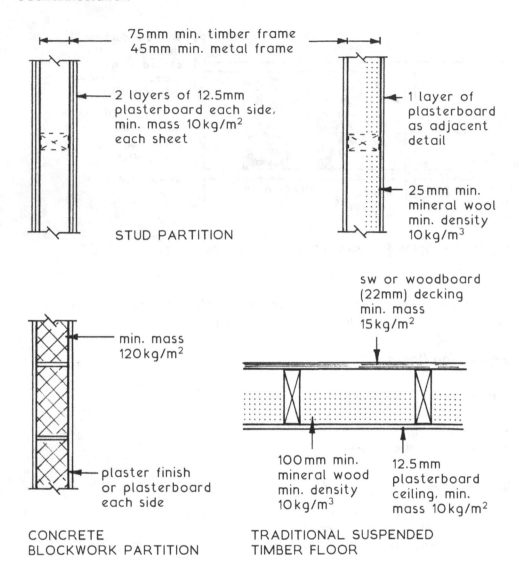

75mm min. timber frame
45mm min. metal frame

← 2 layers of 12.5mm plasterboard each side, min. mass 10kg/m² each sheet

STUD PARTITION

← 1 layer of plasterboard as adjacent detail

← 25mm min. mineral wool min. density 10kg/m³

→ min. mass 120kg/m²

← plaster finish or plasterboard each side

CONCRETE BLOCKWORK PARTITION

sw or woodboard (22mm) decking min. mass 15kg/m²

100mm min. mineral wood min. density 10kg/m³

12.5mm plasterboard ceiling, min. mass 10kg/m²

TRADITIONAL SUSPENDED TIMBER FLOOR

Figure 21.9 Sound insulation methods

Type 1 – relies on mass

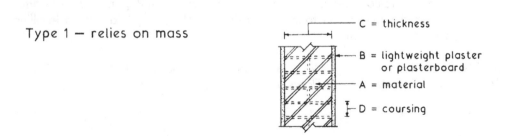

C = thickness

B = lightweight plaster or plasterboard

A = material

D = coursing

Figure 21.10 Type 1 mass

Type 2 — relies on mass and isolation

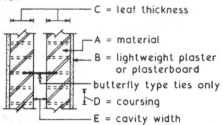

C = leaf thickness

A = material

B = lightweight plaster or plasterboard

butterfly type ties only

D = coursing

E = cavity width

Figure 21.11 Type 2 isolation

Table 21.5 Solid wall mass properties

Material A	Density of A $[kg/m^3]$	Finish B	Combined mass A + B (kg/m^2)	Thickness C [mm]	Coursing D [mm]
Brickwork	1610	13mm lwt. pl. 12.5mm pl. brd.	375	215	75
Concrete block	1840 1840	13mm lwt. pl 12.5mm pl. brd	415		110 150
In-situ concrete	2200	Optional	415	190	n/a

Table 21.6 Cavity wall mass and isolation properties

Material A	Density of A $[kg/m^3]$	Finish B	Mass A + B (kg/m^2)	Thickness C [mm]	Coursing D [mm]	Cavity E[mm]
Brickwork	1970	13mm lwt. pl.	415	102	75	50
Concrete block	1990			100	225	
L/wt. conc. block	1375	or 12.5mm pl. brd.	300	100	225	75

Sound insulation

Type 3 ~ relies on: (a) core material type and mass,
(b) isolation, and
(c) mass of isolated panels.

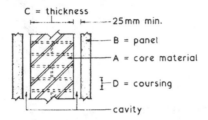

Figure 21.12 Type 3 combined mass and isolation

Note: Panel materials – B: Plasterboard with cellular core plus plaster finish, mass 18kg/m². All joints taped. Fixed floor and ceiling only. Two plasterboard sheets, 12.5mm each, with joints staggered. Frame support of 30mm overall thickness.

Table 21.7 Combined mass and isolation densities

Core material A	Density of A [kg/m³]	Mass A (kg/m²)	Thickness C (mm)	Coursing D (mm)	Cavity (mm)
Brickwork	1290	300	215	75	n/a
Concrete block	2200	300	140	110	n/a
Lwt. conc. block	1400	150	200	225	n/a
Cavity brickwork or block	Any	Any	2 × 100	to suit	50

Timber with floating layer

Airborne resistance varies depending on floor construction, absorbency of materials, extent of pugging and partly on the floating layer. Impact resistance depends mainly on the resilient layer separating the floating floor from the structure.

Platform floor

Note: Minimum mass per unit area = 25kg/m².

Floating layer: 18mm timber or wood-based board, tongue and groove joints glued and spot bonded to a substrate of 19mm plasterboard. Alternative substrate of cement bonded particle board in two 12mm thicknesses, joints staggered, glued and screwed together.

Resilient layer: 25mm mineral fibre, density 60–100kg/m³.

Type 4 — relies on mass, frame separation and absorption of sound.

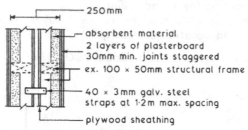

Figure 21.13 Type 4 mass and separation

Table 21.8 Separation zones

Thickness (mm)	Location
25	Suspended in cavity
50	Fixed within one frame
2 × 25	Each quilt fixed within each frame

Type 1. Airborne resistance depends on mass of concrete and ceiling. Impact resistance depends on softness of covering.

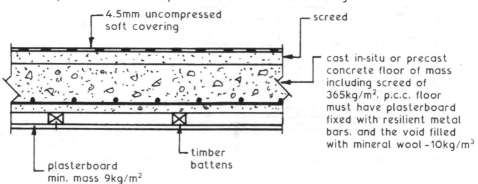

Figure 21.14 Concrete upper floor

Sound insulation

Type 2. Airborne resistance depends mainly on concrete mass and partly on mass of floating layer and ceiling.

Impact resistance depends on resilient layer isolating floating layer from base and isolation of ceiling.

Bases: As type 1. but overall mass minimum 300 kg/m².

Floating layers:

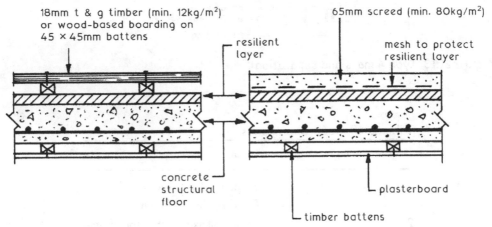

18mm t & g timber (min. 12kg/m²) or wood-based boarding on 45 × 45mm battens

resilient layer

65mm screed (min. 80kg/m²)

mesh to protect resilient layer

concrete structural floor

plasterboard

timber battens

Figure 21.15 Use of resilient layers in concrete floors

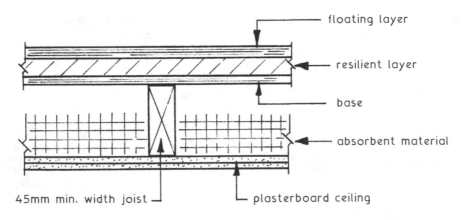

floating layer

resilient layer

base

absorbent material

45mm min. width joist

plasterboard ceiling

Figure 21.16 Use of resilient layer in timber floors

Base: 12mm timber boarding or wood-based board nailed to joists.
Absorbent material: 100mm mineral fibre of minimum density 10kg/m³.
Ceiling: 30mm plasterboard in two layers, joints staggered.

Ribbed or battened floor

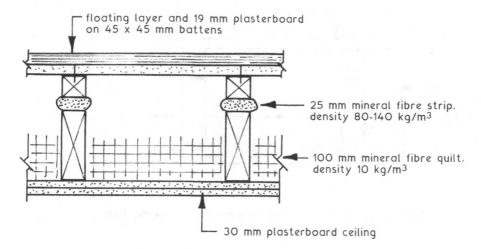

Figure 21.17 Ribbed/battened floor

Note: See Figures 21.19 and 21.20 for other examples of ribbed floors.

Alternative ribbed or battened floor

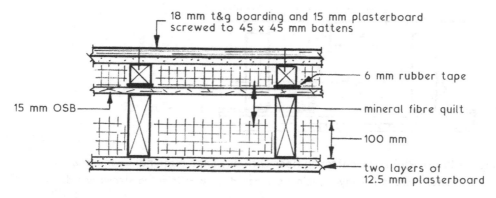

Figure 21.18 Alternative ribbed or battened floor

Sound insulation

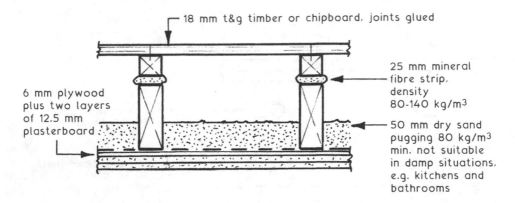

Figure 21.19 Ribbed or battened floor with dry sand pugging: Example 1

18 mm t&g timber or chipboard, joints glued

25 mm mineral fibre strip, density 80-140 kg/m³

6 mm plywood plus two layers of 12.5 mm plasterboard

50 mm dry sand pugging 80 kg/m³ min. not suitable in damp situations, e.g. kitchens and bathrooms

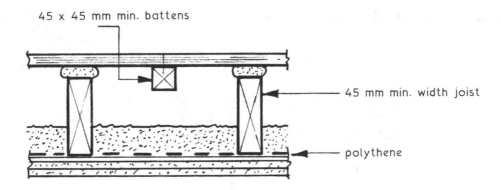

Figure 21.20 Ribbed or battened floor with dry sand pugging: Example 2
Note: Plasterboard in two layers, joints staggered. OSB = Oriented strand board.

45 x 45 mm min. battens

45 mm min. width joist

polythene

21.6 Sound insulation improvements

The intensity of unwanted sound that increases vibration of a separating wall or floor can be reduced to some extent by attaching a lining to the existing structure. Plasterboard or other dense material is suitable but this will have only a limited effect as the lining will also vibrate and radiate sound. The most effective solution is to create another wall or ceiling next to the original but separated from it.

Separating wall

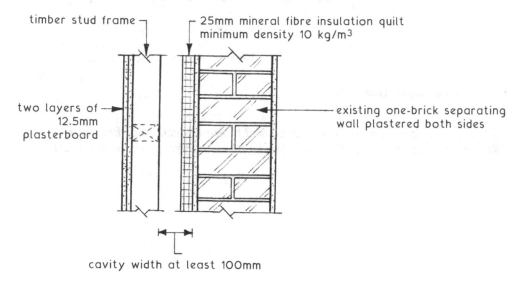

timber stud frame

25mm mineral fibre insulation quilt
minimum density 10 kg/m^3

two layers of
12.5mm
plasterboard

existing one-brick separating
wall plastered both sides

cavity width at least 100mm

Figure 21.21 Internal separating wall

Sound insulation improvements

Separating floor

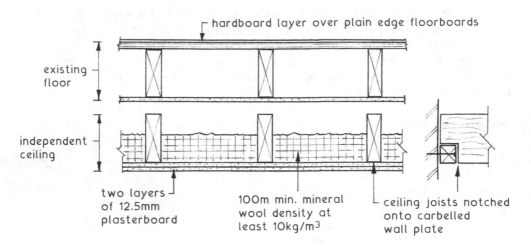

Figure 21.22 Separating floor

BS 8233: *Guidance on Sound Insulation and Noise Reduction for Buildings.*
BRE Digest 293: *Improving the sound insulation of separating walls and floors.*

22 CONSTRUCTION DEFECTS

Opening remarks

The correct application of materials produced to the recommendations of British, European and International Standards authorities, in accordance with local building regulations, bye-laws and the rules of building guarantee companies, i.e. National House Building Council (NHBC) and MD Insurance Services, should ensure a sound and functional structure. However, these controls can be seriously undermined if the human factor of quality workmanship is not fulfilled. The following guidance is designed to promote quality control:

BS 8000: *Workmanship on construction sites.*
Building Regulations, *Approved Document to support Regulation 7 – Materials and workmanship.*

No matter how good the materials, the workmanship and supervision, the unforeseen may still affect a building. This may materialise several years after construction. Some examples of these latent defects include: woodworm emerging from untreated timber, electrolytic decomposition of dissimilar metals inadvertently in contact and chemical decomposition of concrete. Generally, the older a building the more opportunity there is for its components and systems to have deteriorated and malfunctioned – hence the need for regular inspection and maintenance. The profession of facilities management has evolved for this purpose and is represented by the British Institute of Facilities Management (BIFM).

Property values, repairs and replacements are of sufficient magnitude for potential purchasers to engage the professional services of a building surveyor. Surveyors are usually members of the Royal Institution of Chartered Surveyors (RICS). The extent of survey can vary depending on a client's requirements. This may range from no more than a market valuation to secure financial backing, to a full structural survey incorporating specialist reports on electrical installations, drains, heating systems, etc.

BRE Digest No. 268 – *Common defects in low-rise traditional housing.* Available from Building Research Establishment Bookshop – www.brebookshop.com.

22.1 Building survey

The established procedure is for the interested purchaser to engage a building surveyor.

UK government requirement is for the seller to provide an energy performance certificate. This is a fuel use and efficiency appraisal on a numerical scale.

Survey document preliminaries

- Title and address of property.
- Client's name, address and contacts.
- Survey date and time.
- Property status – freehold, leasehold or commonhold.
- Occupancy – occupied or vacant; if vacant, source of keys.
- Extent of survey, e.g. full structural + services reports.
- Specialists in attendance, e.g. electrician, heating engineer, etc.
- Age of property (approx. if very dated or no records).
- Disposition of rooms, i.e. number of bedrooms, etc.
- Floor plans and elevations, if available.
- Elevation (flooding potential) and orientation (solar effect).
- Estate/garden area and disposition, if appropriate.
- Means of access – roads, pedestrian only, rights of way.

Survey tools and equipment

- Drawings + estate agent's particulars if available.
- Notebook and pencil/pen.
- Binoculars and a camera with flash facility.
- Tape measure, spirit level and plumb line.
- Other useful tools to include small hammer, torch, screwdriver and manhole lifting irons.
- Moisture meter.
- Ladders – eaves access and loft access.
- Sealable bags for taking samples, e.g. wood rot, asbestos, etc.

Estate and garden

- Location and establishment of boundaries.
- Fences, gates and hedges – material, condition and suitability.
- Trees – type and height, proximity to building.
- Pathways and drives – material and condition.
- Outbuildings – garages, sheds, greenhouses, barns, etc.
- Proximity of water courses.

Building survey

Roof

- Tile type, treatment at ridge, hips, verge and valleys.
- Age of covering, repairs, replacements, renewals, general condition, defects and growths.
- Eaves finish, type and condition.
- Gutters – material, size, condition, evidence of leakage.
- Rainwater downpipes as above.
- Chimney – DPCs, flashings, flaunching, pointing, signs of movement.
- Flat roofs – materials, repairs, abutments, flashings and drainage.

Walls

- Materials – type of brick, rendering, cladding, etc., condition and evidence of repairs.
- Solid or cavity construction, if cavity extent of insulation and type.
- Pointing of masonry, painting of rendering and cladding.
- Air brick location, function and suitability.
- DPC, material and condition, position relative to ground level.
- Windows and doors, material, signs of rot or damage, original or replacement, frame seal.
- Settlement – signs of cracking, distortion of window and door frames – specialist report.

Drainage

A building surveyor may provide a general report on the condition of the drainage and sanitation installation, however, a full test for leakage and determination of self-cleansing and flow conditions, to include fibre-optic scope examination, is undertaken as a specialist survey.

Roof space

- Access to all parts, construction type – traditional or trussed.
- Evidence of moisture due to condensation – ventilation at eaves, ridge, etc.
- Evidence of water penetration – chimney flashings, abutments and valleys.
- Insulation – type and quantity.
- Party wall in semi-detached and terraced dwellings – suitability as fire barrier.
- Plumbing – adequacy of storage cistern, insulation, overflow function.

Floors

- Construction – timber, precast or cast in-situ concrete? Finish condition?
- Timber ground floor – evidence of dampness, rot, woodworm, ventilation, DPCs.
- Timber upper floor stability, i.e. wall fixing, strutting, joist size, woodworm, span and loading.

Stairs

- Type of construction and method of fixing – built in-situ or preformed.
- Soffit, re fire protection (plasterboard?).
- Balustrading – suitability and stability.
- Safety – adequate screening, balusters, handrail, pitch angle, open tread, tread wear.

Finishes

- General décor, i.e. paint and wallpaper condition – damaged, faded.
- Woodwork/joinery – condition, defects, damage, paintwork.
- Plaster, ceiling (plasterboard or lath and plaster?) – condition and stability.
- Plaster, walls – render and plaster or plasterboard, damage and quality of finish.
- Staining – plumbing leaks (ceiling), moisture penetration (wall openings), rising damp.
- Fittings and ironmongery – adequacy and function, weather exclusion and security.

Supplementary enquiries should determine the extent of additional building work, particularly since the planning threshold of 1948. Check for planning approvals, permitted development and Building Regulation approvals, exemptions and completion certificates.

Services

Apart from a cursory inspection to ascertain location and suitability of system controls, these areas are highly specialised and should be surveyed by those appropriately qualified.

22.2 Cracking in walls

Cracks are caused by applied forces that exceed those that the building can withstand. Most cracking is superficial, occurring as materials dry out and subsequently shrink to reveal minor surface fractures of <2mm. These insignificant cracks can be made good with proprietary fillers.

Severe cracking in walls may result from foundation failure due to inadequate design or physical damage. Further problems could include:

- Structural instability.
- Rain penetration.
- Air infiltration.
- Heat loss.
- Sound insulation reduction.
- Visual depreciation.

A survey should be undertaken to determine the cause of cracking, i.e.:

- Loads applied externally (tree roots, subsoil movement).
- Climate/temperature changes (thermal movement).
- Moisture content change (faulty DPC, building leakage).
- Vibration (adjacent work, traffic).
- Changes in physical composition (salt or ice formation).
- Chemical change (corrosion, sulphate attack).
- Biological change (timber decay).
- The effect on a building's performance (structural and environmental).
- The nature of movement – completed, ongoing or intermittent (seasonal).

Observations over a period of several months, preferably over a full year, will determine whether the cracking is new or established and whether it is progressing.

Simple method for monitoring cracks

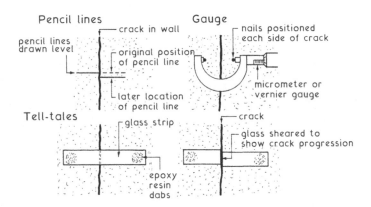

Figure 22.1 Monitoring cracks in walls

BRE Digest 251: *Assessment of damage in low rise buildings.*

22.3 Underpinning

This is a remedial operation to transfer the load carried by a strip foundation from its existing bearing level to a new level at a lower depth or to prop up the foundation.

Underpinning applications

- Uneven settlement: this could be caused by uneven loading of the building, unequal resistance of the soil action of tree roots or cohesive soil settlement.
- Increase in loading: this could be due to the addition of an extra storey or an increase in imposed loadings, such as that which may occur with a change of use.
- Lowering of adjacent ground: usually required when constructing a basement adjacent to existing foundations.

Traditional method wall underpinning

This is a remedial technique used where subsidence or settlement of the foundation has occurred, affecting the walls.

Work is carried out sequentially in short lengths called legs or bays. The length of these bays will depend upon the following factors:

- Total length of wall to be underpinned.
- Wall loading.
- General state of repair and stability of wall and foundation to be underpinned.
- Nature of subsoil beneath existing foundation.
- Estimated spanning ability of existing foundation.

Generally suitable bay lengths are:

- 1–1.5m for mass concrete strip foundations supporting walls of traditional construction.
- 1.5–3m for reinforced concrete strip foundations supporting walls of moderate loading.

In all cases the total sum of the unsupported lengths of wall should not exceed 25% of the total wall length.

The sequence of bays should be arranged so that working in adjoining bays is avoided until one leg of underpinning has been completed, pinned and cured sufficiently to support the wall above.

Needle and pile underpinning

This method of underpinning can be used where the condition of the existing foundation is unsuitable for traditional or jack pile underpinning techniques. The brickwork above the existing foundation must be in a sound condition since this method relies on the 'arching effect' of the brick bonding to transmit the wall loads onto the needles and, ultimately, to the piles. The piles used with this method are usually small diameter bored piles.

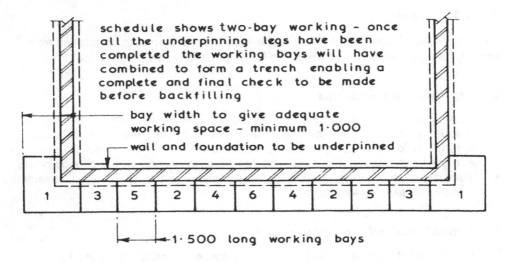

schedule shows two-bay working – once all the underpinning legs have been completed the working bays will have combined to form a trench enabling a complete and final check to be made before backfilling

bay width to give adequate working space – minimum 1·000

wall and foundation to be underpinned

| 1 | 3 | 5 | 2 | 4 | 6 | 4 | 2 | 5 | 3 | 1 |

1·500 long working bays

Figure 22.2 Typical underpinning schedule

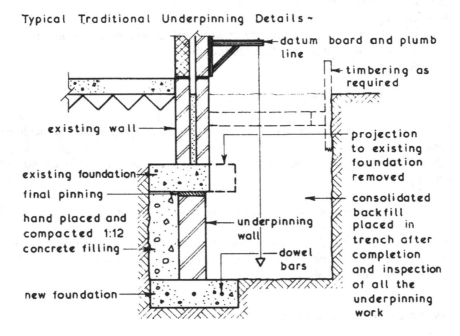

Typical Traditional Underpinning Details ~

datum board and plumb line

timbering as required

existing wall

projection to existing foundation removed

existing foundation

final pinning

consolidated backfill placed in trench after completion and inspection of all the underpinning work

hand placed and compacted 1:12 concrete filling

underpinning wall

dowel bars

new foundation

UNDERPINNING BAY ~ TYPICAL SECTION

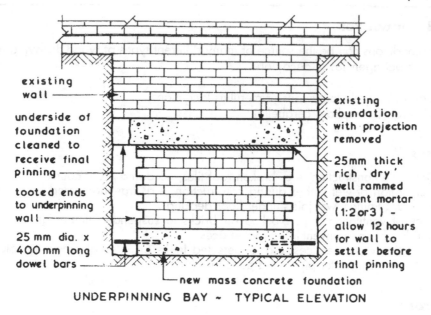

UNDERPINNING BAY ~ TYPICAL ELEVATION

Figure 22.2a Traditional Underpinning Details

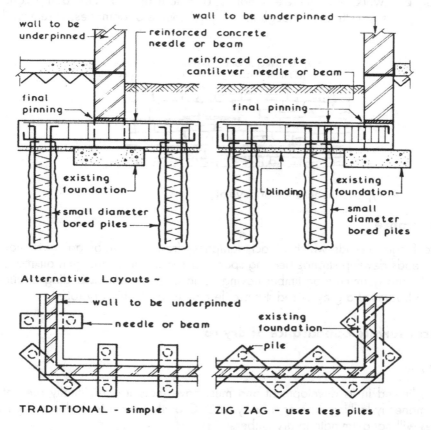

Figure 22.3 Needle or beam pile underpinning

22.4 Timber rot

Damp conditions can be the source of many different types of wood-decaying fungi. The principal agencies of decay are:

- Dry rot (Serpula lacrymans or merulius lacrymans).
- Wet rot (Coniophora cerabella).

Causes

- Defective construction, e.g. broken roof tiles; no damp proof course.
- Installation of wet timber during construction, e.g. framing sealed behind plasterboard linings; wet joists under floor decking.
- Lack of ventilation, e.g. blocked air bricks to suspended timber ground floor; condensation in unventilated roof spaces.
- Defective water services, e.g. undetected leaks on internal pipework; blocked or broken rainwater pipes and guttering.

Dry rot

This is the most difficult to control as its root system can penetrate damp and porous plaster, brickwork and concrete. It can also remain dormant until damp conditions encourage its growth, even though the original source of dampness is removed.

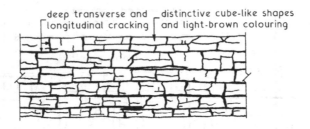

Figure 22.4 Typical appearance of dry rot

Appearance

White fungal threads which attract dampness from the air or adjacent materials. The threads develop strands bearing spores or seeds which drift with air movements to settle and germinate on timber having a moisture content exceeding about 25%. Fruiting bodies of a grey or red flat profile may also identify dry rot.

Typical surface appearance of dry rot

Wet rot

This is limited in its development and must have moisture continually present, e.g. a permanently leaking pipe or a faulty DPC. Growth pattern is similar to dry rot, but spores will not germinate in dry timber.

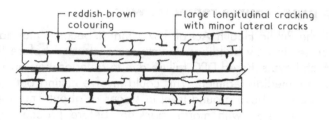

Figure 22.5 Typical surface appearance of wet rot

Appearance

Fungal threads of black or dark-brown colour. Fruiting bodies may be olive-green or dark brown and these are often the first sign of decay.

Treatment and preservation

GENERAL TREATMENT

- Remove source of dampness.
- Allow affected area to dry.
- Remove and burn all affected timber and sound timber within 500mm of rot.
- Remove contaminated plaster and rake out adjacent mortar joints to masonry.

Note: This is normally sufficient treatment where wet rot is identified. However, where dry rot is apparent the following additional treatment is necessary.

STERILISE SURFACE OF CONCRETE AND MASONRY

Heat with a blow torch until the surface is too hot to touch. Apply a proprietary fungicide generously to warm surface. Irrigate badly affected masonry and floors, i.e. provide 12mm diameter boreholes at about 500mm spacing and flood or pressure inject with fungicide: 20:1 dilution of water and sodium pentachlorophenate, sodium orthophenylphate or mercuric chloride. Product manufacturers' safety in handling and use measures must be observed when applying these chemicals.

Replacement work should ensure that new timbers are pressure impregnated with a preservative. Cement and sand mixes for rendering, plastering and screeds should contain a zinc oxychloride fungicide.

BRE: Timber pack (ref. AP 265) – various Digests, Information Papers, Good Repair Guides and Good Building Guides.
Timber Preservation and Repair – Triton Systems.
Building Regulations *Approved Document C, Site preparation and resistance to contaminants and moisture.*

22.5 Damp proof course remedial work

It was not until the Public Health Act of 1875 that it became mandatory to install damp proof courses in new buildings. Structures constructed before that time, and those since, which have suffered DPC failure due to deterioration or incorrect installation will require remedial treatment. This could involve cutting out the mortar bed-joint two brick courses above ground level in stages of about 1m in length. A new DPC can then be inserted with mortar packing, before proceeding to the next length. No two adjacent sections should be worked consecutively. This process is very time consuming and may lead to some structural settlement.

Materials

- Silicone solutions in organic solvent.
- Aluminium stearate solutions.
- Water-soluble silicone formulations (siliconates).

Methods

- High pressure injection (0.70–0.90MPa) solvent based.
- Low pressure injection (0.15–0.30MPa) water based. Gravity feed, water based.
- Insertion/injection, mortar based.

Pressure injection

12mm diameter holes are bored to about two-thirds the depth of masonry, at approximately 150mm horizontal intervals at the appropriate depth above ground (normally two to three brick courses). These holes can incline slightly downwards. With high (low) pressure injection, walls in excess of 120mm (460mm) thickness should be drilled from both sides.

The chemical solution is injected by pressure pump until it exudes from the masonry. Cavity walls are treated as each leaf being a solid wall.

Injection mortars

19mm diameter holes are bored from both sides of a wall, at the appropriate level and no more than 230mm apart horizontally, to a depth equating to three-fifths of the wall thickness. They should be inclined downwards at an angle of 20–30°. The drill holes are flushed out with water before injecting mortar from the base of the hole and outwards. This can be undertaken with a hand operated caulking gun. Special cement mortars contain styrene butadiene resin (SBR) or epoxy resin and must be mixed in accordance with the manufacturer's guidance.

Notes relating to all applications of chemical DPCs

- Before commencing work, old plasterwork and rendered undercoats are removed to expose the masonry. This should be to a height of at least 300mm above the last detectable (moisture meter reading) signs of rising dampness (1 metre min.).
- If the wall is only accessible from one side and both sides need treatment, a second deeper series of holes may be bored from one side, to penetrate the inaccessible side.
- On completion of work, all boreholes are made good with cement mortar. Where dilute chemicals are used for the DPC, the mortar is rammed the full length of the hole with a piece of timber dowelling.
- The chemicals are effective by bonding to and lining the masonry pores by curing and solvent evaporation.
- The process is intended to provide an acceptable measure of control over rising dampness. A limited amount of water vapour may still rise but this should be dispersed by evaporation in a heated building.
- Replastering should be undertaken with lightweight renovating plaster.

BS 6576: *Code of Practice for Diagnosis of Rising Damp in Walls of Buildings and Installation of Chemical Damp-Proof Courses.*
BRE Digest 245: *Rising damp in walls: diagnosis and treatment.*
BRE Digest 380: *Damp-proof courses.*
BRE Good Repair Guide 6: *Treating rising damp in houses.*

INDEX

Index

Index

Index

Index

Index

Index

Index

Index

Index

Printed in the United States
by Baker & Taylor Publisher Services